Digital Picture Processing

Second Edition

Volume 2

Digital Picture Processing

Second Edition

Volume 2

AZRIEL ROSENFELD

Computer Science Center
University of Maryland
College Park, Maryland

AVINASH C. KAK

School of Electrical Engineering
Purdue University
West Lafayette, Indiana

ACADEMIC PRESS, INC.
Harcourt Brace Jovanovich, Publishers
San Diego   New York   Berkeley   Boston
London   Sydney   Tokyo   Toronto

ACADEMIC PRESS, INC.
San Diego, California 92101

United Kingdom Edition published by
ACADEMIC PRESS LIMITED
24-28 Oval Road, London NW1 7DX

Library of Congress Cataloging in Publication Data

Rosenfeld, Azriel.
    Digital picture processing.

    (Computer science and applied mathematics)
    Includes bibliographical references and index.
    1. Optical data processing. I. Kak, Avinash C.
II. Title. III. Series.
TA1630.R67 1981   621.3819'598   81-17611
  ISBN—0—12—597302—0   AACR2

PRINTED IN THE UNITED STATES OF AMERICA
89 90 91 92 93   9 8 7 6 5

Contents

v

Chapter 11   **Representation**

Chapter 12   **Description**

Preface

The size of this second edition has made it necessary to publish this book in two volumes. However, a single chapter numbering has been maintained. Volume 1 contains Chapters 1–8, covering digitization, compression, enhancement, restoration, and reconstruction; and Volume 2 contains Chapters 9–12, covering matching, segmentation, representation, and description. The material in Volume 2 is not strongly dependent on that in Volume 1; and to make it even more self-contained, the Preface and Introduction (called Chapter 1 in Volume 1) are reproduced at the beginning of Volume 2.

The rapid rate at which the field of digital picture processing has grown in the past five years has necessitated extensive revisions and the introduction of topics not found in the original edition.

Two new chapters have been added: Chapter 8 (by A. C. K.) on reconstruction from projections and Chapter 9 (by A. R.) on matching. The latter includes material from Chapters 6 and 8 of the first edition on geometric transformations and matching, but it consists primarily of new material on imaging geometry, rectification, and stereomapping, as well as an appendix on the analysis of time-varying imagery.

Chapter 2 incorporates a new section on vector space representation of images. Chapter 5 on compression has undergone a major expansion. It includes a new section on interpolative representation of images and fast implementation of the Karhunen–Loève transforms based thereon. Also included in this chapter is image compression using discrete cosine transforms—a technique that has attracted considerable attention in recent years. New sections on block quantization, the recently discovered technique of block truncation compression, and error-free compression have also been added to Chapter 5. New material has been added to Chapter 6 on gray level and histogram transformation and on smoothing. Chapter 7 has also been considerably expanded and includes many new restoration techniques. This chapter incorporates a new frequency domain derivation of the constrained least squares filter. The treatment of Markov representation has been expanded with a section on vector–matrix formulation of such representations. Chapters 10, 11, and 12 are major expansions of the first edition's Chapters 8–10, dealing with segmentation of pictures into parts, representations of the parts (formerly "geometry"), and description of pictures in terms of parts. Chapter 10 incorporates much new material on pixel classification, edge detection, Hough transforms, and picture partitioning, reflecting recent developments in these areas; it also contains an entirely new section (10.5) on iterative "relaxation" methods for fuzzy or probabilistic segmentation. Chapter 11 is now organized according to types of representations (runs, maximal blocks, quadtrees, border codes), and discusses how to convert between these representations and how to use them to compute geometrical properties of picture subsets. Chapter 12 treats picture properties as well as descriptions of pictures at various levels (numerical arrays, region representations, relational structures). It also discusses models for classes of pictures, as defined, in particular, by constraints that must be satisfied at a given level of description ("declarative models") or by grammars that generate or accept the classes. It considers how to construct a model consistent with a given set of descriptions and how to extract a description that matches a given model; and it also contains an appendix on the extraction of three-dimensional information about a scene from pictures.

The authors of the chapters are as follows: Chapters 2, 4, 5, 7, and 8 are by A. C. K, whereas Chapters 1, 3, 6, and 9 through 12 are by A. R.

Acknowledgments

A. C. K. would like to express appreciation to his colleague O. Robert Mitchell and his graduate students David Nahamoo, Carl Crawford, and Kai-Wah Chan for many stimulating discussions. A number of our colleagues helped us by providing figures (see the individual figure credits). In addition, many of the figures were produced by Andrew Pilipchuk and others at the University of Maryland; and by Carl Crawford, Doug Morton, and Kai-Wah Chan at Purdue University. Kathryn Riley, Dawn Shifflett, and Mellanie Boes, among others, did an excellent job of preparing the text. To these, and to others too numerous to mention, our sincerest thanks.

The authors wish to express their indebtedness to the following individuals and organizations for their permission to reproduce the figures listed below.

Chapter 9: Figure 10, from D. A. O'Handley and W. B. Green, "Recent developments in digital image processing at the Image Processing Laboratory of the Jet Propulsion Laboratory," *Proceedings of the IEEE* **60,** 1972, 821–828. Figure 11, from R. Bernstein, "Digital image processing of earth observation sensor data," *IBM Journal of Research and Development* **20,** 1976, 40–57.

Contents of Volume 1

Chapter 7 **Restoration**

Chapter 8 **Reconstruction**

Introduction

PICTURE PROCESSING

Picture processing or *image processing* is concerned with the manipulation and analysis of pictures by computer. Its major subareas include

(a) *Digitization and compression*: Converting pictures to discrete (digital) form; efficient coding or approximation of pictures so as to save storage space or channel capacity.

(b) *Enhancement, restoration, and reconstruction*: Improving degraded (low-contrast, blurred, noisy) pictures; reconstructing pictures from sets of projections.

(c) *Matching, description, and recognition*: Comparing and registering pictures to one another; segmenting pictures into parts, measuring properties of and relationships among the parts, and comparing the resulting descriptions to models that define classes of pictures.

In this chapter we introduce some basic concepts about pictures and digital pictures, and also give a bibliography of general references on picture processing and recognition. (References on specific topics are given at the end of each chapter.) Chapter 2 reviews some of the mathematical tools used in later chapters, including linear systems, transforms, and random fields, while Chapter 3 briefly discusses the psychology of visual perception. The

1

remaining chapters deal with the theory of digitization (4); coding and compression (5); enhancement (6); restoration and estimation (7); reconstruction from projections (8); registration and matching (9); segmentation into parts (10); representation of parts and geometric property measurement (11); and nongeometric properties, picture descriptions, and models for classes of pictures (12).

The level of treatment emphasizes concepts, algorithms, and (when necessary) the underlying theory. We do not cover hardware devices for picture input (scanners), processing, or output (displays); nondigital (e.g., optical) processing; or picture processing software.

SCENES, IMAGES, AND DIGITAL PICTURES

Scenes and Images

When a scene is viewed from a given point, the light received by the observer varies in brightness and color as a function of direction. Thus the information received from the scene can be expressed as a function of two variables, i.e., of two angular coordinates that determine a direction. (The scene brightness and color themselves are resultants of the illumination, reflectivity, and geometry of the scene; see Section 6.2.2.)

In an optical image of the scene, say produced by a lens, light rays from each scene point in the field of view are collected by the lens and brought together at the corresponding point of the image. Scene points at different distances from the lens give rise to image points at different distances; the basic equation is

$$\frac{1}{u} + \frac{1}{v} = \frac{1}{f}$$

where u, v are the distances of the object and image points from the lens (on opposite sides), and f is a constant called the focal length of the lens. If u is large, i.e., the scene points are all relatively far from the lens, $1/u$ is negligible, and we have $v \approx f$, so that the image points all lie at approximately the same distance from the lens, near its "focal plane." Thus the imaging process converts the scene information into an illumination pattern in the image plane; this is still a function of two variables, but they are now coordinates in the plane. (Image formation by optical systems will not be further discussed here. On the geometry of the mapping from three-dimensional scene coordinates to two-dimensional image coordinates, see Section 9.1.2.)

We can now record or measure the pattern of light from the scene by placing some type of sensor in the image-plane. (Some commonly used sensors will be mentioned in the next paragraph.) Any given sensor has a characteristic spectral sensitivity, i.e., its response varies with the color of the light; thus its total response to the light at a given point can be expressed by an integral of the form $\int S(\lambda)I(\lambda)\,d\lambda$, where $I(\lambda)$ is light intensity and $S(\lambda)$ is sensitivity as functions of wavelength. This means that if we use only a single sensor, we can only measure (weighted) light intensity. If we want to measure color, we must use several sensors having different spectral responses; or we must split the light into a set of spectral bands, using color filters, and measure the light intensity in each band. (Knowing the intensities in three suitably chosen bands, e.g., in the red, green, and blue regions of the spectrum, is enough to characterize any color; see Section 3.3.) In other words, when we use only one sensor, we are representing the scene information by a scalar-valued function of position in the image, representing scene brightness. To represent color, we use a k-tuple (usually a triple) of such functions, or equivalently, a vector-valued function, representing the brightnesses in a set of spectral bands. We will usually assume in this book that we are dealing with a single scalar-valued brightness function. Photometric concepts and terminology will not be further discussed here; we use terms such as "brightness" and "intensity" in an informal sense.

Image sensors will not be discussed in detail in this book, but we briefly mention here some of the most common types.

(a) We can put an *array* of photosensitive devices in the image plane; each of them measures the scene brightness at a particular point (or rather, the total scene brightness in a small patch).

(b) We need only a single photosensor in the image plane if we can illuminate the scene one point (or small patch) at a time; this is the principle of the *flying-spot scanner*. Similarly, we need only one photosensor if we can view the scene through a moving aperture so that, at any given time, the light from only one point of the scene can reach the sensor.[§]

(c) In a *TV camera*, the pattern of brightnesses in the scene is converted into an electrical charge pattern on a grid; this pattern can then be scanned by an electron beam, yielding a *video signal* whose value at any given time corresponds to the brightness at a given image point.

In all of these schemes, the image brightness is converted into a pattern of electrical signals, or into a time-varying signal corresponding to a sequential scan of the image or scene. Thus the sensor provides an electrical or electronic analog of the scene brightness function, which is proportional to it,

---

[§] As a compromise between (a) and (b), we can use a one-dimensional array of sensors in the image plane, say in the horizontal direction, and scan in the vertical direction, so that light from only one "row" of the scene reaches the sensors at any given time.

if the sensors are linear. More precisely, an array sensor provides a discrete array of samples of this function; while scanning sensors provide a set of cross sections of the function along the lines of the scanning pattern.

If, instead of using a sensor, we put a piece of photographic film (or some other light-sensitive recording medium) in the image plane, the brightness pattern gives rise to a pattern of variations in the optical properties of the film. (Color film is composed of layers having different spectral sensitivities; we will discuss here only the black-and-white case.) In a film transparency, the optical transmittance t (i.e., the fraction of the light transmitted by the film) varies from point to point; in an opaque print, the reflectance r ($=$ the fraction of light reflected) varies. Evidently we have $0 \leqslant t \leqslant 1$ and $0 \leqslant r \leqslant 1$. The quantity $-\log t$ or $-\log r$ is called optical *density*; thus a density close to zero corresponds to almost perfect transmission or reflection, while a very high density, say 3 or 4, corresponds to almost perfect opaqueness or dullness (i.e., only 10^{-3} or 10^{-4} of the incident light is transmitted or reflected). For ordinary photographic processes, the density is roughly a linear function of the log of the amount of incident light (the log of the "exposure") over a range of exposures; the slope of this line is called photographic *gamma*. Photographic processes will not be discussed further in this book. A photograph of a scene can be converted into signal form by optically imaging it onto a sensor.

Pictures and Digital Pictures

We saw in the preceding paragraphs that the light received from a scene by an optical system produces a two-dimensional image. This image can be directly converted into electrical signal form by a sensor, or it can be recorded photographically as a picture and subsequently converted. Mathematically, a *picture* is defined by a function $f(x, y)$ of two variables (coordinates in the image plane, corresponding to spatial directions). The function values are brightnesses, or k-tuples of brightness values in several spectral bands. In the black-and-white case, the values will be called *gray levels*. These values are real, nonnegative (brightness cannot be negative), and bounded (brightness cannot be arbitrarily great). They are zero outside a finite region, since an optical system has a bounded field of view, so that the image is of finite size; without loss of generality, we can assume that this region is rectangular. Whenever necessary, we will assume that picture functions are analytically well-behaved, e.g., that they are integrable, have invertible Fourier transforms, etc.

When a picture is digitized (see Chapter 4), a *sampling* process is used to extract from the picture a discrete set of real numbers. These *samples* are usually the gray levels at a regularly spaced array of points, or, more realistically, average gray levels taken over small neighborhoods of such points. (On other methods of sampling see Section 4.1.) The array is almost always taken to be Cartesian or rectangular, i.e., it is a set of points of the form (md, nd), where m and n are integers and d is some unit distance. (Other types of regular arrays, e.g., hexagonal or triangular, could also be used; see Section 11.1.7, Exercise 4, on a method of defining a hexagonal array by regarding alternate rows of a rectangular array as shifted $d/2$ to the right.) Thus the samples can be regarded as having integer coordinates, e.g., $0 \leqslant m < M, 0 \leqslant n < N$.

The picture samples are usually *quantized* to a set of gray level values which are often taken to be equally spaced (but see Section 4.3). In other words, the *gray scale* is divided into equal intervals, say $I_0, \ldots, I_K$, and the gray level $f(x, y)$ of each sample is changed into the level of the midpoint of the interval I_i in which $f(x, y)$ falls. The resulting quantized gray levels can be represented by their interval numbers $0, \ldots, K$, i.e., they can be regarded as integers.

The result of sampling and quantizing is a *digital picture*. As just seen, we can assume that a digital picture is a rectangular array of integer values. An element of a digital picture is called a *picture element* (often abbreviated *pixel* or *pel*); we shall usually just call it a *point*. The value of a pixel will still be called its gray level. If there are just two values, e.g., "black" and "white," we will usually represent them by 0 and 1; such pictures are called two-valued or binary-valued.

Digital pictures are often very large. For example, suppose we want to sample and quantize an ordinary (500-line) television picture finely enough so that it can be redisplayed without noticeable degradation. Then we must use an array of about 500 by 500 samples, and we should quantize each sample to about 50 discrete gray levels, i.e., to about a 6-bit number. This gives us an array of 250,000 6-bit numbers, for a total of $1\frac{1}{2}$ million bits. In many cases, even finer sampling is necessary; and it has become standard to use 8-bit quantization, i.e., 256 gray levels.

Except on the borders of the array, any point (x, y) of a digital picture has four horizontal and vertical neighbors and four diagonal neighbors, i.e.,

$$
\begin{array}{lll}
(x-1, y+1) & (x, y+1) & (x+1, y+1) \\
(x-1, y) & (x, y) & (x+1, y) \\
(x-1, y-1) & (x, y-1) & (x+1, y-1)
\end{array}
$$

In this illustration of the 3×3 *neighborhood* of a point we have used Cartesian coordinates (x, y) with x increasing to the right and y increasing upward. There are other possibilities; for example, one could use matrix coordinates (m, n), in which m increases downward and n to the right. Note that the diagonal neighbors are $\sqrt{2}$ units away from (x, y), while the horizontal and vertical neighbors are only one unit away. If we think of a pixel as a unit square, the horizontal and vertical neighbors of (x, y) share a side with (x, y), while its diagonal neighbors only touch it at a corner. Some of the complications introduced by the existence of these two types of neighbors will be discussed in Chapter 11. Neighborhoods larger than 3×3 are sometimes used; in this case, a point may have many types of neighbors.

If (x, y) is on the picture border, i.e., $x = 0$ or $M - 1$, $y = 0$ or $N - 1$, some of its neighbors do not exist, or rather are not in the picture. When we perform operations on the picture, the new value of (x, y) often depends on the old values of (x, y) and its neighbors. To handle cases where (x, y) is on the border, we have several possible approaches:

(a)   We might give the operation a complex definition that covers these special cases. However, this may not be easy, and in any case it is computationally costly.

(b)   We can regard the picture as cyclically closed, i.e., assume that column $M - 1$ is adjacent to column 0 and row $N - 1$ to row 0; in other words, we take the coordinates (x, y) modulo (M, N). This is equivalent to regarding the picture as an infinite periodic array with an $M \times N$ period. We will sometimes use this approach, but it is usually not natural, since the opposite rows and columns represent parts of the scene that are not close together.

(c)   We can assume that all values outside the picture are zero. This is a realistic way of representing the image (see the first paragraph of this section), but not the scene.

(d)   The simplest approach is to apply the operation only to a *subpicture*, chosen so that for all (x, y) in the subpicture, the required neighbors exist in the picture. This yields results all of which are meaningful; but note that the output picture produced by the operation is smaller than the input picture.

Operations on Pictures

In this book we shall study many different types of operations that can be performed on digital pictures to produce new pictures. The following are some of the important types of picture operations:

(a)   *Point operations*: The output gray level at a point depends only on the input gray level at the same point. Such operations are extensively used for gray scale manipulations (Section 6.2) and for segmentation by pixel classification (Section 10.1). There may be more than one input picture;

for example, we may want to take the difference or product of two pictures, point by point. In this case, the output level at a point depends only on the set of input levels at the same point.

(b) *Local operations*: The output level at a point depends only on the input levels in a neighborhood of that point. Such operations are used for deblurring (Section 6.3), noise cleaning (Section 6.4), and edge and local feature detection (Sections 10.2 and 10.3), among other applications.

(c) *Geometric operations*: The output level at a point depends only on the input levels at some other point, defined by a geometrical transformation (e.g., translation, rotation, scale change, etc.), or in a neighborhood of that point. On such operations see Section 9.3.

An operation $\mathcal{O}$ is called *linear* if we get the same output whether we apply $\mathcal{O}$ to a linear combination of pictures (i.e., we take $\mathcal{O}(af + bg)$) or we apply $\mathcal{O}$ to each of the pictures and then form the same linear combination of the results (i.e., $a\mathcal{O}(f) + b\mathcal{O}(g)$). Linear operations on pictures will be discussed further in Section 2.1.1. Point and local operations may or may not be linear. For example, simple stretching of the gray scale ($\mathcal{O}(f) = cf$) is linear, but thresholding ($\mathcal{O}(f) = 1$ if $f \geqslant t$, $= 0$ otherwise) is not; local averaging is linear, but local absolute differencing is not. Geometric operations are linear, if we ignore the need to redigitize the picture after they are performed (Section 9.3).

$\mathcal{O}$ is called *shift invariant* if we get the same output whether we apply $\mathcal{O}$ to a picture and then shift the result, or first shift the picture and then apply $\mathcal{O}$. Such operations will be discussed further in Section 2.1.2. The examples of point and local operations given in the preceding paragraph are all shift invariant, but we can also define shift-variant operations of these types, e.g., modifying the gray level of a point differently, or taking a different weighted average, as a function of position in the picture. The only shift-invariant geometric operations are the shifts, i.e., the translations. It is shown in Section 2.1.2 that an operation is linear and shift invariant iff it is a *convolution*; this is an operation in which the output gray level at a point is a linear combination of the input gray levels, with coefficients that depend only on their positions relative to the given point, but not on their absolute positions.

In Chapters 11 and 12 we will discuss *picture properties*, i.e., operations that can be performed on pictures to produce numerical values. In particular, we will deal with *point* and *local properties* (whose values depend only on one point, or on a small part, of the picture); *geometric properties* of picture subsets (whose values depend only on the set of points belonging to the given subset, but not on their gray levels); and *linear properties* (which give the same value whether we apply them to a linear combination of pictures, or apply them to each picture and then form the same linear combination of the results). It will be shown in Section 12.1.1a that a property is linear and bounded (in a certain sense) iff it is a linear combination of the picture's gray levels.

We will also be interested in certain types of *transforms* of pictures, particularly in their *Fourier transforms*. These are of interest because they make it easier to measure certain types of picture properties, or to perform certain types of operations on pictures, as we will see throughout this book. Basic concepts about (continuous) Fourier transforms are reviewed in Sections 2.1.3 and 2.1.4, and various types of discrete transforms are discussed in Section 2.2.

A GUIDE TO THE LITERATURE

Papers on picture processing and its various applications are being published at a rate of more than a thousand a year. Regular meetings are held on many aspects of the subject, and there are many survey articles, paper collections, meeting proceedings, and journal special issues. Picture processing is also extensively represented in the literature on (two-dimensional) signal processing and pattern recognition (and, to a lesser extent, artificial intelligence).

No attempt has been made here to give a comprehensive bibliography. Selected references are given at the end of each chapter on the subject matter of that chapter, but their purpose is only to cite material that provides further details about the ideas treated in the chapter – it is not practical to cite references on every idea mentioned in the text. Annual bibliographies [9-19] covering some of the non-application-oriented U.S. literature contain several thousand additional references, arranged by subject; they may be consulted for further information.

The principal textbooks on the subject through 1979, including the predecessors to this book, are [1-8, 20]. The following are some of the journals that frequently publish papers on the subject:

IEEE Transactions on Acoustics, Speech and Signal Processing
IEEE Transactions on Communications
IEEE Transactions on Computers
IEEE Transactions on Information Theory
IEEE Transactions on Pattern Analysis and Machine Intelligence
IEEE Transactions on Systems, Man, and Cybernetics
Computer Graphics and Image Processing
Pattern Recognition

We shall not attempt to list the numerous meeting proceedings or paper collections here; see [9-19] for further information.

REFERENCES

1. H. C. Andrews, "Computer Techniques in Image Processing." Academic Press, New York, 1970.
2. K. R. Castleman, "Digital Image Processing." Prentice-Hall, Englewood Cliffs, New Jersey, 1979.
3. R. O. Duda and P. E. Hart, "Pattern Classification and Scene Analysis." Wiley, New York, 1973.
4. R. C. Gonzalez and P. Wintz, "Digital Image Processing." Addison-Wesley, Reading, Massachusetts, 1977.
5. E. L. Hall, "Computer Image Processing and Recognition." Academic Press, New York, 1979.
6. T. Pavlidis, "Structural Pattern Recognition." Springer, New York, 1977.
7. W. K. Pratt, "Digital Image Processing." Wiley, New York, 1978.
8. A. Rosenfeld, "Picture Processing by Computer." Academic Press, New York, 1969.
9. A. Rosenfeld, Picture processing by computer, *Comput. Surveys* **1**, 1969, 147–176.
10. A. Rosenfeld, Progress in picture processing: 1969–71, *Comput. Surveys* **5**, 1973, 81–108.
11. A. Rosenfeld, Picture processing: 1972, *Comput. Graphics Image Processing* **1**, 1972, 394–416.
12. A. Rosenfeld, Picture processing: 1973, *Comput. Graphics Image Processing* **3**, 1974, 178–194.
13. A. Rosenfeld, Picture processing: 1974, *Comput. Graphics Image Processing* **4**, 1975, 133–155.
14. A. Rosenfeld, Picture processing: 1975, *Comput. Graphics Image Processing* **5**, 1976, 215–237.
15. A. Rosenfeld, Picture processing: 1976, *Comput. Graphics Image Processing* **5**, 1977, 157–183.
16. A. Rosenfeld, Picture processing: 1977, *Comput. Graphics Image Processing* **7**, 1978, 211–242.
17. A. Rosenfeld, Picture processing: 1978, *Comput. Graphics Image Processing* **9**, 1979, 354–393.
18. A Rosenfeld, Picture processing: 1979, *Comput. Graphics Image Processing* **12**, 1980, 46–79.
19. A. Rosenfeld, Picture processing: 1980, *Comput. Graphics Image Processing* **16**, 1981, 52–89.
20. A. Rosenfeld and A. C. Kak, "Digital Picture Processing," Academic Press, New York, 1976.

Chapter 9

Matching

There are many situations in which we want to match or register two pictures with one another, or match some given pattern (i.e., piece of picture) with a picture. The following are some common examples:

(a) Given two or more pictures of the same scene taken by different sensors, if we bring them into registration with one another we can determine the characteristics of each pixel with respect to all of the sensors; this can be very useful for classifying the pixels, as discussed in Section 10.1.3.

(b) Given two pictures of a scene taken at different times, if they are registered we can determine the points at which they differ, and thus analyze the changes that have taken place.

(c) Given two pictures of a scene taken from different positions, if we can identify corresponding points in the pictures, we can determine their distances from the camera and thus obtain three-dimensional information about the scene. This process of *stereomapping* will be discussed in Section 9.2.3.

(d) We often want to find places in a picture where it matches a given pattern. The pattern may be very simple (e.g., a spot, a streak, etc.; see Section 10.3 on local feature detection), or it may be relatively complex (e.g., a printed character on a page, a landmark or target on the terrain, etc.). This concept has many applications in pattern recognition ("template matching" for detecting specified patterns), as well as in navigation (landmark or map matching, star pattern matching, etc., for determining the location at which a picture was taken).

Registration of two pictures taken from different positions requires an understanding of imaging geometry, i.e., of how three-dimensional scenes are mapped into two-dimensional pictures. This subject is reviewed in Section 9.1.

In some cases, two pictures of a scene can be registered by applying an overall perspective transformation. More commonly, a simple transformation cannot be used; rather, a transformation must be defined piecewise, based on comparing the positions of corresponding "landmarks" in the two pictures. Position comparisons can also be used to determine the three-dimensional positions of points in the pictures. These topics are treated in Section 9.2.

The problems involved in applying geometrical transformations to digital pictures are discussed in Section 9.3. Finally, measurement of the match or mismatch between two given pictures is treated in Section 9.4.

9.1 IMAGING GEOMETRY

This section discusses the mapping of a three-dimensional scene into a two-dimensional picture from the standpoint of the coordinate transformations that are involved. Subsection 9.1.1 reviews the transformations that map one Cartesian coordinate system into another—translation, rotation, scale change, and reflection. Using such transformations, we can convert any given system of scene coordinates to a standardized system of "camera coordinates," with respect to which the mapping of the scene into the picture has an especially simple form. Some of the properties of this mapping are discussed in Subsection 9.1.2; in particular, we show which scene points map into a given image point. Subsection 9.1.3 discusses some additional properties of the imaging transformation, involving images of straight lines and families of lines.

The treatment of imaging geometry in this section assumes that we are dealing with a simple optical imaging system, such as a camera, that is free of optical aberrations, so that a plane at constant distance from the camera is imaged at a constant scale factor. Other types of sensors may give rise to other imaging geometries; for example, in uncorrected images obtained from radar or from mechanically scanning sensors, distance in the image may be a quadratic or trigonometric function, rather than a linear function, of distance in the scene. Such nonlinear imaging geometries will not be discussed further here.

9.1.1 Coordinate Transformations

a. Two dimensions

For simplicity, we first review two-dimensional coordinate transformations. Let (x, y) and (x', y') be any two Cartesian coordinate systems in the plane, as shown in Fig. 1. Given the coordinates (x_p, y_p) of a point P in the (x, y) system, we want to find the coordinates (x_p', y_p') of P in the (x', y') system. It can be shown that this can always be done by applying a sequence of elementary transformations to (x_p, y_p); by an elementary transformation we mean a translation, a rotation, or a scale change (or reflection).

A *translation* by (a, b), illustrated in Fig. 2, has the equations

$$x_p' = x_p - a, \qquad y_p' = y_p - b \tag{1}$$

A *rotation* (counterclockwise about the origin) through the angle θ, shown in Fig. 3, has the equations

$$x_p' = x_p \cos \theta + y_p \sin \theta, \qquad y_p' = -x_p \sin \theta + y_p \cos \theta \tag{2}$$

A *scale change* of r in x and of s in y, illustrated in Fig. 4, has the equations

$$x_p' = rx_p, \qquad y_p' = sy_p \tag{3}$$

A *reflection* in the x-axis has the equations $x_p' = x_p$, $y_p' = -y_p$; a reflection in the y-axis has the equations $x_p' = -x_p$, $y_p' = y_p$. We can thus regard such reflections as special cases of scale changes, in which r or s is -1.

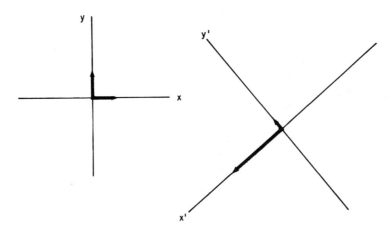

Fig. 1 Cartesian coordinate systems in two dimensions, showing unit vectors in the positive x- and y- (or x'- and y'-) directions. These two systems differ by translation, rotation, scale change, and reflection.

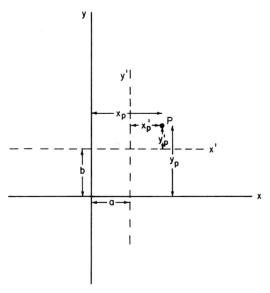

Fig. 2   Translation by (a,b). $x_p' = x_p - a$; $y_p' = y_p - b$.

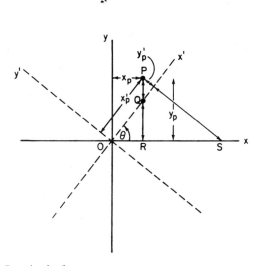

Fig. 3   Rotation by θ:

$$\overline{OS} = x_p + \overline{RS},$$
$$x_p' = \overline{OS} \cos \theta,$$
$$\overline{RS} = y_p \tan \theta,$$
$$\therefore x_p' = x_p \cos \theta + y_p \tan \theta \cos \theta$$
$$= x_p \cos \theta + y_p \sin \theta$$

$$\overline{PQ} = y_p - \overline{QR},$$
$$y_p' = \overline{PQ} \cos \theta,$$
$$\overline{QR} = x_p \tan \theta,$$
$$\therefore y_p' = y_p \cos \theta - x_p \tan \theta \cos \theta$$
$$= -x_p \sin \theta + y_p \cos \theta$$

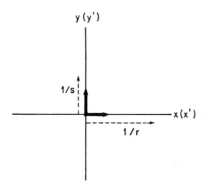

Fig. 4   Scale change by (r, s). $x_p' = rx_p$;
$y_p' = sy_p$.

Suppose that we know the position, orientation, and scale of one coordinate system relative to the other. Then we can easily determine the transformation that takes the coordinates of a point P in the first system into the coordinates of P in the second system. This is done as follows: Suppose that the origin of the (x', y') system is at position (A, B) relative to the (x, y) system; that the scale factors of x' and y' are λ and μ times those of x and y, respectively; and that the x'-axis makes counterclockwise angle θ with the x-axis. Let (x_p, y_p) be the coordinates of P in the unprimed system. Let (x_1, y_1) be the coordinate system in which the scale and orientation are the same as in the (x, y) system, but the origin is at (A, B); then the coordinates (x_{1p}, y_{1p}) of P in the (x_1, y_1) system are evidently $(x_p - A, y_p - B)$. Let (x_2, y_2) be the coordinate system obtained from (x_1, y_1) by rescaling by λ and μ, respectively, while leaving the position and orientation unchanged; then the coordinates (x_{2p}, y_{2p}) of P in the (x_2, y_2) system are evidently $(\lambda x_{1p}, \mu y_{1p})$. The coordinate system (x', y') is clearly obtained from (x_2, y_2) by a counterclockwise rotation through θ around the origin; hence

$$(x_p', y_p') = (x_{2p} \cos \theta + y_{2p} \sin \theta, \, -x_{2p} \sin \theta + y_{2p} \cos \theta).$$

In summary, we have

$$
\begin{aligned}
x_p' &= x_{2p} \cos \theta + y_{2p} \sin \theta \\
     &= \lambda x_{1p} \cos \theta + \mu y_{1p} \sin \theta \\
     &= \lambda(x_p - A) \cos \theta + \mu(y_p - B) \sin \theta \\
y_p' &= -x_{2p} \sin \theta + y_{2p} \cos \theta \\
     &= -\lambda x_{1p} \sin \theta + \mu y_{1p} \cos \theta \\
     &= -\lambda(x_p - A) \sin \theta + \mu(y_p - B) \cos \theta
\end{aligned}
\tag{4}
$$

Exercise 9.1.   Show by example that translation and rotation do not commute—in other words, the result of a given translation followed by a given rotation is not the same as the result of performing the same operations in the opposite order. Similarly, show that translation and scale change do not commute; but prove that rotation and scale change do commute. ∎

b. Three dimensions

Analogously, let (x, y, z) and (x', y', z') be any two Cartesian coordinate systems in three-dimensional space (Fig. 5). Here again, the coordinates (x_p', y_p', z_p') of any point P in the primed system can be computed in terms of those in the unprimed system by applying a sequence of elementary transformations: translations, rotations, and rescalings (or reflections). The equations of these transformations are as follows:

Translation by (a, b, c):

$$x_p' = x_p - a$$
$$y_p' = y_p - b \qquad (5)$$
$$z_p' = z_p - c$$

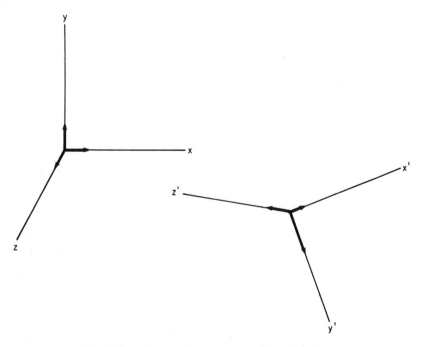

Fig. 5 Cartesian coordinate systems in three dimensions.

Rotation through angle θ:

(1)   About the *z*-axis:

$$x_p' = x_p \cos \theta + y_p \sin \theta$$
$$y_p' = -x_p \sin \theta + y_p \cos \theta \tag{6a}$$
$$z_p' = z_p.$$

(2)   About the *x*-axis:

$$x_p' = x_p$$
$$y_p' = y_p \cos \theta + z_p \sin \theta \tag{6b}$$
$$z_p' = -y_p \sin \theta + z_p \cos \theta$$

(3)   About the *y*-axis:

$$x_p' = x_p \cos \theta - z_p \sin \theta$$
$$y_p' = y_p \tag{6c}$$
$$z_p' = x_p \sin \theta + z_p \cos \theta$$

[In each case, the rotation is assumed to be counterclockwise from the viewpoint of an observer who is looking along the positive *z*- (or *x*-, or *y*-) axis toward the origin.]

Scale change by (r, s, t):

$$x_p' = rx_p$$
$$y_p' = sy_p \tag{7}$$
$$z_p' = tz_p$$

Reflection in the *xy*-plane:

$$x_p' = x_p, \qquad y_p' = y_p, \qquad z_p' = -z_p$$

in the *yz*-plane:

$$x_p' = -x_p, \qquad y_p' = y_p, \qquad z_p' = z_p$$

in the *zx*-plane:

$$x_p' = x_p, \qquad y_p' = -y_p, \qquad z_p' = z_p$$

As in the two-dimensional case, we can determine how to transform the coordinates of *P* from the unprimed to the primed system if we know the position, scale, and orientation of the latter system relative to the former. Specifically, suppose that this position is (A, B, C); that the scale factors are (λ, μ, ν); and that the direction cosines of the *z'*-axis relative to the unprimed axes are $\cos \alpha$, $\cos \beta$, $\cos \gamma$ (Fig. 6). We then proceed as follows:

(1)   Translate the (x, y, z) system by (A, B, C) to obtain system (x_1, y_1, z_1); this has the same origin as (x', y', z').

(2)   Rescale the (x_1, y_1, z_1) system by (λ, μ, v) to obtain system (x_2, y_2, z_2); this has the same scale as (x', y', z').

(3)   Rotate the (x_2, y_2, z_2) system counterclockwise about the z_2-axis to obtain system (x_3, y_3, z_3) having the z'-axis in the $x_3 z_3$-plane. The rotation angle φ can be derived as follows (see Fig. 6): Let z'' be the projection of the z'-axis onto the xy- (or $x_2 y_2$-) plane. Then the angle φ between x and z'' is given by $\cos \varphi = \cos \alpha / \sin \gamma$.

(4)   Finally, if we rotate the (x_3, y_3, z_3) system counterclockwise through angle γ about the y_3-axis, we obtain the (x', y', z') system.

In summary, using (5), (6a), (6c), and (7), we have

$$
\begin{aligned}
x_p' &= x_{3p} \cos \gamma - z_{3p} \sin \gamma \\
&= (x_{2p}\cos \varphi + y_{2p} \sin \varphi) \cos \gamma - z_{2p} \sin \gamma \\
&= (\lambda x_{1p} \cos \varphi + \mu y_{1p} \sin \varphi) \cos \gamma - v z_{1p} \sin \gamma \\
&= \lambda(x_p - A) \cos \varphi \cos \gamma + \mu(y_p - B) \sin \varphi \cos \gamma - v(z_p - C) \sin \gamma \\
y_p' &= y_{3p} = -x_{2p} \sin \varphi + y_{2p} \cos \varphi \\
&= -\lambda x_{1p} \sin \varphi + \mu y_{1p} \cos \varphi \\
&= -\lambda(x_p - A) \sin \varphi + \mu(y_p - B) \cos \varphi \\
z_p' &= x_{3p} \sin \gamma + z_{3p} \cos \gamma \\
&= (x_{2p} \cos \varphi + y_{2p} \sin \varphi) \sin \gamma + z_{2p} \cos \gamma \\
&= \lambda x_{1p} \cos \alpha + \mu y_{1p} \sin \varphi \sin \gamma + v z_{1p} \cos \gamma \\
&= \lambda(x_p - A) \cos \alpha + \mu(y_p - B) \sin \varphi \sin \gamma + v(z_p - C) \cos \gamma
\end{aligned}
\tag{8}
$$

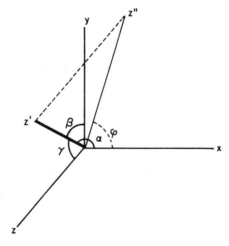

Fig. 6   Angular relationships between the z'-axis and the x-, y-, and z-axes.

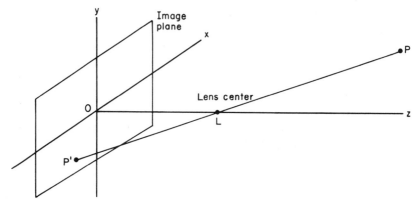

Fig. 7 Camera coordinates.

Exercise 9.2. Show by example that rotations about different axes do not commute. ∎

9.1.2 The Imaging Transformation

In the process of optical imaging, the given scene is mapped onto the image plane by projection through a point, called the optical center of the imaging lens. Let us define the *camera coordinate system* as having xy-plane in the image plane, and z-axis ("optical axis") through the lens center, as shown in Fig. 7. Thus the center of the film is at the origin O, and the lens center L is at $(0, 0, f)$, say. (If the camera is in focus for distant objects, f is the focal length of the lens.)[§]

Let P be any point of the scene, say with coordinates (x_0, y_0, z_0) in the camera system. We will assume that $z_0 > f$; this corresponds to the assumption that P is in front of the camera, rather than alongside or behind it. Then P', the image of P, is the point at which line PL hits the image plane. Let the coordinates of P' be $(x_i, y_i, 0)$. Then by similar triangles we have

$$\frac{x_i}{f} = -\frac{x_0}{z_0 - f}, \qquad \frac{y_i}{f} = -\frac{y_0}{z_0 - f} \tag{9}$$

(The negative signs indicate that the image is inverted.) If we want to use a different coordinate system for the scene points, and we know the position and orientation of the camera in the scene coordinate system, we can use (8) to find the camera coordinates of any scene point P (given its scene coordinates), and then use (9) to find the coordinates of P's image on the film.

[§] In Sections 9.1 and 9.2, f denotes focal length; elsewhere in this book, f is a picture.

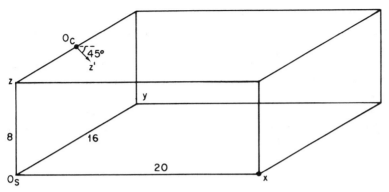

Fig. 8 Example illustrating computation of image coordinates from scene coordinates when camera position and orientation in the scene are known.

To illustrate this, suppose that the scene is a room $20' \times 16' \times 8'$ with the floor as xy-plane, and that the camera is located at the center of a $16'$ wall where it meets the ceiling, and points downward at 45°, parallel to the xz-plane, as shown in Fig. 8. Thus the origin O_c of the camera coordinate system is at point $(0, 8, 8)$ in the scene coordinate system, and the direction cosines of the optical axis relative to the scene axes are $(\sqrt{2}/2, 0, -\sqrt{2}/2)$ (i.e., the optical axis is perpendicular to the y-axis and makes angles of 45° and 135° with the x- and z-axes, respectively). We thus have $\cos \gamma = -\sqrt{2}/2$, $\cos \alpha = \sin \gamma = \sqrt{2}/2$, $\cos \varphi = 1$, $\sin \varphi = 0$. Given a point P in the room, we can thus use (8) to find its camera coordinates:

$$x_p' = -x_p \frac{\sqrt{2}}{2} - (z_p - 8) \frac{\sqrt{2}}{2}, \qquad y_p' = (y_p - 8),$$

$$z_p' = x_p \frac{\sqrt{2}}{2} - (z_p - 8) \frac{\sqrt{2}}{2}$$

For example, the room corner $(20, 0, 0)$ has camera coordinates $(-6\sqrt{2}, -8, 14\sqrt{2})$. Let $f = \sqrt{2}$; then using (9), we find that the film coordinates (x_i, y_i) of this point are

$$x_i = \frac{12}{13\sqrt{2}} = \frac{6\sqrt{2}}{13}, \qquad y_i = \frac{8\sqrt{2}}{13\sqrt{2}} = \frac{8}{13}$$

Conversely, suppose we know the image coordinates of three points P_1, P_2, P_3 whose scene coordinates are known; then we can find the position (A, B, C) and orientation $(\cos \alpha, \cos \beta, \cos \gamma)$ of the camera in the scene coordinate system. In fact, we can use (8) and (9) to find the image coordinates of the P_i's in terms of their scene coordinates and of the parameters

A, B, C, α, β, γ, and f. Each P_i thus gives us two equations in these parameters, so that three P's yield a total of six equations in the six independent parameters (α, β, and γ are not independent, since $\cos^2 \alpha + \cos^2 \beta + \cos^2 \gamma = 1$). If we have more than three P's, we can find a set of values for the parameters that satisfies the six equations in a least squares sense. The details will not be given here.

The mapping from scene onto image is many-to-one. The image point (x_i, y_i) corresponds to a set of collinear points in the scene—namely, all the points that lie on the line through $(x_i, y_i, 0)$ amd $(0, 0, f)$. The equations of this line are

$$x = -\frac{x_i}{f}(z - f), \qquad y = -\frac{y_i}{f}(z - f) \tag{10}$$

If we know nothing about the scene coordinates of a given image point, we can only say that it lies somewhere along this line. On the other hand, if we do have additional information, e.g., that the scene point is at a given distance from the camera (i.e., z is known), then the position of the point is uniquely determined.

Exercise 9.3. Let the scene contain a plane whose equation, in camera coordinates, is $ax + by + cz + d = 0$. Which point of this plane corresponds to a given image point? ∎

9.1.3 Properties of Imaging

a. Straight lines map into straight lines

Any straight line in the scene maps into a straight line in the image. To see this, consider the line l whose equations are $(x - x_0)/a = (y - y_0)/b = (z - z_0)/c$, which passes through the point (x_0, y_0, z_0). Then l and the point $(0, 0, f)$ determine a plane Π, and the image l' of l must lie in the intersection of Π with the image plane, so that l' must be a straight line [or a point, if l passes through $(0, 0, f)$].

To find the equation of l', let the equation of Π be $Ax + By + Cz = D$. Since $(0, 0, f)$ is in Π, we have $D = Cf$; and since (x_0, y_0, z_0) is in Π, we have $Ax_0 + By_0 + Cz_0 = Cf$. Finally, since Π contains the line l, we have $Aa + Bb + Cc = 0$. We can solve these last two equations for A and B in terms of C to obtain

$$A = -C \begin{vmatrix} z_0 - f & y_0 \\ c & b \end{vmatrix} \bigg/ \begin{vmatrix} x_0 & y_0 \\ a & b \end{vmatrix}, \qquad B = -C \begin{vmatrix} x_0 & z_0 - f \\ a & c \end{vmatrix} \bigg/ \begin{vmatrix} x_0 & y_0 \\ a & b \end{vmatrix}$$

Then l' is the line $Ax + By = Cf$ in the image plane; in other words, it is the line

$$\begin{vmatrix} z_0 - f & y_0 \\ c & b \end{vmatrix} x + \begin{vmatrix} x_0 & z_0 - f \\ a & c \end{vmatrix} y + f \begin{vmatrix} x_0 & y_0 \\ a & b \end{vmatrix} = 0 \qquad (11)$$

Conversely, let l' be a line in the image, and let (x_1, y_1), (x_2, y_2) be any points on l'. Then l' is the image of the plane through the three points $(x_1, y_1, 0)$, $(x_2, y_2, 0)$, and $(0, 0, f)$; readily, the equation of this plane is

$$(y_2 - y_1)x - (x_2 - x_1)y + (x_1 y_2 - x_2 y_1)z/f = (x_1 y_2 - x_2 y_1)$$

Moreover, l' is the image of any line in this plane except for lines that pass through $(0, 0, f)$, which may map into single points of l'.

Exercise 9.4. Let l be a line parallel to the image plane; prove that the image of l is parallel to l. ∎

b. Vanishing points

Let $\mathscr{L}$ be a family of straight lines all of which pass through a common point P. It follows that the images of the lines in $\mathscr{L}$ must evidently all pass through the image P' of P. If P is in the plane $z = f$, then by (11) with $(x_0, y_0, z_0) = P$, the images of the lines are all of the form

$$-y_0 cx + x_0 cy + f \begin{vmatrix} x_0 & y_0 \\ a & b \end{vmatrix} = 0$$

so that they all have slope y_0/x_0, independent of a, b, and c; thus in this special case, the images are all parallel (provided x_0 and y_0 are not both zero). We thus see that a family of concurrent straight lines maps into a family of concurrent or parallel straight lines under the imaging mapping.

Now let $\mathscr{L}$ be a family of *parallel* straight lines, say all of the form $(x - x_0)/a = (y - y_0)/b = (z - z_0)/c$. The line l_0 of this form that passes through $(0, 0, f)$ is $x/a = y/b = (z - f)/c$; it intersects the image plane $z = 0$ at $x = -af/c, y = -bf/c$. (Indeed, the entire line l_0 maps into this point P_0 under the imaging transformation.) In fact, P_0 is on the image of *every* line in $\mathscr{L}$;[§] we can verify this by setting $x = -af/c, y = -bf/c$ in the left side of (11) to obtain

$$\begin{vmatrix} z_0 - f & y_0 \\ c & b \end{vmatrix} a + \begin{vmatrix} x_0 & z_0 - f \\ a & c \end{vmatrix} b - \begin{vmatrix} x_0 & y_0 \\ a & b \end{vmatrix} c = a[bz_0 - bf - cy_0]$$

$$+ b[cx_0 - az_0 + af] - c[bx_0 - ay_0] = 0$$

[§] It is not hard to verify that, on any line in $\mathscr{L}$, as the coordinates of a point become very large, the image of the point approaches P_0.

Thus a family of parallel straight lines in the scene maps into a family of concurrent straight lines in the image. We call P_0 the *vanishing point* for $\mathcal{L}$. This argument breaks down if P_0 is not defined, which implies that the lines are all parallel to the xy-plane; in this case it is not hard to see that they map into a family of parallel lines in the image. In fact, the lines of the form $(x - x_0)/a = (y - y_0)/b$ (where $z = z_0$) map into the lines parallel to the line $x/a = y/b$ in the image plane.

To illustrate the results just obtained about images of families of parallel lines, let us first consider the family of lines parallel to the y-axis, for which we have $a = c = 0, b = 1$. These vertical lines evidently all map into the vertical lines ($x = $ constant) in the image plane. On the other hand, consider any family of lines perpendicular to the y-axis, i.e., parallel to the xz-plane, so that $b = 0$. Any such family of horizontal lines has its vanishing point on the x-axis of the image ($x = -af/c, y = 0$). As the lines become parallel to the z-axis ($a \to 0$), the vanishing point approaches the origin; as they become parallel to the x-axis ($c \to 0$), it recedes to infinity.

These remarks can be summarized as follows: if the camera is in a room pointing parallel to the floor, all vertical straight lines appear vertical on the image. On the other hand, any family of parallel straight lines in a horizontal plane maps into a family of concurrent lines on the image, with vanishing point on the image's x-axis.

Exercise 9.5. Let l_1 and l_2 be two perpendicular horizontal lines, and let their vanishing points be x_1 and x_2. Prove that $x_1 x_2 = -f^2$. ∎

c. Cross ratio

Let $(x_k, y_k, z_k), k = 1, 2, 3, 4$ be four points on the line $l: (x - x_0)/a = (y - y_0)/b = (z - z_0)/c$. Let $(x_{ik}, y_{ik}, z_{ik}), k = 1, 2, 3, 4$ be the images of these points. We will show in the next paragraph that the imaging transformation preserves the *cross ratio* of the coordinates of these points, i.e., that

$$\frac{(x_{i3} - x_{i1})(x_{i4} - x_{i2})}{(x_{i3} - x_{i2})(x_{i4} - x_{i1})} = \frac{(x_3 - x_1)(x_4 - x_2)}{(x_3 - x_2)(x_4 - x_1)}$$

and similarly for the y's. If we know the coordinates of three collinear points, this formula can be used to compute the coordinates of any other point on the same line from its coordinates in the image.

To see this, note first that for each k we have $z_k = (c/a)(x_k - x_0) + z_0$. But from (9) we know that $f/x_{ik} + (z_k - f)/x_k = 0$; hence

$$\frac{f}{x_{ik}} + \frac{z_0 - f - cx_0/a}{x_k} + \frac{c}{a} = 0$$

which we can write in the form $u/x_{ik} + v/x_k + w = 0$.

Subtracting these equations for two different k's, say 1 and 3, we get

$$\frac{u(x_{i3} - x_{i1})}{x_{i3} x_{i1}} = -\frac{v(x_3 - x_1)}{x_3 x_1}$$

and similarly for $k = 2$ and 4, 2 and 3, and 1 and 4. Dividing the product of the first two equations by the product of the last two cancels u^2, v^2, $x_{i1} x_{i2} x_{i3} x_{i4}$, and $x_1 x_2 x_3 x_4$ and leaves us with the desired cross-ratio expression.

9.2 REGISTRATION

If a scene is relatively flat, two pictures of it taken from different positions can be registered by applying a perspective transformation. This method can also be used to remove perspective distortion from a picture of such a scene. The transformations involved in this process are derived in Subsection 9.2.1.

In general, two pictures of the same scene cannot be registered, and the distortion in a single picture cannot be corrected, by applying a simple transformation to the entire picture(s), since objects at different distances transform differently under the imaging process. Rather, it is necessary to pair off corresponding "landmark" points in the two pictures, or to identify points in the single picture whose positions in the scene are known; based on this information, the registration or distortion correction transformation can be defined piecewise. Arbitrary (nonperspective) distortions can also be corrected in this way. This subject is discussed in Subsection 9.2.2.

If we can identify corresponding points in two pictures of a given scene taken from different positions, we can determine the three-dimensional positions of those points, and can correct their perspective distortion. These processes of *stereomapping* and *orthophoto* construction are treated briefly in Subsection 9.2.3.

9.2.1 Rectification

Suppose that the camera is pointed at a flat surface S; this is approximately true in remote sensing, where we may regard the terrain as essentially flat if the sensor is at a very high altitude. If S is perpendicular to the camera axis (i.e., parallel to the image plane), it is mapped onto the image by a simple scale change, without distortion. In fact, if S is the plane $z = h$, by (9) we have

$$x_i = \frac{-x_0 f}{h - f}, \qquad y_i = \frac{-y_0 f}{h - f}$$

so that the image is a rescaling of S by the factor $-f/(h - f)$. On the other hand, if S is tilted (equivalently: if the camera is not pointing directly toward S), the image is distorted.

A simple way of describing this *perspective distortion* is to consider S as composed of lines, each of which is parallel to the image plane. Specifically, let S be the plane

$$x \cos \alpha + y \cos \beta + z = h$$

so that the intersection of S with the camera axis ($x = y = 0$) is at distance $z = h$ from the camera. Then the line $x \cos \alpha + y \cos \beta = 0$, $z = h$ lies in S and is at constant distance h from the camera, so that its image is scaled by the factor $-f/(h - f)$. Similarly, the lines in S parallel to this line are all at constant distances from the camera; in particular, the line $x \cos \alpha + y \cos \beta = h'$, $z = h - h'$ lies in S and is at distance $h - h'$ from the camera, so that its image is scaled by the factor $-f/(h - h' - f)$.

Let l be the line $x \cos \alpha + y \cos \beta = h'$, $z = h - h'$. Then the image P' of an arbitrary point $P = (x, y)$ on l has coordinates

$$(x', y') = \left(\frac{-xf}{h - h' - f}, \frac{-yf}{h - h' - f} \right)$$

Since $x \cos \alpha + y \cos \beta = h'$, the coordinates of P' satisfy $x' \cos \alpha + y' \cos \beta = -h'f/(h - h' - f)$; thus the image l' of l is the line $x \cos \alpha + y \cos \beta = -h'f/(h - h' - f)$ in the image plane. In particular, taking $h' = 0$, we find that the line $l_0 : (x \cos \alpha + y \cos \beta = 0, z = h)$ has as image the line $l_0' : x \cos \alpha + y \cos \beta = 0$.

Suppose that we want to *rectify* the image of S, e.g., so that it becomes a simple rescaling of S by the constant factor $-f/(h - f)$. For any point (x_i, y_i) in the image, the corresponding point (x_0, y_0, z_0) on S satisfies

$$fx_0 = (f - z_0)x_i, \qquad fy_0 = (f - z_0)y_i$$
$$x_0 \cos \alpha + y_0 \cos \beta + z_0 = h$$

If we let $Z = f - x_i \cos \alpha - y_i \cos \beta$, then readily we have

$$x_0 = (f - h)x_i/Z, \qquad y_0 = (f - h)y_i/Z, \qquad z_0 = f - f(f - h)/Z$$

Rescaling by the factor $-f/(h - f) = f/(f - h)$ thus maps x_0 and y_0 into $fx_i/Z, fy_i/Z$. Hence to rectify the image, we should map (x_i, y_i) into $(fx_i/Z, fy_i/Z)$, where $Z = f - x_i \cos \alpha - y_i \cos \beta$. Note that this mapping is independent of h; only the resulting scale factor depends on h. An example of perspective distortion is shown in Fig. 9.

Fig. 9 Oblique photograph illustrating perspective distortion.

Suppose now that we take two pictures of S from different positions and orientations. Since there are two camera positions, it is convenient to use a coordinate system related to S, and to express both positions in terms of this system. Specifically, we shall use the system in which S is the xy-plane $(z = 0)$. Let the two camera positions in this (x, y, z) system be (A_1, B_1, C_1) and (A_2, B_2, C_2), and let the orientations of the camera axis $(z_1$ or $z_2)$ relative to the (x, y, z)-axes be $(\cos \alpha_1, \cos \beta_1, \cos \gamma_1)$ and $(\cos \alpha_2, \cos \beta_2, \cos \gamma_2)$. Then the camera coordinates (x_1, y_1, z_1) and (x_2, y_2, z_2) of any given point $P = (x_0, y_0, 0)$ of S can be found by using (8) with $\lambda = \mu = \nu = 1$. They are

$$x_1 = (x_0 - A) \cos \varphi_1 \cos \gamma_1 + (y_0 - B) \sin \varphi_1 \cos \gamma_1 + C \sin \gamma_1$$
$$y_1 = -(x_0 - A) \sin \varphi_1 + (y_0 - B) \cos \varphi_1 \qquad (12a)$$
$$z_1 = (x_0 - A) \cos \alpha_1 + (y_0 - B) \sin \varphi_1 \sin \gamma_1 - C \cos \gamma_1$$

where $\cos \varphi_1 = \cos \alpha_1 / \sin \gamma_1$, and analogously for (x_2, y_2, z_2). $\qquad (12b)$

To register the first picture of S with the second, we proceed as follows:

(a) Let (x_{i2}, y_{i2}) be the image plane coordinates of any point in the second picture.

(b) By (10), this point is the image of the point P of S whose coordinates (x_2, y_2, z_2) satisfy

$$x_2 = -x_{i2}(z_2 - f)/f, \qquad y_2 = -y_{i2}(z_2 - f)/f \qquad (13)$$

(c)   To find the coordinates (x_0, y_0, z_0) of P in the (x, y, z) system, we eliminate x_2, y_2, z_2 from the five equations (12b) and (13), and solve for (x_0, y_0).

(d)   We then obtain the coordinates of P in the (x_1, y_1, z_1) system from (12a).

(e)   Finally, using (9), we obtain the image plane coordinates of the image of P in the first picture.

Note that the rectification transformations derived in this section do not, in general, map digital picture points into digital picture points. The application of geometrical transformations to digital pictures will be discussed in Section 9.3.

9.2.2   Landmark Matching

Pictures are subject to many different types of geometrical distortion, aside from perspective distortion. When the scene is not flat, perspective distortion itself varies from point to point; its removal under these conditions will be discussed in Section 9.2.3. Other types of distortion are due to limitations of optical imaging or electronic scanning systems. In *pincushion* distortion, the scale of the picture increases (slightly) with distance from the center; as a result, a line that does not pass through the center bends inward toward the midpoint since the displacement of its center is less than that of its ends. In *barrel* distortion, the scale decreases with distance from the center, so that lines bend outward. An example of barrel distortion is shown in Fig. 10a.

We can remove geometrical distortion from a picture, at least approximately, if the picture contains a known pattern, such as a regular grid, or a set of landmarks whose positions in the scene are known. To do this, we must find a transformation that maps the grid points (or landmarks) into their observed (distorted) positions. This can be done piecewise, by taking a few of the grid points at a time, and solving for the coefficients of a transformation of some given form that takes the known (ideal) coordinates of these points into their observed coordinates.

As a simple illustration of this idea, suppose we are given a distorted picture f that contains an (ideally) square grid, where we know that the horizontal and vertical spacing of the grid is supposed to be k. Let us take three grid points at a time, say with ideal coordinates (x, y), $(x + k, y)$, and $(x, y + k)$. Let the observed coordinates of these points in f be (u_1, v_1), (u_2, v_2), and (u_3, v_3). Then we can find a linear transformation of the form

$$x' = Ax + By + C, \qquad y' = Dx + Ey + F \qquad (14)$$

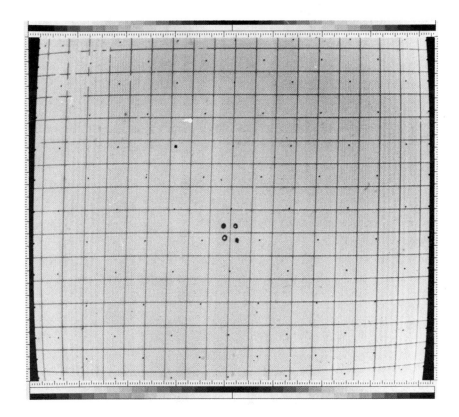

(a)

Fig. 10a Geometric correction. (a) Mariner 9 grid target, showing geometrical distortion.

that takes the ideal coordinates into the observed ones, by solving the six linear equations

$$u_1 = Ax + By + C, \qquad v_1 = Dx + Ey + F,$$
$$u_2 = A(x + k) + By + C, \qquad v_2 = D(x + k) + Ey + F,$$
$$u_3 = Ax + B(y + k) + C, \qquad v_3 = Dx + E(y + k) + F$$

for the six coefficients A, B, C, D, E, F. Readily, the solution is

$$A = (u_2 - u_1)/k, \qquad B = (u_3 - u_1)/k,$$
$$C = [u_1(x + y + k) - xu_2 - yu_3]/k, \qquad D = (v_2 - v_1)/k, \qquad (15)$$
$$E = (v_3 - v_1)/k, \qquad F = [v_1(x + y + k) - xv_2 - yv_2]/k$$

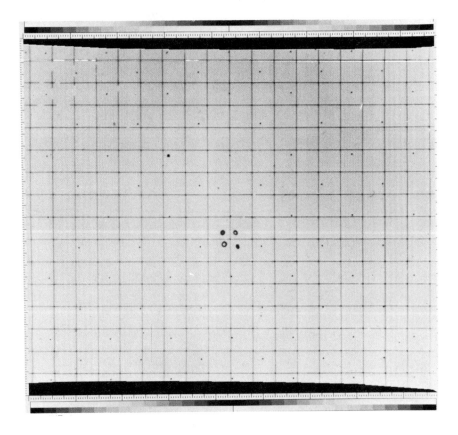

(b)

Fig. 10b Geometric correction. (b) Results of distortion removal. (From [18].)

The transformation (14), with coefficients given by (15), is an approximation to the distorting transformation for the triangular portion T of f bounded by the three points (u_1, v_1), (u_2, v_2), and (u_3, v_3). Thus it is the inverse of the transformation that is needed to correct the distortion of T. We shall see in Section 9.3 how to use this transformation to compute the gray levels in order to construct an undistorted version of T.

Analogous transformations can be applied to other triangular pieces of the grid. Note that since these transformations are linear, they take straight lines into straight lines, and so map the triangles into triangles. Thus for two adjacent triangles T_1 and T_2, having a common side, the transformations will agree exactly all along that side, since they agree at the end points.

Rather than the piecewise linear correction scheme just described, we can use a higher-order scheme involving local polynomial transformations of the form $x' = h_1(x, y)$, $y' = h_2(x, y)$, where h_1 and h_2 are polynomials. Given that such a transformation takes the ideal points $(x_1, y_1), \ldots, (x_n, y_n)$ into the observed points $(u_1, v_1), \ldots, (u_n, v_n)$, we can solve the $2n$ linear equations $u_i = h_1(x_i, y_i)$, $v_i = h_2(x_i, y_i)$, $1 \leq i \leq n$, for the coefficients of h_1 and h_2. If there are fewer than $2n$ of these coefficients, we can still find a best-fitting solution in the least squares sense.

An example of a distorted picture that has been corrected by piecewise transformation is shown in Fig. 10b.

9.2.3   Stereomapping

Let P be a scene point, and let P_1, P_2 be the images of P in two pictures obtained from two different (known) camera positions. Given P_1 and P_2, we know by (10) that P lies on a specific line in each camera coordinate system. We may assume, as in Section 9.2.1, that we know how to express the camera coordinates in terms of a common coordinate system (x, y, z), and vice versa. Thus we can find the equations of the two lines in the (x, y, z) system, and P is then at their intersection. In other words, given that P_1 and P_2 are images of the same point P, we can find the coordinates of P in the (x, y, z) system in terms of the image plane coordinates of P_1 and P_2 in the two pictures, provided the camera positions are known. [If they are only known relative to one another, we can take one of them as defining the (x, y, z) system.]

We shall not go through the details of this computation in the general case, but shall only treat a special case which is often encountered—namely, that in which the two camera positions differ only in x and y, but not in z or orientation. [For example, in remote sensing these conditions are satisfied when the camera remains at the same altitude and pointing (say) vertically downward.] In this case the two camera coordinate systems (x_1, y_1, z_1) and (x_2, y_2, z_2) differ only by a translation of the form

$$x_2 = x_1 - A, \qquad y_2 = y_1 - B, \qquad z_2 = z_1$$

Let $P_1 = (x_{i1}, y_{i1})$, $P_2 = (x_{i2}, y_{i2})$. Thus in the (x_1, y_1, z_1) system, P lies on the line

$$x_1 = -x_{i1}(z_1 - f)/f, \qquad y_1 = -y_{i1}(z_1 - f)/f \tag{16}$$

and in the (x_2, y_2, z_2) system, it lies on the line

$$x_2 = -x_{i2}(z_2 - f)/f, \qquad y_2 = -y_{i2}(z_2 - f)/f$$

The equations of the latter line in the (x_1, y_1, z_1) system are

$$x_1 = A - x_{i2}(z_1 - f)/f, \qquad y_1 = B - y_{i2}(z_1 - f)/f \qquad (17)$$

We can solve either the first or the second equations in (16) and (17) to find z_1, the "depth" of P from the camera (as measured parallel to the camera axis):

$$z_1 = f\left[1 + \frac{A}{x_{i2} - x_{i1}}\right] = f\left[1 + \frac{B}{y_{i2} - y_{i1}}\right] \qquad (18)$$

As an example, let $f = 1$, let the coordinates of P be $(0, 0, 10)$, and let the camera positions be $(0, 0, 0)$ and $(-2, 0, 0)$, so that $A = 2, B = 0$. Then readily $(x_{i1}, y_{i1}) = (0, 0)$ and $(x_{i2}, y_{i2}) = (\frac{2}{9}, 0)$. In this case we use the first equation of (18) to verify that

$$z_1 = \left[1 + \frac{2}{\frac{2}{9}}\right] = 1 + 9 = 10$$

The relative displacement of P_1 and P_2 on the two pictures is called *stereoscopic parallax*. If every scene point had the same z value, we see from (18) that this displacement would be constant; in fact, we would have

$$x_{i2} - x_{i1} = Af/(z - f), \qquad y_{i2} - y_{i1} = Bf/(z - f)$$

It is the variation in z that causes the displacement to vary. Note that the greater z, the smaller the displacement, and vice versa.

If we have two pictures of a scene, and know which pairs of points in the two pictures are the images of the same scene point, we can in principle construct a complete three-dimensional model of the scene. (In practice, the difficulty arises that some scene points are visible in one picture but not in the other, due to occlusion; but we can still obtain three-dimensional coordinates for many of the scene points.) As an application of this, we can construct a "map" of the scene in which each point is displayed as though it were viewed from directly above. In other words, given an (x, y, z) scene coordinate system, we can display each point at a position corresponding to its x- and y-coordinates. Such a display is sometimes called an *orthophoto*.

For the case treated in this section, we have from (16) and (18)

$$z_1 = f\left[1 + \frac{A}{x_{i2} - x_{i1}}\right] = f\left[1 + \frac{B}{y_{i2} - y_{i1}}\right]$$

$$(z_1 - f)/f = \frac{A}{x_{i2} - x_{i1}} = \frac{B}{y_{i2} - y_{i1}}$$

$$x_1 = \frac{-Ax_{i1}}{x_{i2} - x_{i1}}, \qquad y_1 = \frac{-By_{i1}}{y_{i2} - y_{i1}}$$

Thus we can compute x_1 and y_1 in terms of $x_{i1}, y_{i1}, x_{i2}, y_{i2}, A,$ and B.

Given a set of pictures that show overlapping parts of a scene, we can register them and combine them into a *mosaic* picture of the entire scene, provided the overlapping pairs of pictures have consistent perspective. To avoid visible "seams" where adjacent pictures meet, it may be necessary to adjust their gray scales relative to one another (see Section 6.2.4), and to stagger the points of transition from one picture to the other within the overlap area [14, 15].

9.3 GEOMETRIC TRANSFORMATION

A *geometrical transformation* of the plane is defined by a pair of equations of the form

$$x' = h_1(x, y), \qquad y' = h_2(x, y)$$

which specify the new coordinates of each point as functions of the old coordinates. It is nontrivial to apply such a transformation to a digital picture, since the new coordinates are not necessarily integers. To make the results of the transformation into a digital picture, they must be resampled or interpolated. Methods of applying geometric transformations to digital pictures are described in Section 9.3.1, and methods of interpolation are discussed in Section 9.3.2.

9.3.1 Applying a Transformation

Suppose that we want to apply to the picture f the geometrical transformation defined by

$$x' = h_1(x, y), \qquad y' = h_2(x, y) \tag{19}$$

A straightforward way of doing this would be as follows: For each point (x, y) of f, compute the new coordinates (x', y') as given by (19); round these coordinates to the nearest integers, say $[x'], [y']$; and assign gray level $f(x, y)$ to point $([x'], [y'])$ of the new picture. However, this simple approach will usually lead to unacceptable results, since some points of the new picture will have no gray levels assigned to them by this method, while others will have more than one gray level assigned. To see this, let us consider a simple example, namely that of applying a 45° counterclockwise rotation to the 3 × 3 picture whose gray levels are

A B C

D E F

G H I

This rotation takes (x, y) into the point having coordinates

$$x' = (x - y)\sqrt{2}/2, \qquad y' = (x + y)\sqrt{2}/2$$

Let the center of rotation and the origin be at point G; then the new coordinates of the nine points, and their rounded values, are as follows:

Point	Original coordinates	New coordinates	Rounded new coordinates
A	$(0, 2)$	$(-\sqrt{2}, \sqrt{2})$	$(-1, 1)$
B	$(1, 2)$	$(-\sqrt{2}/2, 3\sqrt{2}/2)$	$(-1, 2)$
C	$(2, 2)$	$(0, 2\sqrt{2})$	$(0, 3)$
D	$(0, 1)$	$(-\sqrt{2}/2, \sqrt{2}/2)$	$(-1, 1)$
E	$(1, 1)$	$(0, \sqrt{2})$	$(0, 1)$
F	$(2, 1)$	$(\sqrt{2}/2, 3\sqrt{2}/2)$	$(1, 2)$
G	$(0, 0)$	$(0, 0)$	$(0, 0)$
H	$(1, 0)$	$(\sqrt{2}/2, \sqrt{2}/2)$	$(1, 1)$
I	$(2, 0)$	$(\sqrt{2}, \sqrt{2})$	$(1, 1)$

Thus points $(-1, 1)$ and $(1, 1)$ each get two gray levels (A and D, H and I), while point $(0, 2)$ gets no gray level, even though it is surrounded by points that do get a gray level. Schematically, the new picture looks like this:

C

B   —   F

A, D   E   H, I

G

We can avoid this problem by applying the transformation (19) "backwards." Let the inverse of (19) be

$$x = H_1(x', y'), \qquad y = H_2(x', y') \tag{20}$$

which specifies the old coordinates as functions of the new ones; this is well defined provided that (19) is a one-to-one transformation. We apply (20) to each point (x', y') of the desired *new* picture; this maps it into the plane of the old picture, say into (x'', y''). We then assign a gray level to (x', y') by interpolating between the gray levels of the old picture at the points that surround (x'', y''). Alternative methods of interpolation will be discussed later; for the moment, let us assume that we use *zero-order interpolation*, in which (x', y') is given the gray level of the old picture point closest to (x'', y'').

Under this assumption, the 45° rotation example can be redone as follows: The inverse transformation is a clockwise 45° rotation which takes (x', y') into the point having coordinates

$$x = (y' + x')\sqrt{2}/2, \qquad y = (y' - x')\sqrt{2}/2$$

We apply this transformation to find gray levels for the points of the new picture as follows:

Point	Inverse transformed coordinates	Nearest old point	Gray level
$(0, 0)$	$(0, 0)$	$(0, 0)$	G
$(0, 1)$	$(\sqrt{2}/2, \sqrt{2}/2)$	$(1, 1)$	E
$(0, 2)$	$(\sqrt{2}, \sqrt{2})$	$(1, 1)$	E
$(0, 3)$	$(3\sqrt{2}/2, 3\sqrt{2}/2)$	$(2, 2)$	C
$(1, 1)$	$(\sqrt{2}, 0)$	$(1, 0)$	H
$(1, 2)$	$(3\sqrt{2}/2, \sqrt{2}/2)$	$(2, 1)$	F
$(-1, 1)$	$(0, \sqrt{2})$	$(0, 1)$	D
$(-1, 2)$	$(\sqrt{2}/2, 3\sqrt{2}/2)$	$(1, 2)$	B

Thus the new picture now looks like

$$\begin{array}{ccc} & C & \\ B & E & F \\ D & E & H \\ & G & \end{array}$$

Note that the same gray level is assigned to more than one point [both $(0, 1)$ and $(0, 2)$ get E], and some of the gray levels are not assigned to any point (A and I); but every point in the new picture does get a unique gray level. Of course, the new picture is not square; it is a sampled version of a 45° rotated square. In general, a geometrically transformed picture will not be square; if we want to display it as a square array, we must either leave the corners of that array blank, or else lose the information corresponding to the corners of the old picture. If the transformation is not continuous, the new picture may not even be connected!

9.3.2 Interpolating

Let us now consider other methods of assigning gray levels to the points of the new picture. The simplest such method is *bilinear interpolation*, which is defined as follows: Let the integer parts $\lfloor x'' \rfloor$, $\lfloor y'' \rfloor$ of x'' and y'' be x and y, so that the point (x'', y'') is surrounded by the four integer-coordinate points

$$(x, y + 1) \qquad (x + 1, y + 1)$$
$$(x'', y'')$$
$$(x, y) \qquad (x + 1, y)$$

Let the fractional parts of x'' and y'' be $\alpha = x'' - \lfloor x'' \rfloor$ and $\beta = y'' - \lfloor y'' \rfloor$; thus $0 \leqslant \alpha, \beta < 1$. Then the gray level that we assign to (x', y') is given by

$$(1 - \alpha)(1 - \beta)f(x, y) + (1 - \alpha)\beta f(x, y + 1) + \alpha(1 - \beta)f(x + 1, y)$$
$$+ \alpha\beta f(x + 1, y + 1) \tag{21}$$

Note that if x'' is an integer (i.e., $\alpha = 0$), (x'', y'') is on the line segment between (x, y) and $(x, y + 1)$, and its gray level is assigned by linear interpolation between their levels, namely $(1 - \beta)f(x, y) + \beta f(x, y + 1)$. Similarly, if y'' is an integer, $\beta = 0$; here (x'', y'') is collinear with (x, y) and $(x + 1, y)$, and gets gray level $(1 - \alpha)f(x, y) + \alpha f(x + 1, y)$. Finally, if x'' and y'' are both integers, we have $\alpha = \beta = 0$, and $(x'', y'') = (x, y)$ gets gray level $f(x, y)$, as would be expected.

To illustrate this method, let us again consider our 45° rotation example. Here the gray levels to be assigned to the points of the new picture are determined as follows:

Point	Inverse transformed coordinates	α, β	Gray levels of surrounding old points	Assigned gray level (rounded to one decimal place)
$(0, 0)$	$(0, 0)$	$0, 0$	G	G
$(0, 1)$	$(\sqrt{2}/2, \sqrt{2}/2)$	$0.7, 0.7$	D, E, G, H	$0.1G + 0.2(D + H) + 0.5E$
$(0, 2)$	$(\sqrt{2}, \sqrt{2})$	$0.4, 0.4$	B, C, E, F	$0.4E + 0.2(B + F) + 0.2C$
$(0, 3)$	$(3\sqrt{2}/2, 3\sqrt{2}/2)$	$0.1, 0.1$	C, —, —, —	$0.8C$
$(1, 1)$	$(\sqrt{2}, 0)$	$0.4, 0$	H, I	$0.6H + 0.4I$
$(1, 2)$	$(3\sqrt{2}/2, \sqrt{2}/2)$	$0.1, 0.7$	F, I, —, —	$0.3I + 0.6F$
$(-1, 1)$	$(0, \sqrt{2})$	$0, 0.4$	D, A	$0.6D + 0.4A$
$(-1, 2)$	$(\sqrt{2}/2, 3\sqrt{2}/2)$	$0.7, 0.1$	A, B, —, —	$0.3A + 0.6B$

Now all nine of the original gray levels contribute to the gray levels of the new points, though two of them (A and I) do not make the largest contribution to any point, while E makes the largest contribution to two different points. Thus no gray levels get entirely discarded; on the other hand, only one of the old gray levels (G) is preserved exactly, while the others are all blurred or attenuated. Note that points $(0, 3)$, $(1, 2)$, and $(-1, 2)$, when inverse transformed, lie slightly outside the old 3×3 array, so that some of the integer-coordinate points that surround their preimages (x'', y'') are blanks (indicated above by dashes). In computing the gray levels for these points, we have treated the blanks as having gray level 0.

One can also use higher-order interpolation schemes; these generally yield better-appearing results. For example, one can use *bicubic spline interpolation*, in which the picture is approximated by a linear combination of products of cubic polynomials, $\sum \sum c_{ij} g_i(x) g_j(y)$. The coefficients of these polynomials can be chosen so that the approximation has the same values and same first partial derivatives ($\partial/\partial x$ and $\partial/\partial y$) at the sample points. The details of this method of interpolation will not be given here.

The preferred interpolation scheme depends on the nature of the transformation and also on the type of picture being transformed. For example, if we are transforming a two-valued (black-and-white) picture, we might prefer to use zero-order interpolation, since the higher-order schemes introduce intermediate gray levels, which is presumably undesirable. As another example, suppose we are magnifying a picture, i.e., applying a transformation of the form

$$x' = kx, \qquad y' = ky$$

where $k \gg 1$. If we use zero-order interpolation, the same gray level will be assigned to large blocks of points in the new picture; e.g., in our earlier example, if $k = 3$, the new picture will be

```
A A A B B B C C C
A A A B B B C C C
A A A B B B C C C
D D D E E E F F F
D D D E E E F F F
D D D E E E F F F
G G G H H H I I I
G G G H H H I I I
G G G H H H I I I
```

This blocky appearance will ordinarily be objectionable. On the other hand, if we use bilinear interpolation, the new gray levels will be (in part)

$$\vdots$$

$$0.3G + 0.7D \qquad\qquad \vdots$$

$$0.7G + 0.3D \quad 0.4G + 0.2D + 0.2H + 0.1E \qquad \vdots$$

$$G \qquad\qquad 0.7G + 0.3H \qquad\qquad 0.3G + 0.7H \cdots$$

which will have a much smoother appearance. Bicubic interpolation yields even smoother results. An example comparing the results of distortion correction using bicubic and bilinear interpolation is shown in Fig. 11.

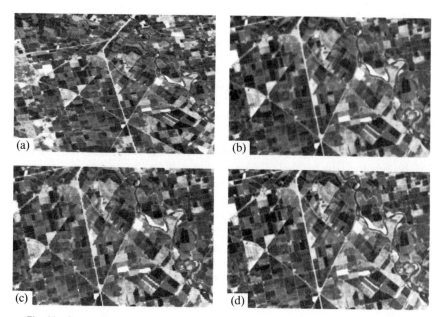

Fig. 11 Comparison of interpolation techniques: (a) Portion of a LANDSAT image prior to geometric correction; (b) corrected using zero-order interpolation; (c) corrected using bilinear interpolation; (d) corrected using cubic interpolation. (From [6].)

Conversely, when we demagnify a picture (i.e., $k \ll 1$), if we use low-order interpolation, the new gray levels will depend only on some of the old gray levels, namely those of the old points in the vicinities of the preimages of the new points; many of the old points will have no influence on the new picture. If we do not want to discard these old gray levels completely, we should use an interpolation scheme which allows a large neighborhood of the preimage of each new point to contribute to its gray level.

9.4   MATCH MEASUREMENT

This section discusses picture matching, with emphasis on the problem of matching a pattern (e.g., a template) to a picture, or a piece of one picture to another. Subsection 9.4.1 discusses measures of match or mismatch, including the use of cross correlation in matching. Subsection 9.4.2 treats optimal linear operators for detecting a given pattern in a picture. (Of course, nonlinear operators may give still better results; see Section 10.3 on the use of such operators in local feature detection.) Subsection 9.4.3 considers distortion-tolerant approaches to matching, while Subsection 9.4.4 describes methods of speeding up the matching process.

9.4.1 Match and Mismatch Measures

There are many possible ways of measuring the degree of match or mismatch between two functions f and g over a region $\mathscr{A}$. For example, one can use as mismatch measures such expressions as

$$\max_{\mathscr{A}} |f - g| \qquad \text{or} \qquad \iint_{\mathscr{A}} |f - g| \qquad \text{or} \qquad \iint_{\mathscr{A}} (f - g)^2$$

and so on (where the integrals become sums in the digital case).

If we use $\iint (f - g)^2$ as a measure of mismatch, we can derive an important measure of match from it. Note, in fact, that

$$\iint (f - g)^2 = \iint f^2 + \iint g^2 - 2 \iint fg$$

Thus if $\iint f^2$ and $\iint g^2$ are fixed, the mismatch measure $\iint (f - g)^2$ is large if and only if $\iint fg$ is small. In other words, for given $\iint f^2$ and $\iint g^2$, we can use $\iint fg$ as a measure of match.

The same conclusion can be reached by making use of the well-known Cauchy–Schwarz inequality, which states that for f and g nonnegative we always have

$$\iint fg \leqslant \sqrt{\iint f^2 \iint g^2}$$

with equality holding if and only if $g = cf$ for some constant c.[§] [The analogous result in the digital case is

$$\sum_{i,j} f(i,j)g(i,j) \leqslant \sqrt{\sum_{i,j} f(i,j)^2 \cdot \sum_{i,j} g(i,j)^2}$$

[§] To prove the Cauchy–Schwarz inequality, consider the polynomial

$$P(z) = \left(\iint f^2\right)z^2 + 2\left(\iint fg\right)z + \iint g^2$$

Now $P(z) = \iint (fz + g)^2 \geqslant 0$ for all z. Since P is a quadratic polynomial in z, this implies that it either has no real roots or has just one ("double") root; in other words the discriminant of P must be negative or zero, i.e., $(\iint fg)^2 - (\iint f^2)(\iint g^2) \leqslant 0$, which immediately gives the desired inequality. Moreover, if equality holds, P does have a root, say, z_0; but then $P(z_0) = \iint (fz_0 + g)^2 = 0$ evidently implies $fz_0 + g = 0$ for all x, y, so that $g = (-z_0)f$.

with equality holding if and only if $g(i, j) = cf(i, j)$ for all i, j.] Thus when $\iint f^2$ and $\iint g^2$ are given, the value of $\iint fg$ is a measure of the degree of match between f and g (up to a constant factor).

Let us now apply these ideas to the problem of matching a pattern f with a picture g, i.e., of finding places where g matches f. (We are tacitly assuming that f is small compared to g, i.e., f is zero outside a small region $\mathscr{A}$, and we are interested only in matching the nonzero part of f against g.) We can do this by shifting f into all possible positions relative to g, and computing $\iint fg$ for each such shift (u, v). By the Cauchy–Schwarz inequality, we have

$$\iint_{\mathscr{A}} f(x, y)g(x + u, y + v) \, dx \, dy$$

$$\leqslant \left[\iint_{\mathscr{A}} f^2(x, y) \, dx \, dy \iint_{\mathscr{A}} g^2(x + u, y + v) \, dx \, dy \right]^{1/2}$$

Since f is zero outside $\mathscr{A}$, the left-hand side is equal to

$$\iint_{-\infty}^{\infty} f(x, y)g(x + u, y + v) \, dx \, dy$$

which is just the *cross correlation* C_{fg} of f and g (Section 2.1.4). Note that on the right-hand side, while $\iint f^2$ is a constant, $\iint g^2$ is not, since it depends on u and v. Thus we cannot simply use C_{fg} as a measure of match; but we can use instead the *normalized cross correlation*

$$C_{fg} \bigg/ \left[\iint_{\mathscr{A}} g^2(x + u, y + v) \, dx \, dy \right]^{1/2}$$

This quotient takes on its maximum possible value (namely, $\sqrt{\iint_{\mathscr{A}} f^2}$) for displacements (u, v) at which $g = cf$. Some simple examples of the use of normalized cross correlation to find places where a template matches a picture are shown in Fig. 12; a real example is shown in Fig. 13.

In many cases of interest (character recognition, for example), the templates and picture are—at least approximately—two-valued, i.e., "black and white." In such cases, the process of finding matches can be greatly simplified. Specifically, let the two values be 0 and 1; then $\iint fg$ is just the area of the set of points where f and g are both 1. Let f' be the "negative template" which is 0 when f is 1, and vice versa; then $\iint f'g$ is just the area of the set of points where f' and g are both 1, i.e., where f is 0 and g is 1. Now consider

$$\iint (f - f')g = \iint fg - \iint f'g$$

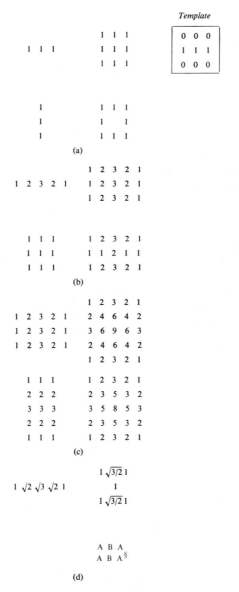

Fig. 12 Matching by normalized cross correlation: some simple examples. (a) Picture g and template f. (b) Cross correlation C_{fg}. (c) $\sum\sum g^2$. (d) $C_{fg}/\sqrt{\sum\sum g^2}$; values less than 1 have been discarded. Note that the perfect match value ($\sqrt{3}$) is not much better than the near misses in position ($\sqrt{2}$'s) and in shape ($3/\sqrt{5}$'s). All points not shown are 0's.

§ $A = 2/\sqrt{3}$; $B = 3/\sqrt{5}$.

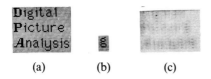

(a)   (b)   (c)

Fig. 13 Matching by normalized cross correlation: a real example. (a) Picture, (b) Template. (c) Their normalized cross correlation. Note the many near misses.

This takes on its greatest possible value when

(a) $g = 1$ wherever $f = 1$, so that the positive term is as large as possible, namely, equal to the area of the set of points where $f = 1$;

(b) $g = 0$ wherever $f = 0$ (i.e., g is never 1 when $f' = 1$), so that the negative term is zero.

Thus $\iint (f - f')g$ is maximized when f and g match exactly.

It follows from these remarks that we can find matches to f in g by cross-correlating $f - f'$ with g and looking for maxima. Note that $f - f'$ is just the two-valued "picture" that has value 1 where $f = 1$, and value -1 where $f = 0$. Note also that in this two-valued case we can detect matches without the need to "normalize" the cross correlation. Simple examples of this method are shown in Fig. 14.

Exercise 9.6 Prove that $\iint (f - f')g = 2 \iint fg - \iint g.$ ∎

Template

1	1	1
0	0	0
1	1	1

(a)

```
              1 2 3 2 1
1 2 3 2 1     1 2 3 2 1
              2 4 6 4 2
1 2 3 2 1     1 2 3 2 1
              1 2 3 2 1
```

```
    1 1 1     1 2 3 2 1
    1 1 1     1 1 2 1 1
    2 2 2     2 4 6 4 2
    1 1 1     1 1 2 1 1
    1 1 1     1 2 3 2 1
```
(b)

```
1 2 3 2 1
```

```
    1 1 1

    1 1 1
```
(c)

Fig. 14 Matching by subtracting cross correlations in the two-valued case. (a) Negative template f'; the picture and positive template are the same as in Fig. 12a. (b) $C_{f'g}$. (c) $C_{fg} - C_{f'g}$; negative values have been discarded.

For digital pictures, cross correlations would normally be computed point by point; for each relative shift (u, v) of f and g, we multiply them point-wise and sum the results to obtain $C_{fg}(u, v)$. If f is $m \times m$ and g is $n \times n$, where m is much smaller than n, this requires m^2 multiplications to be performed for each of n^2 relative shifts. If a parallel cellular array computer were available, the sets of multiplications could presumably be done as single operations, but we would still have to go through all n^2 shifts, and for each shift, we must add up the m^2 terms of the resulting product of arrays, which may not be a natural operation for an array processing machine. For the parallel computer, a different method of computing the cross correlation might be more efficient: We multiply g by each of the m^2 elements in the array f; shift the resulting m^2 arrays relative to one another, by amounts corresponding to the positions of the elements in f; and add them pointwise to obtain the array C_{fg}. This requires only m^2 multiplications and shifts (each carried out in parallel on the array g), and addition of the resulting m^2 arrays; note that this is pointwise addition of arrays, rather than addition of the elements of an array.

By the convolution theorem, cross-correlating a template f with a picture g is equivalent to pointwise multiplying the Fourier transforms F^* and G and then taking the inverse transform. This method can be faster than direct cross correlation, if an efficient Fourier transform algorithm is used. Recall that convolutions obtained using the discrete Fourier transform are *cyclic*, since the transform treats the pictures as though they were periodic (see Section 2.2.1). Thus these convolutions will have values even for shifts such that the template is no longer entirely inside the picture. For most purposes, such values would not be useful. Note also that the Fourier transform arrays that we multiply pointwise should be of the same size; to achieve this when the template is smaller than the picture, we can fill out the template to the size of the picture by adding rows and columns of zeros, or of some other constant value (the average gray level of the picture is sometimes a good choice).

It is sometimes advantageous to work with pictures whose gray scales have been normalized (see Section 12.1.1c) by subtracting the average gray level of the picture from the gray level of each point. Given two pictures f and g, whose average gray levels are μ and v, respectively, the normalized pictures are then $f - \mu$ and $g - v$. Our normalized match measure for $f - \mu$ and $g - v$ is then

$$\iint (f - \mu)(g - v) \bigg/ \sqrt{\iint (f - \mu)^2 \iint (g - v)^2} = \left(\left(\iint fg/A \right) - \mu v \right) \bigg/ \sigma \tau$$

where σ and τ are the standard deviations of the gray levels of f and g, respectively, and A is the area of the region of integration.

9.4.2 Linear Matched Filtering

Cross-correlating f and g is a linear shift-invariant operation, since it is the same as convolving g with f rotated $180°$. In this section we consider the question of finding matches between a template f and a picture g by cross-correlating an arbitrary filter with g. We shall show that, under certain assumptions, the best filter to use for this purpose is f itself; this result is known as the *matched filter* theorem. In what follows we use boldface to denote random fields, as in Section 2.4.

Let $\mathbf{g} = f + \mathbf{n}$, where the noise field $\mathbf{n}$ is homogeneous, and suppose that we convolve the filter h with $\mathbf{g}$ to obtain $h * \mathbf{g}$. Let $\mathbf{g}' = h * \mathbf{g}$ and $\mathbf{n}' = h * \mathbf{n}$. We have seen in Section 2.4.5 that $S_{n'n'} = S_{nn}|H|^2$. Now the expected noise power, i.e., the expected value $E\{|h * \mathbf{n}|^2\}$ at any point, by homogeneity, is $R_{n'n'}(0, 0)$. This, however, is just the inverse Fourier transform of $S_{n'n'}$ evaluated at $(0, 0)$, i.e., it is $\iint S_{n'n'} = \iint S_{nn}|H|^2$. If the noise is "white," i.e., S_{nn} is constant, the expected noise power is thus $S_{nn} \iint |H|^2$.

On the other hand, the signal power $|h * f|^2$ at a point (x, y) is equal to

$$|\mathscr{F}^{-1}(HF)|^2 = \left| \iint HF \exp[2\pi j(ux + vy)] \, du \, dv \right|^2$$

By the Cauchy–Schwarz inequality,[§] the right-hand side is less than or equal to $\iint |H|^2 \iint |F|^2$. Hence the ratio of signal power to expected noise power in the white noise case is

$$\frac{|h * f|^2}{E\{|h * \mathbf{n}|^2\}} \leqslant \frac{\iint |H|^2 \iint |F|^2}{S_{nn} \iint |H|^2} = \frac{1}{S_{nn}} \iint |F|^2$$

Moreover, the upper bound is achieved only when

$$H = F^* \exp[-2\pi j(ux + vy)]$$

This is equivalent to

$$h(\alpha, \beta) = f(x - \alpha, y - \beta)$$

In other words, h is f rotated $180°$ and shifted by the amount (x, y). Thus we see that in this case, the linear shift-invariant filter that maximizes the ratio

---

[§] We use here the complex form of the Cauchy–Schwarz inequality:

$$\left| \iint fg \right|^2 \leqslant \iint |f|^2 \iint |g|^2$$

with equality holding only when $g = cf^*$.

of signal power to expected noise power at (x, y) is just the template f, rotated 180°. Convolving this filter with $\mathbf{g}$ is thus the same as cross-correlating f with $\mathbf{g}$.

Other signal-to-noise criteria can also be used to define optimum filters. Suppose, for instance, that we want to choose h so as to maximize the ratio

$$(E\{h * \mathbf{g}\})^2 / \text{var}(h * \mathbf{g})$$

where the numerator is the expected value of $h * \mathbf{g}$ at the point (x, y), and the denominator is the variance of its values taken over all points. It turns out that this can be done by solving the integral equation

$$R_{gg} * h = f$$

where we assume that f is the expected value of $\mathbf{g}$. If $\mathbf{g}$ is highly uncorrelated, so that R_{gg} approximates a δ function, this gives $h = f$; but in other cases, $h = f$ is no longer the optimum solution. For example, in the one-dimensional case, if R_{gg} is exponential, say of the form $R_{gg}(x - u) = e^{-|x-u|/L}$, it can be shown that the solution is

$$h = \frac{1}{2L} f - \frac{L}{2} \frac{\partial^2 f}{\partial x^2}$$

Thus for L small ($\mathbf{g}$ highly uncorrelated) the f term dominates; but for L large, the term $\partial^2 f/\partial x^2$ dominates.

The use of derivatives of f as filters can be justified on other grounds as well. If we are interested only in finding the shape of the template f in g, but we do not care about its gray level, it makes sense to convert f and g into outline form, say by differentiation or high-pass filtering, before we cross-correlate them. This type of filtering can be used to maximize the ratio of power in the derivative (gradient, Laplacian, etc.) of the signal to expected power in the derivative of the noise. It should also be noted that matching of outlines tends to yield sharper matches than does matching of solid objects. Some simple examples of this last phenomenon are shown in Fig. 15. Fig. 14 showed the advantages of matching a "second-difference" template to the original picture; the result in Fig. 14c corresponds to correlating the template

$$
\begin{array}{ccc}
-1 & -1 & -1 \\
1 & 1 & 1 \\
-1 & -1 & -1
\end{array}
$$

with the picture in Fig. 12a, and discarding negative values.

A straight piece of outline (i.e., a straight edge in the picture) yields a sharp match in the direction across it, but the match falls off slowly in the

Template

0	0	0	0	0
0	1	1	1	0
0	1	1	1	0
0	1	1	1	0
0	0	0	0	0

1	1	1
1	1	1
1	1	1

(a)

1	2	3	2	1
2	4	6	4	2
3	6	9	6	3
2	4	6	4	2
1	2	3	2	1

(b)

1	2	3	3	3	2	1
2	4	6	6	6	4	2
3	6	9	9	9	6	3
3	6	9	9	9	6	3
3	6	9	9	9	6	3
2	4	6	6	6	4	2
1	2	3	3	3	2	1

(c)

$$
\begin{array}{ccccc}
 & & \sqrt{3/2} & & \\
 & 4/3 & 2 & 4/3 & \\
\sqrt{3/2} & 2 & 3 & 2 & \sqrt{3/2} \\
 & 4/3 & 2 & 4/3 & \\
 & & \sqrt{3/2} & & \\
\end{array}
$$

(d)

Fig. 15 Matching of outline figures yields sharper matches than matching of solid figures. (a) Templates f: a solid square and a hollow square. The pictures g are the same as the templates, surrounded by many 0's on all sides. (b) C_{fg}. (c) $\sum\sum g^2$. (d) $C_{fg}/\sqrt{\sum\sum g^2}$; values less than 1 have been discarded.

direction along it. However, if there are many edges in a variety of directions, the overall match will fall off relatively sharply in any direction. Very sharp match peaks are obtained if we match patterns of sharply localized features (e.g., "branch" points where curves intersect, or "corner" points where edges make abrupt turns).

For digital pictures, the role of cross correlation in picture matching can be further justified on grounds of minimum-error decision making. Suppose that, in a given region $\mathscr{A}$ of the digital picture g, we are trying to decide between the two equally likely alternatives $g = f_1 + n$ and $g = f_2 + n$, where n is signal-independent Gaussian noise with mean 0 and standard deviation σ. If f_1 is present, the probability of obtaining the observed pattern of gray levels g over $\mathscr{A}$ is proportional to $\prod_{\mathscr{A}} \exp[-(g(i,j) - f_1(i,j))^2/2\sigma^2]$ (where

Template

0	0	0	0	0
0	1	1	1	0
0	1	0	1	0
0	1	1	1	0
0	0	0	0	0

1	1	1
1		1
1	1	1

(a′)

1	2	3	2	1
2	2	4	2	2
3	4	8	4	3
2	2	4	2	2
1	2	3	2	1

(b′)

1	2	3	3	3	2	1
2	3	5	5	5	3	2
3	5	8	8	8	5	3
3	5	8	8	8	5	3
3	5	8	8	8	5	3
2	3	5	5	5	3	2
1	2	3	3	3	2	1

(c′)

$$
\begin{array}{ccccc}
 & & 3/\sqrt{5} & & \\
 & & \sqrt{2} & & \\
3/\sqrt{5} & \sqrt{2} & 2\sqrt{2} & \sqrt{2} & 3/\sqrt{5} \\
 & & \sqrt{2} & & \\
 & & 3/\sqrt{5} & & \\
\end{array}
$$

(d′)

Fig. 15   (*Continued*)

we have ignored a factor which is a power of $1/\sqrt{2\pi}\sigma$). This is because, for each (i, j) in $\mathscr{A}$, the probability of having $n(i, j) = g(i, j) - f_1(i, j)$ is $\exp[-n(i, j)^2/2\sigma^2]$, and these events are independent for the different points of $\mathscr{A}$, so that their probabilities multiply. Similarly, if f_2 is present, the probability of observing g is proportional to $\prod_{\mathscr{A}} \exp[-(g(i, j) - f_2(i, j))^2/2\sigma^2]$. To determine which of f_1 and f_2 is more likely, we can compare these probabilities and decide in favor of the case having the larger probability. Now comparing the probabilities is equivalent to comparing their logarithms

$$-\sum_{\mathscr{A}} [g(i, j) - f_1(i, j)]^2/2\sigma^2 \quad \text{and} \quad -\sum_{\mathscr{A}} [g(i, j) - f_2(i, j)]^2/2\sigma^2$$

Here we can cancel the common factor $1/2\sigma^2$, and can also cancel $\sum_{\mathscr{A}} g(i, j)^2$ from both sides, so that it remains to compare

$$\sum [2g(i, j)f_1(i, j) - f_1(i, j)^2] \quad \text{and} \quad \sum [2g(i, j)f_2(i, j) - f_2(i, j)^2]$$

If $\sum f_1(i, j)^2 = \sum f_2(i, j)^2$ (i.e., the two patterns have the same "power"), this is equivalent to comparing $\sum g(i, j) f_1(i, j)$ and $\sum g(i, j) f_2(i, j)$; doing this for all regions $\mathscr{A}$ of g is tantamount to cross-correlating f_1 and f_2 with g, and deciding in favor of whichever one yields the highest cross correlation at a given point. Note that if $\sum f_1(i, j)^2 \neq \sum f_2(i, j)^2$, it is no longer sufficient to compare the cross correlations.

9.4.3   Distortion-Tolerant Matching

When we match a template with a picture, e.g., by cross correlation, we may expect to find any exact copies of the template that are present in the picture. These copies must be of the same size, however, and have the same orientation, as the template; we will not be able to find a rotated, enlarged, or perspective-transformed version of the template in the picture, since such a version will not, in general, give rise to high values of the cross correlation. In general, if a geometrically distorted copy of the template is present, we will have difficulty detecting it by matching unless the distortion is quite small.

If we want to use template matching to recognize a pattern that is subject to rotation, scale change, or other geometrical transformations, a very large number of templates would ordinarily be needed. Another possibility, however, is to use a single template, and search through the space of permissible distortions or transformations of the template, in an attempt to optimize its degree of match with the picture. This "rubber mask" approach may be practical if the space of possibilities is relatively small, and if it is possible to make a good initial guess as to the correct transform of the template that should be used. In some cases, searching may not be necessary, since it may be possible to "normalize" the pictures being matched so as to undo the effects of the transformations; see Section 12.1.1c.

If unlimited amounts of distortion (in gray scale and/or geometry) are allowed, any two pictures can be made to match. Evidently, the "value" of a match obtained in this way should depend both on the goodness of the match and on how little distortion was required to achieve it. To illustrate how such a match evaluation measure could be defined, imagine that we have a template composed of subtemplates that are connected by springs. As we search for a match between this template and some part of a picture, we may find many good matches to the various subtemplates, but these partial matches may not be in the right relative positions, so that to achieve them all simultaneously, we may have to stretch or compress the springs. Our goal is to find a combination of partial matches which (a) are as good as possible and (b) require as little tension in the springs as possible. This can be done, in principle, using mathematical programming techniques; but in practice this would usually be computationally costly. A useful shortcut is to first find good matches for the individual subtemplates, and then try to build up a low-tension combination of these matches stepwise, using heuristic search techniques. Another approach is to find all the matches to the subtemplates, and then iteratively strengthen or weaken each of these matches in accordance with the strengths of the matches to the other subtemplates and the closeness of these matches to their ideal relative positions. This "relaxation" approach will be considered further in Section 10.5.3.

The use of subtemplates makes the matching process more tolerant to geometric distortion because the subtemplates, being smaller, will match relatively well even under distortion. Various types of subtemplates can be used, e.g., subtemplates representing distinctive local features. A different approach to distortion-tolerant matching is to segment the picture and template into regions, and try to pair off corresponding regions by comparing their properties (size, shape, texture, etc.). The success of this approach depends on the reliability of the segmentation; if a region splits apart, or several regions merge, it becomes difficult or impossible to match them with the corresponding template regions.

When matching pictures containing sets of local features, it may be advantageous to work with the coordinates of the features explicitly rather than matching the pictures as arrays. To do this, we pair off each feature in the first set with every feature in the second set and determine the displacement under which they would coincide. If there are m features in each set, this gives us m^2 displacements; m of these should be the same, corresponding to the correct displacement, while the others should be "noise," so that we should have a detectable match peak. The counts of point pairs having a given displacement correspond to the cross correlation of two binary pictures with 1's at the positions of the features. The computational cost of this process is proportional to the square of the number of features: if they are sparse, this may be much less than the cost of matching the pictures. We can allow for geometric distortion by introducing a tolerance into the displacements – i.e., by mapping each pair of features into a disk in displacement space; we can give this disk a peak value at the ideal displacement, and taper its values away from the ideal. This corresponds to cross-correlating a binary picture of one point pattern with a blurred picture of the other.

Feature matching techniques can also be used to match patterns that differ by transformations other than translation. For example, suppose that we are given two sets of edge points with associated orientations (see Section 10.2 on edge detection). We can determine the orientation difference of each pair of edge points; this gives us m^2 differences, m of which should be the same, yielding a match peak. This corresponds to one-dimensionally cross-correlating the orientation histograms corresponding to the two sets of edge points. In general, given a space of allowable transformations, each pair of feature points determines a (subspace of) transformation(s), and there should be a match peak in transformation space where m of these subspaces meet. This approach is closely related to the Hough transform methods of detecting curves that have given shapes; see Section 10.3.3.

9.4.4 Fast Matching

Even in the absence of distortions, looking for matches to a template in all possible positions in a picture is a relatively time-consuming process. This process can be shortened if we can develop fast, inexpensive tests for determining positions or regions in which matches are likely to be found, so that it is no longer necessary to look for a match in every possible position. It can be further shortened if, when we are testing for a match in a given position, we can develop methods of rejecting mismatches rapidly, without having to go through the entire point-by-point comparison of the template and picture. These two approaches will be discussed in the following paragraphs.

If the template f is not too highly uncorrelated, there will be fairly good matches to it in the picture g in the vicinity of any ideal match—in other words, the peak(s) in the (normalized) cross correlation C_{fg} will be relatively smooth and broad, rather than spikelike. Hence in this case, we can attempt to find ideal matches by first computing C_{fg} at a coarsely spaced grid of points, and then searching for better matches in the vicinity of those grid points at which C_{fg} is high. In these searches, it may be possible to use hill-climbing techniques, if the correlation peaks are smooth enough.

Another possible approach to reducing the number of match positions that need to be examined is to compute some simple picture property, say a textural property, in the vicinity of every point of g, and to attempt matches first at those points where the value of the property is close to its value for the template f. (Of course, this makes sense only if computing the property at a point is much cheaper than measuring the degree of match between g and f at that point.) This approach should make it possible to find a match sooner, on the average, than if we just tested match positions in some arbitrary order. Incidentally, when we are matching pieces of one picture with another picture (e.g., to determine stereoscopic depth), the sizes of the pieces that should be used also depend on the local texture; when the picture contains fine detail, relatively small pieces can be used.

Matching with a subtemplate is a particularly useful way of eliminating match positions; the rest of the template needs to be matched only at those positions where the subtemplate has matched well. (Note that if the subtemplate is too small, there will be many false alarms, and the computational saving will be reduced; while if it is too large, the saving is small to begin with.) Another possibility is to match a reduced-resolution template to a reduced-resolution version of the picture, and use the full-resolution template only in the vicinity of positions where the crude template has matched; this coarse–fine approach is closely related to the hill-climbing idea mentioned earlier. (Here again, if the reduction in resolution is too great, there will be many false alarms, while if it is too little, the saving may not be significant.)

When we measure the degree of match in a given position, we would like to be able to reject mismatches rapidly, since most positions will be mismatches. For this purpose, it is convenient to use a sum of absolute differences measure of mismatch $\sum |f - g|$. Given a mismatch threshold t, we want to compute $\sum |f - g|$ in such a way that, in a mismatch position, the sum can be expected to rise above t as rapidly as possible. One way to do this is to measure $|f - g|$ first for those points of f whose gray levels have high expected absolute differences from the gray level of a randomly selected point of g. The expected contributions of such points to $\sum |f - g|$, when we are not in a match position, are large; hence when we measure $|f - g|$ for these points first, the sum $\sum |f - g|$ should tend to rise more rapidly than if we measure $|f - g|$ for the points of f in some arbitrary order.

9.5 BIBLIOGRAPHICAL NOTES

Imaging geometry is treated in books on photogrammetry (e.g., Thompson [23]) or on computer graphics. A treatment from the standpoint of scene analysis can be found in Duda and Hart [8]. Geometrical operations on digital pictures are discussed in Johnston and Rosenfeld [12]. For examples of their application to geometrical distortion correction see O'Handley and Green [18] and Bernstein [6].

Picture matching techniques, particularly those employing correlation measures of match, have a long history; as indicated at the beginning of the chapter, they have been used extensively for character recognition, target recognition, change detection, map-matching navigation, and stereomapping. An early reference on the two-valued case is Horowitz [11]. On matched filtering see, e.g., Turin [24]; on the importance of derivative filtering in the case of a correlated scene see Arcese et al. [2], as well as Andrus et al. [1]. On the minimum-error motivation of the cross-correlation match measure see Harris [10].

"Rubber masks" are discussed by Widrow [27]. The spring-loaded sub-template model is treated in Fischler and Elschlager [9]; the relaxation approach is described in Davis and Rosenfeld [7]. "Symbolic matching" of regions based on their properties was investigated by Price [19]. On matching of local features and its advantages see Barrow et al. [5] and Stockman and Kopstein [22].

On methods of finding good match positions see Barnea and Silverman [4] and Nagel and Rosenfeld [17]. The use of subtemplates or coarse templates for preliminary screening is treated by VanderBrug and Rosenfeld [21, 26]; another multilevel approach is described by Ramapriyan [20].

APPENDIX: ANALYSIS OF TIME-VARYING IMAGERY[§]

During the past few years there has been a rapid growth of interest in the analysis of time sequences of pictures. Tasks that involve such sequences include interframe TV image coding, real-time object tracking, change detection, and dynamic scene analysis. Research on these tasks has taken many different approaches, and the subject is so new that no generally accepted scheme for classifying these approaches has emerged as yet. Most of the work involves the comparison between two successive pictures of a sequence; this work may therefore be regarded as an extension to the subject of image matching. For brevity, we will refer to the successive pictures as "frames."

This section briefly discusses, on a conceptual level, some of the many approaches to picture sequence analysis. For further details and an extensive bibliography, the reader is referred to two 1978 surveys [13, 16] and to the proceedings of a conference held in 1979 [3]. References on individual approaches will not be given here.

The methods that are appropriate in processing a sequence of frames depend on the nature of the changes that are expected to have taken place from one frame to the next. For example, suppose that the camera moves relative to a distant scene, or that cloud cover moves uniformly against a featureless background. In such cases we expect that the entire frame shifts more or less as a unit, with the possible exception of some areas near its borders. Here straightforward correlation of the two frames, as in Section 9.4, can be used to estimate the shift. Note that if a scene contains objects at significantly different distances, they will shift by different amounts (motion

[§] The authors are grateful to Professor H. H. Nagel for his detailed and insightful comments on an earlier version of this appendix, most of which have been incorporated in the present version. The responsibility for all omissions and misrepresentations remains with the authors.

parallax) when the camera moves, so that simple correlation will no longer work. The situation is also more complicated if the motion is not a pure translation.

In most situations, the frame-to-frame changes are very nonuniform, and must be described piecewise. This is true whether the frames are far apart in time (the standard "change detection" scenario) or are in close succession [so that the observed changes are likely to be due to motion of the camera or object(s)].

In the former case, we usually lack *a priori* knowledge about where in the frames changes are to be expected, and what kinds of changes to expect. Thus we commonly attempt to detect the changes, by systematic comparison of the frames, and try to identify their causes, on the basis of general knowledge about the given class of scenes.

In the latter case, on the other hand, we may be able to use local matching techniques to track displaced objects from frame to frame, and thus develop a model for the object or camera motion. Note that if the object displacements are due to camera motion, they must all be consistent with the requirements of imaging geometry (Section 9.1), and if we can identify corresponding pairs of objects, we can determine the camera motion. On the other hand, if the objects themselves move, their displacements relative to each other can be arbitrary.

The approaches discussed in the following paragraphs will generally assume that we are dealing with situations involving arbitrary object motion. We assume that the frames are close enough together in time for most object displacements to be small (or at least not very large) relative to object size.

a. Feature or object tracking

If the objects (or features) that one wants to track from frame to frame are easily distinguishable from their background, it may not be necessary to compare the entire frames. Instead, one can independently extract the objects from each frame, and analyze the resulting object sequences. Segmentation and feature detection techniques are treated in Chapter 10.

If only one object candidate is present in each frame, tracking is trivial. In this case we can also use the object's position in one frame to predict its approximate position in the next frame, so that once it has been found in the first frame, we only need to examine portions of the subsequent frames. The same is true if there are several objects which are far apart or easily distinguishable from one another.

If only one object is expected, but several object candidates are present in each frame, we must choose one candidate from each frame so as to obtain a best sequence of candidates. Sequences can be evaluated based on both similarity of the successive objects and consistency of the successive positions. Techniques such as mathematical programming and relaxation (see Section 10.5.3) can be used to find best sequences.

If many objects are expected, and they are not easily distinguishable, the object candidates in successive frames must be matched with one another, based on similarity and relative position. If there are more than a few candidates, it may be impractical to find best correspondences by exhaustive comparison; instead, here again one may need to use mathematical programming or relaxation methods. Note that the correspondence will not, in general, be one-to-one, since "objects" may appear and disappear.

More generally, one can segment each frame completely into regions, as represented by a "map" or "sketch," and compare successive maps (see Section 9.4.3); this is a higher-level analog of direct frame comparison.§ Note that here again, the correspondence will usually not be one-to-one.

b.   Frame comparison

We next consider the more general case where candidate objects or features are not easily extracted from the individual frames. In this case, comparison of entire frames will usually be necessary, at least initially. (Once objects are being tracked, we should be able to predict where to look for them on the next frame, and we may also know at what places in a frame new objects are likely to appear, e.g., at the borders.) We shall first assume that successive frames are in register except for the effects of object motion, and that other changes from frame to frame are insignificant.

One possible approach is to estimate the motion directly on a pointwise basis. We know from calculus that $\partial g/\partial t = \partial g/\partial x \; \partial x/\partial t + \partial g/\partial y \; \partial y/\partial t$. If we approximate the derivatives of g by differences, and assume that the unit of time is the interframe interval and the unit of x or y is the pixel, then $\Delta g/\Delta t$ is the frame-to-frame gray level difference, while $\Delta g/\Delta x$ and $\Delta g/\Delta y$ are the gray level differences in the horizontal and vertical directions. Thus by measuring these temporal and spatial differences at a given pixel we obtain a constraint on the velocity vector $(\partial x/\partial t, \partial y/\partial t)$ of that pixel. In particular, at places where $\Delta g/\Delta x$ is zero, we can solve for $\partial y/\partial t$ $(\doteq (\Delta g/\Delta t)/(\Delta g/\Delta y))$, and vice versa. This approach is appropriate when the objects are strongly patterned and contain many internal edges in different orientations.

Another possibility is to detect moving objects by using local correlation, i.e., by matching pieces of one frame to nearby pieces of the other frame. This will work best if the objects are strongly patterned, and the pieces are comparable in size to the expected objects, so that strong, sharp match peaks

§ If desired, the segmentation of the previous frame can be used as an initial estimate of the segmentation of the current frame, and can then be adjusted as necessary.

will be obtained. If we know in which places on the frames motion has probably occurred (e.g., by crude pointwise comparison), we need only do the correlation in the vicinity of such places.

Given a set of velocity estimates at different points, obtained pointwise or by local correlation, we can cluster them to obtain estimates of object or region motion. The use of iterative estimation techniques or relaxation methods may be desirable in order to improve the reliability or local consistency of the estimates.

Yet another possibility is to use frame comparison as a basis for extracting moving objects. Suppose that the frames are composed of homogeneous patches, some of which are objects. If we compare the frames pointwise, we can classify the points as changed or unchanged. Recalling our assumption that the frame to frame displacement is small relative to the object size, we can expect a moving object to give rise to two types of connected components of changed points: Pieces of background that have been covered by the object (i.e., regions that are background on the first frame and object on the second), and pieces that have been uncovered (i.e., vice versa); these will be separated by region(s) that are in the object on both frames. If we can distinguish these types of regions, we have detected the object and have also tracked it from frame to frame.

If the frames are not in register, some form of local correlation must first be used to determine corresponding parts of the frames. This yields a set of disparity vectors which contain information about the relative distortion of the frames as well as about object motion. For example, a local cluster of disparities that differ sharply from the surrounding disparities may arise from a moving object. (Note that if the camera has been moved, it could also correspond to an object that is at a significantly different distance than its surround from the camera.) Ullman [25] has shown that under certain circumstances the three-dimensional motion of a rigid object is uniquely determined if we can correctly track four noncoplanar points of the object through three successive frames.

Situations in which frame comparison is required, but the frames are out of register, give rise to the most difficult problems in time-varying image analysis. However, research on this subject is quite active, and rapid progress is being made on the development of real-time techniques for analyzing picture sequences.

REFERENCES

1. J. F. Andrus, C. W. Campbell, and R. R. Jayroe, Digital image registration using boundary maps, *IEEE Trans. Comput.* **C-24**, 1975, 935–940.
2. A. Arcese, P. H. Mengert, and E. W. Trombini, Image detection through bipolar correlation, *IEEE Trans. Informat. Theory* **IT-16**, 1970, 534–541.

3. N. I. Badler and J. K. Aggarwal (eds.), Abstracts of the workshop on computer analysis of time-varying imagery (April 5–6, 1979, Philadelphia, PA).

4. D. I. Barnea and H. F. Silverman, A class of algorithms for fast digital image registration, *IEEE Trans. Comput.* **C-21**, 1972, 179–186.

5. H. G. Barrow, J. M. Tenenbaum, R. C. Bolles, and H. C. Wolf, Parametric correspondence and chamfer matching: two new techniques for image matching, *Proc. Internat. Joint Conf. on Artificial Intelligence, 5th* 1977, 659–663.

6. R. Bernstein, Digital image processing of earth observation sensor data, *IBM J. Res. Develop.* **20**, 1976, 40–57.

7. L. S. Davis and A. Rosenfeld, An application of relaxation labelling to spring-loaded template matching, *Proc. Internat. Joint Conf. Pattern Recognition, 3rd* 1976, 591–597.

8. R. O. Duda and P. E. Hart, "Pattern Classification and Scene Analysis." Wiley, New York, 1973.

9. M. A. Fischler and R. A. Elschlager, The representation and matching of pictorial structures, *IEEE Trans. Comput.* **C-22**, 1973, 67–92.

10. J. L. Harris, Resolving power and decision theory, *J. Opt. Soc. Amer.* **54**, 1964, 606–611.

11. M. Horowitz, Efficient use of a picture correlator, *J. Opt. Soc. Amer.* **47**, 1957, 327.

12. E. G. Johnston and A. Rosenfeld, Geometrical operations on digital pictures, *in* "Picture Processing and Psychopictorics" (B. S. Lipkin and A. Rosenfeld, eds.), pp. 217–240. Academic Press, New York, 1970.

13. W. N. Martin and J. K. Aggarwal, Dynamic scene analysis, *Comput. Graphics Image Processing* **7**, 1978, 356–374.

14. D. L. Milgram, Computer methods for creating photomosaics, *IEEE Trans. Comput.* **C-24**, 1975, 1113–1119.

15. D. L. Milgram, Adaptive techniques for photomosaicking, *IEEE Trans. Comput.* **C-26**, 1977, 1175–1180.

16. H. H. Nagel, Analysis techniques for image sequences, *Proc. Internat. Joint Conf. Pattern Recognition, 4th* 1978, 186–211.

17. R. N. Nagel and A. Rosenfeld, Ordered search techniques in template matching, *Proc. IEEE* **60**, 1972, 242–244.

18. D. A. O'Handley and W. B. Green, Recent developments in digital image processing at the Image Processing Laboratory of the Jet Propulsion Laboratory, *Proc. IEEE* **60**, 1972, 821–828.

19. K. Price and D. R. Reddy, Matching segments of images, *IEEE Trans. Pattern Anal. Machine Intelligence* **1**, 1979, 110–116.

20. H. K. Ramapriyan, A multilevel approach to sequential detection of pictorial features, *IEEE Trans. Comput.* **C-25**, 1976, 66–78.

21. A. Rosenfeld and G. J. VanderBrug, Coarse-fine template matching, *IEEE Trans. Systems Man Cybernet.* **SMC-7**, 1977, 104–107.

22. G. Stockman and S. Kopstein, Image registration from edge content, *Proc. Symp. Automatic Imagery Pattern Recognition, 8th* 1978, 139–157.

23. M. M. Thompson (ed.), "Manual of Photogrammetry." American Society of Photogrammetry, Falls Church, Virginia, 1966.

24. G. L. Turin, An introduction to digital matched filters, *Proc. IEEE* **64**, 1976, 1092–1112.

25. S. Ullman, "The Interpretation of Visual Motion." MIT Press, Cambridge, Massachusetts, 1979.

26. G. J. VanderBrug and A. Rosenfeld, Two-stage template matching, *IEEE Trans. Comput.* **C-26**, 1977, 384–393.

27. B. Widrow, The "rubber-mask" technique, *Pattern Recognition* **5**, 1973, 175–211.

Chapter 10

Segmentation

In image compression or enhancement, the desired output is a picture—an approximation to, or an improved version of, the input picture. Another major branch of picture processing deals with *image analysis* or *scene analysis*; here the input is still pictorial, but the desired output is a *description* of the given picture or scene. The following are examples of image analysis problems which have been extensively studied:

(1)   The input is text (machine printed or handwritten), and it is desired to read the text; here the desired description of the input consists of a sequence of names of characters.

(2)   The input is a nuclear bubble chamber picture, and it is desired to detect and locate certain types of "events" (e.g., particle collisions); here the description consists of a set of coordinates and names of event types.

(3)   The input is a photomicrograph of a mitotic cell; the desired output is a karyotype, or "map" showing the chromosomes arranged in a standard order. Note that here the output is pictorial, but its construction requires location and identification of the chromosomes.

(4)   The input is an aerial photograph of terrain; the desired output is a map showing specific types of terrain features (forests, urban areas, bodies of water, roads, etc.). Here again the output is pictorial, and is even in registration with the input; but construction of the output requires location and identification of the desired terrain types.

(5)   The input is a television image of a pile of parts; the desired output is a plan of action that can be used by a robot to assemble a device out of the parts. This evidently requires identification and location of individual parts in the scene.

In all of these examples, the description refers to specific *parts* (regions or objects) in the picture or scene; to generate the description, it is necessary to *segment* the picture into these parts. Thus to identify the individual characters in text, they must first be singled out; to locate bubble chamber events, the bubble tracks and their ends or branches must be found; to make a karyotype, the individual chromosomes must be "scissored out"; and so on. This chapter discusses picture and scene segmentation techniques. The remaining chapters of the book discuss how picture segments—regions or objects—are used in picture descriptions, which involve properties of objects and relationships among objects.

Some segmentation operations can be applied directly to any picture; others can only be applied to a picture that has already been partially segmented, since they depend on the geometry of the parts that have already been extracted from the picture. For example, a chromosome picture can be (crudely) segmented by thresholding its gray level—dark points probably belong to chromosomes, while light points are probably background. Once this has been done, further segmentation into individual chromosomes can be attempted, based on connectedness, size, and shape criteria. This chapter will deal primarily with the initial segmentation of pictures; methods of further segmenting an already segmented picture will be discussed in Chapter 11.

It should be emphasized that there is no single standard approach to segmentation. Many different types of picture or scene parts can serve as the segments on which descriptions are based, and there are many different ways in which one can attempt to extract these parts from the picture. The perceptual processes involved in segmentation of a scene by the human visual system, e.g., the Gestalt laws of organization, are not yet well understood (Section 3.4). For this reason, no attempt will be made here to define criteria for successful segmentation; success must be judged by the utility of the description that is obtained using the resulting objects.

In order to "extract" an object from a picture explicitly, we must somehow mark the points that belong to the object in a special way. This marking process can be thought of as creating a "mask" or "overlay," congruent with the picture, in which there are marks at positions corresponding to object points. We can regard this overlay as a two-valued picture (e.g., 1's at object points, 0's elsewhere); the overlay thus represents the "characteristic function" of the object, i.e., the function that has the value 1 for points belonging to the object, and value 0 elsewhere. There are many other ways of representing objects or regions, e.g., using their boundaries or their "skeletons" (see Chapter 11).

In practice, the objects that we try to extract from pictures are not always clearly defined, and it is not always wise to attempt to map these objects into clear-cut, two-valued overlays. Very often, our objects are defined only in a "fuzzy" sense, and it would be more appropriate to represent them by continuous-valued "membership functions." These are functions that can take on any value between 0 and 1; a point that has value 1 definitely belongs to the object, a point with value 0 definitely does not belong, while points with intermediate values have intermediate "degrees of membership" in the object. For an introduction to the theory of fuzzy sets see Kaufmann [31]. We shall indicate in the course of these chapters how some of the concepts discussed can be extended from clear-cut objects to fuzzy objects.

In Section 10.1 we review some basic concepts of pattern classification and clustering, and then discuss segmenting a picture by classifying the pixels on the basis of their gray levels or their spectral or spatial signatures.§ Sections 10.2–10.3 treat detection of local features (edges, lines, etc.) in pictures based on the values of appropriate local operators. Section 10.4 deals with sequential methods of segmentation in which the classification decision for a given pixel depends on the decisions made for previously examined pixels. Section 10.5 discusses iterative techniques in which pixels are "classified" fuzzily or probabilistically, and the degrees of class membership are then iteratively adjusted based on their mutual compatibilities.

10.1 PIXEL CLASSIFICATION

Segmentation is basically a process of *pixel classification*; the picture is segmented into subsets by assigning the individual pixels to classes.¶ For example, when we segment a picture by thresholding its gray level, we are classifying the pixels into "dark" and "light" classes, in an attempt to distinguish (e.g.) dark objects from their light background. Similarly, in edge

§ It should be pointed out that if range or velocity information is available for each pixel, obtained by special sensors or derived from stereopairs or image sequences as in Chapter 9, we can use this information as a basis for segmenting the picture. All of the segmentation methods discussed in this chapter—pixel classification, edge detection, region growing, etc.—can be used with range or velocity data; the details will not be given here.

¶ For an alternative definition of segmentation, in terms of partitioning the picture into homogeneous parts, see the beginning of Section 10.6.

detection, we are classifying pixels into "edge" and "not edge" by threshold-ing the response of some difference operator that has high values when the rate of change of gray level is high (see Section 10.2).

In Section 10.1.1 we review some basic concepts of classification and clustering. Sections 10.1.2 and 10.1.3 apply these concepts to picture segmen-tation by gray level thresholding or by classification of spectral or spatial signatures.

10.1.1   Pattern Classification and Clustering

Suppose that we want to classify a collection of objects (e.g. pixels) on the basis of the value of some property or properties (e.g., gray level, dif-ference value, etc.) that can be measured for each object. If we know the prob-ability density of the property values for each class, and the probabilities with which the classes occur, we can derive classification criteria that minimize the expected classification error. This is a standard problem in pattern recognition; see, e.g., [18] for an introduction to this subject. For simplicity, we shall first consider the case where there are only two classes (e.g., light/dark, edge/not-edge), and where classification is done on the basis of a single property value; other cases will be discussed later. Since we are interested in picture segmentation, we will assume that the objects are pixels.

a.   One property, two classes

Let the property value be denoted by z, and let the probability densities of the values of z for the two classes of pixels be $p(z|1)$ and $p(z|2)$, respectively. Moreover, let the *a priori* probabilities of the classes be $p(1)$ and $p(2)$, so that $p(1) + p(2) = 1$; then the overall probability density of the values of z for the entire picture is

$$p(1)p(z|1) + p(2)p(z|2) \tag{1}$$

Exercise 10.1.   Let μ, v be the means, and σ^2, τ^2 the variances of $p(z|1)$ and $p(z|2)$. Prove that (1) has mean $p(1)\mu + p(2)v$, and variance $p(1)\sigma^2 + p(2)\tau^2 + p(1)p(2)(\mu - v)^2$.   ∎

Suppose that we classify the pixels by thresholding z at t; in other words, pixels for which $z < t$ are assigned to class 1, and those for which $z \geq t$, to class 2. Then the probability of misclassifying a class 2 point as class 1 is $P(t|2) \equiv \int_{-\infty}^{t} p(z|2)\, dz$; and the probability of misclassifying a class 1 point as class 2 is $1 - P(t|1) \equiv \int_{t}^{\infty} p(z|1)\, dz$. The overall misclassification prob-ability is thus

$$p(1)[1 - P(t|1)] + p(2)P(t|2) \tag{2}$$

To find the value of t for which this probability is a minimum, we simply differentiate (2) with respect to t and set the result equal to zero, obtaining

$$p(2)p(t|2) = p(1)p(t|1) \tag{3}$$

In general, (3) can be solved for t numerically. If we know the mathematical forms of the probability densities for $p(z|1)$ and $p(z|2)$, it may also be possible to solve (3) analytically. For example, suppose that $p(z|1)$ and $p(z|2)$ are Gaussian with means and variances as in Exercise 10.1, so that

$$p(z|1) = \frac{1}{\sqrt{2\pi}\sigma} \exp[-(z-\mu)^2/2\sigma^2], \quad p(z|2) = \frac{1}{\sqrt{2\pi}\tau} \exp[-(z-v)^2/2\tau^2] \tag{4}$$

Of course, gray level probabilities cannot really be Gaussian, since they are nonnegative and lie in a bounded range; but if μ, v are well away from the ends of this range, and σ, τ are small, the probabilities can be approximately Gaussian. If we set $z = t$ in (4), substitute in (3), and take logarithms of both sides, we obtain

$$\ln \sigma + \ln p(2) + (t-\mu)^2/2\sigma^2 = \ln \tau + \ln p(1) + (t-v)^2/2\tau^2$$

or

$$\tau^2(t-\mu)^2 - \sigma^2(t-v)^2 = 2\sigma^2\tau^2 \ln[\tau p(1)/\sigma p(2)] \tag{5}$$

This quadratic equation can be solved for t in terms of μ, v, σ, τ, $p(1)$, and $p(2)$. For example, if $p(1) = p(2) = \frac{1}{2}$ and $\sigma = \tau$, then (5) simplifies to $(t-\mu)^2 = (t-v)^2$, so that $t = (\mu + v)/2$.

Exercise 10.2.   Verify that in the Gaussian case, if $\sigma = \tau$ but $p(1)$ is arbitrary, the solution of (5) is

$$t = \frac{\mu + v}{2} + \frac{\sigma^2}{v - \mu} \ln[p(1)/p(2)] \quad \blacksquare$$

More generally, if we know $p(z|1)$, $p(z|2)$, $p(1)$, and $p(2)$, then for any given z we can determine whether it is more likely to arise from a pixel in class 1 or in class 2. In fact, let $p(1|z)$ and $p(2|z)$ be the probabilities that a pixel having value z is in class 1 or class 2. Then the joint probability $p(1, z)$ that a pixel belongs to class 1 and has value z satisfies

$$p(1, z) = p(1)p(z|1) = p(1|z)p(z) \tag{6a}$$

where $p(z)$ is the *a priori* probability that a pixel has value z; this is evidently given by (1). Similarly, the joint probability $p(2, z)$ that a pixel belongs to class 2 and has value z satisfies

$$p(2, z) = p(2)p(z|2) = p(2|z)p(z) \tag{6b}$$

It follows that

$$p(1|z) = p(1)p(z|1)/p(z), \qquad p(2|z) = p(2)p(z|2)/p(z) \qquad (7)$$

where $p(z)$ is given by (1). Thus we can decide which class is more likely, for a given value of z, by determining which of $p(1|z)$ and $p(2|z)$, as given by (7), is larger—or, equivalently, which of $p(1)p(z|1)$ and $p(2)p(z|2)$ is larger.

b. Several properties, several classes

This discussion is easily generalized to the case where there are $k \geqslant 2$ classes of pixels. Let $p(z), p(i), p(i, z), p(i|z)$, and $p(z|i)$ be defined just as in (a) for $i = 1, ..., k$. Then we have

$$p(z) = \sum_{i=1}^{k} p(i)p(z|i) \qquad (1')$$

and

$$p(i, z) = p(i)p(z|i) = p(z)p(i|z), \qquad 1 \leqslant i \leqslant k \qquad (6')$$

so that

$$p(i|z) = p(i)p(z|i)/p(z) \qquad (7')$$

Thus if the $p(i)$'s and $p(z|i)$'s are known, we can determine the i for which $p(i|z)$, as given by (7'), is greatest for any given z. In practice, the $p(i)$'s and $p(z|i)$'s are not known; but they can be estimated if we are given a set of pictures in which the pixels have been correctly classified.

Still more generally, suppose that we want to classify pixels on the basis of an m-tuple of properties (e.g., color components, or a set of local property values; see Section 10.1.3). Let $\vec{z}$ be the m-dimensional vector whose components are the given property values. Then we can define $p(\vec{z}), p(i, \vec{z})$, $p(i|\vec{z})$, and $p(\vec{z}|i)$ just as above, and can analogously determine the i for which $p(i|\vec{z})$ is greatest for any given $\vec{z}$; in fact, we use equations exactly like (6') and (7'), except that the z's are now vectors rather than scalars. Unfortunately, the probability densities involved in these equations are now much harder to estimate, since $\vec{z}$ can take on many possible m-tuples of values, and we need a very large number of classified samples in order to estimate the probabilities of all these m-tuples for each class.

The approach discussed in this section minimizes the expected classification error, since it assigns each pixel to the most probable class. More generally, suppose that we are given a *cost function* which specifies, for each pair of classes i, j, the cost $\lambda(i|j)$ of assigning a pixel to class i when it really belongs to class j. These costs may not all be equal; for example, if the classes are "target" and "background," the cost of a false alarm (background classified as target) may be very different from that of a false dismissal (target classified as background). Thus we may really want to minimize the expected cost $c(i|\vec{z})$, which is given by $\sum_j p(j|\vec{z})\lambda(i|j)$. Note that if $\lambda(i|i) = 0$ and the $\lambda(i|j)$'s are all equal for $j \neq i$, this is equivalent to minimizing $\sum_{j \neq i} p(j|\vec{z})$, or to maximizing $p(i|\vec{z})$, as above.

c. Clustering

As a final generalization, suppose that the set of classes is not known *a priori*. In this situation we can look for *clusters* in the set of z (or $\vec{z}$) values, i.e., for ranges of values that occur relatively often, separated by ranges that occur less often. These clusters intuitively correspond to natural subpopulations of the pixels, i.e., to natural classes.

In the case of a single feature z, the clusters should show up as peaks on the histogram of z values. These peaks should be at least some minimum distance apart, to insure that they represent distinctive subpopulations; we might also require that there be relatively deep valleys between them. Under these circumstances, we can segment the picture by choosing thresholds so as to separate the peaks, e.g., at the bottoms of the valleys between the peaks. This method of thresholding will be discussed further in Section 10.1.2.

For two (or more) features, say z_1 and z_2, the clusters should show up as dense regions on the scatter plot of z_1 versus z_2, and we should be able to separate them by partitioning the (z_1, z_2) space into appropriate pieces. For example, we can select a set of local maxima of density on the scatter plot, and expand a rectangular neighborhood around each of them; we stop the expansion if the rate at which points are taken into the rectangle starts increasing, since this presumably means that the rectangle has hit another cluster. When this has been done for all the chosen maxima, the resulting rectangles contain a set of clusters. Points not in any of the rectangles can be classified as belonging to the nearest cluster. The procedure is analogous for more than two features. There are many different methods of cluster detection and separation; we will not attempt to review this subject here.

10.1.2 Gray Level Thresholding

We now discuss segmentation techniques which classify pixels based on their gray levels. In Section 10.1.3 we will consider pixel classification on the basis of color or local property values.

a. Thresholding

Many types of pictures are composed of two kinds of regions that occupy different gray level ranges. For example, in a picture of printing or writing, the characters are generally darker than the paper. In pictures of mitotic cells or of bubble chambers, the chromosomes or tracks are darker (lighter) than the background; in pictures of the earth taken from a satellite, the clouds are lighter than most types of terrain.

In all of these cases, the gray level histogram of the picture should display peaks corresponding to the two gray level ranges (characters and paper, chromosomes or tracks and background, terrain and clouds). The picture can thus be segmented by choosing a threshold that separates these peaks, as discussed in Section 10.1.1c. A histogram having two peaks is called *bimodal*.

To illustrate this process, consider the pictures of writing, chromosomes, and clouds shown in Figs. 1a–1c. The histograms of these pictures are shown in Fig. 2; note that the histogram of the cloud picture has no well-defined valley. Results of applying various thresholds to these pictures are shown in Figs. 1d–1l; the pixels whose gray levels are darker than the threshold are displayed as black, and those lighter than the threshold as white. In parts (d)–(f) there is too much black; in parts (g)–(i), too much white; while in parts (j)–(l), the results seem to be about right. Note that these last thresholds correspond to histogram valley bottoms, except in the case of the cloud picture, where there is no clear-cut bottom; in this case the good threshold is just adjacent to the peak.

As a useful variation on the display of thresholding results, we can display the pixels that are lighter than threshold as white, and leave those darker than threshold with their original gray levels, or vice versa. This process, which might be called *semithresholding*, is illustrated in Fig. 3.

b. Threshold selection

A threshold that separates the peaks on a bimodal gray level histogram can be automatically chosen by a method such as the following: Let the histogram be $p(z)$. We find two local maxima on the histogram that are at least some minimum distance apart; let these be at z_i and z_j. Let z_k be the point between z_i and z_j at which $p(z)$ has its lowest value. If $p(z_k)/[\min(p(z_i),\ p(z_j))]$ is small, then $p(z_k)$ is much smaller than either $p(z_i)$ or $p(z_j)$, so that the histogram has a significant valley between the peaks, and z_k should thus be a useful threshold for segmenting the picture.

Using the lowest point between two peaks as a threshold is intuitively plausible, but if we know the probability densities of the two subpopulations, we may find that this threshold is not optimal. For example, let $p(z|1)$ and $p(z|2)$ be given by (4), and let the overall probability density for the picture be given by (1); then for $\sigma = \tau$ [but $p(1)$ arbitrary], the threshold $z = t$ that minimizes misclassification probability is given in Exercise 10.2. On the other hand, (1) does not, in general, have derivative equal to zero at $z = t$, so that this value of z is not a valley bottom.

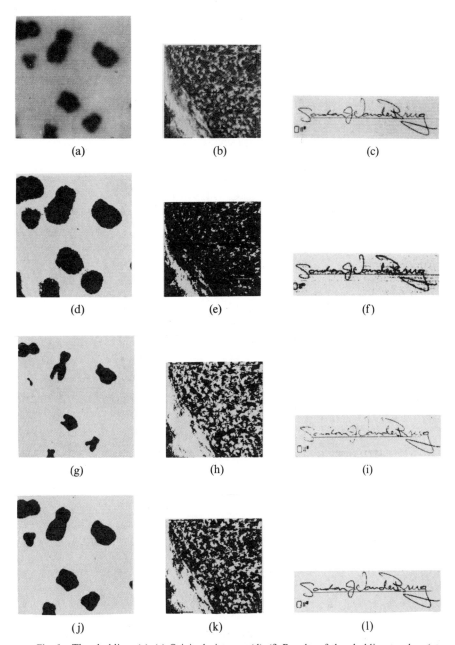

Fig. 1 Thresholding. (a)–(c) Original pictures. (d)–(f) Results of thresholding too low (at 11, 25, and 10, respectively, on a scale of 0–63). (g)–(i) Results of thresholding too high (at 45, 52, and 35). (j)–(l) Results of using good threshold levels (28, 37, and 29).

```
 0     0  0
 2     0  0
 4     0  0
 6     0  0
 8   106  0 +
10   892  1 ++++++++
12  3438  6 +++++++++++++++++++++++++++++++++++
14  9773 21 ++++++++++++++++++++++++++++++++++++++++++++++++++++++++++++++++++++++++++++++++
16  8514 34 ++++++++++++++++++++++++++++++++++++++++++++++++++++++++++++++++++++++
18 11039 51 ++++++++++++++++++++++++++++++++++++++++++++++++++++++++++++++++++++++++++++++++++++++
20  8345 64 +++++++++++++++++++++++++++++++++++++++++++++++++++++++++++++++++
22  4574 71 ++++++++++++++++++++++++++++++++++++
24  2796 76 ++++++++++++++++++++++++
26  1656 78 ++++++++++++++++
28  1249 80 +++++++++++++
30   936 81 ++++++++
32   787 83 +++++++
34   664 84 ++++++
36   594 85 +++++
38   572 86 +++++
40   563 86 +++++
42   551 87 +++++
44   628 88 ++++++
46   661 89 ++++++
48   798 90 +++++++
50   996 92 +++++++++
52  1365 94 ++++++++++++
54  1680 97 +++++++++++++++
56  1402 99 ++++++++++++
58   442 99 ++++
60     4 100
                                                                            (a)
```

```
 0  2048  4 ++++++++++++++++++++
 2   949  6 +++++++++
 4  1052  9 +++++++++
 6   968 11 ++++++++
 8   976 13 +++++++++
10   960 15 ++++++++
12  1025 18 +++++++++
14   986 20 ++++++++
16  1034 22 +++++++++
18  1117 25 +++++++++++
20  1164 27 ++++++++++
22  1136 30 ++++++++++
24  1146 33 ++++++++++
26  1228 35 +++++++++++
28  1324 39 ++++++++++++
30  1560 42 ++++++++++++++
32  1724 46 +++++++++++++++++
34  1992 51 ++++++++++++++++++
36  2692 57 ++++++++++++++++++++++++++
38  3557 65 ++++++++++++++++++++++++++++++++++
40  4220 74 +++++++++++++++++++++++++++++++++++++++
42  4660 85 +++++++++++++++++++++++++++++++++++++++++++
44  3437 93 +++++++++++++++++++++++++++++++++
46  1589 96 +++++++++++++++
48   622 98 ++++++
50   249 98 ++
52   153 99 +
54   126 99 +
56    97 99
58    46 99
60    21 99
62     2 100
                                                                            (b)
```

```
 0  3818 12 ++++++++++++++++++++++++++++++++++++
 2  5104 28 ++++++++++++++++++++++++++++++++++++++++++++++++
 4  5900 47 +++++++++++++++++++++++++++++++++++++++++++++++++++++++++
 6  6020 67 +++++++++++++++++++++++++++++++++++++++++++++++++++++++++++
 8  4312 80 ++++++++++++++++++++++++++++++++++++++++++
10  1827 86 +++++++++++++++++++
12   763 89 +++++++
14   288 90 ++
16   173 90 +
18   123 91 +
20    85 91
22    83 91
24    82 91
26    74 92
28    69 92
30   106 92 +
32    85 92
34    68 93
36    93 93
38   121 93 +
40   134 94 +
42   154 94 +
44   212 95 ++
46   252 96 ++
48   295 97 ++
50   313 98 +++
52   249 99 ++
54   161 99 +
56    97 99
58    36 99
60     7 100
                                                                            (c)
```

Fig. 2 Gray level histograms for the pictures in Figs. 1a–1c. Each value i (0, 2, 4, . . .) in the first column represents gray levels i and $i + 1$. The second column shows the number of pixels having these gray levels (proportional to the bar height), and the third column shows the percentile.

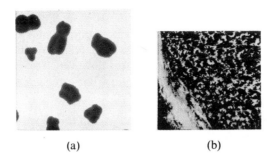

(a)   (b)

Fig. 3 Semithresholding. (a) Upper semithresholding: In Fig. 1a, gray levels below 37 have been mapped into white, while the remaining gray levels are preserved intact. (b) Lower semithresholding: In Fig. 1b, gray levels 29 and above have been mapped into black.

Exercise 10.3. Prove that if $\sigma = \tau$ and $p(1) = p(2) = \frac{1}{2}$, then the derivative of (1) is zero at $z = (\mu + \nu)/2$ (which is the minimum-error threshold; see just prior to Exercise 10.2). On the other hand, prove that this point is a local minimum only if $|\nu - \mu| > 2\sigma$, while if $|\nu - \mu| < 2\sigma$ it is a local maximum. ▌

This discussion suggests that an alternative method of threshold selection might be to approximate the given gray level histogram by a linear combination of standard probability densities, e.g., Gaussians, each of which is *unimodal* (i.e., has only one peak), and choose the minimum-error threshold defined by these Gaussians.

Still another approach to threshold selection is to try a range of thresholds, and choose the one for which the thresholded picture has some desired property. The following are some examples of this approach:

(1) For any given threshold t, we can measure the "busyness" of the thresholded picture, i.e., the number of adjacencies between above-threshold and below-threshold pixels. Intuitively, a good choice of t would ordinarily be one that minimizes this busyness—provided, of course, that t does segment the picture nontrivially, and does not make it completely blank.

(2) If we know what fraction of the pixels should be above threshold, we can choose t accordingly; in other words, if the fraction is $1 - p$, we choose t at the *p-tile* of the picture's histogram. This method is applicable to, e.g., pictures of printed pages, where we know approximately what fraction of the page should be occupied by characters.

(3) If we know something about the individual sizes of the above-threshold objects (e.g., that printed characters are composed of strokes having specified widths), we can choose t to give these sizes the desired value —as measured, e.g., by the lengths of runs of above-threshold points on the rows of the picture.

Exercise 10.4.   Let f_t be the result of thresholding the picture f at t, and let
the two gray levels in f_t be b and w. One might think that a possible basis for
choosing t would be to minimize the mean-squared error between f and f_t,
i.e., to minimize $\iint (f - f_t)^2 \, dx \, dy$. Prove, however, that no matter what
the probability density of the gray levels in f, this error is always minimized
by taking $t = (b + w)/2$.   ∎

One can also choose a threshold so as to minimize some measure of the
homogeneity of the two resulting populations (above and below the threshold), and/or maximize some measure of the difference between the populations [37, 55].

c. Multilevel thresholding

If a picture contains more than two types of regions, it may still be possible
to segment it by applying several thresholds. For example, in pictures of
white blood cells the nucleus is generally darker than the cytoplasm, which
is in turn darker than the background; thus the histogram has three peaks,
and the picture can be segmented using two thresholds that separate these
peaks, as illustrated in Fig. 4. The two thresholds divide the gray scale into
three ranges, which we have displayed as black, gray, and white. A histogram
having several peaks is called *multimodal*.

As the number of region types increases, the peaks become harder to
distinguish, and segmentation by thresholding becomes more difficult.
Figure 5, for example, shows a picture of a house; if we threshold so as to
separate the peaks on its histogram, we can segment out the sky, but we

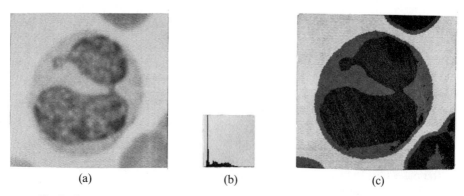

| (a) | (b) | (c) |

Fig. 4 Segmentation by multilevel thresholding. (a) Picture of blood cells. (b) Gray level
histogram of (a), showing three peaks, corresponding (for the white blood cell in the center) to
background, cytoplasm, and nucleus. (c) Result of applying two thesholds (9 and 19); gray
levels in the two extreme ranges are displayed as white and black, and those in the midrange as
gray.

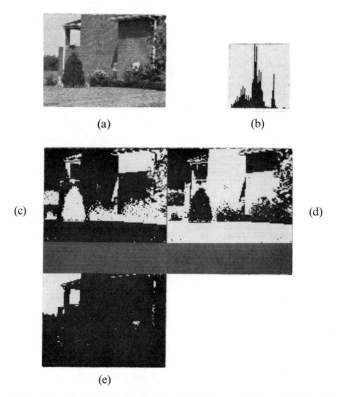

Fig. 5   Limitations of multilevel thresholding. (a) Picture of a house. (b) Gray level histogram of (a), showing three principal peaks. (c)–(e) Result of applying two thresholds to (a); in each part, gray levels in a given range are displayed as white, and levels outside that range as black, for ranges [0, 23], [24, 44], and [45, 56].

cannot distinguish the sunlit brick from the grass, or the shadowed brick from the bushes.

All of the methods of threshold selection described in (b) can be generalized, in principle, to the multimodal case. However, the results will often be less reliable, since the gray level subpopulations may be less clearly separable.

It is sometimes useful to select a gray level range that corresponds to a valley, rather than a peak, on the histogram; this will extract the pixels that lie on borders between light and dark regions, and so will produce outlines of the regions, provided that the transitions between regions are not too abrupt. As an example, consider the chromosome picture in Fig. 1a. If we map the valley gray levels of this picture into white, and all other levels into black, we obtain white outlines of the chromosomes, as shown in Fig. 6.

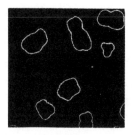

Fig. 6 Outlining by thresholding. Using Fig. 1a, gray levels 35–39 have been mapped into white, all other levels into black.

(Other, more conventional methods of edge detection will be discussed in Section 10.2.)

If we choose thresholds that slice the gray scale into many small ranges, and display every kth range as black and the other ranges as white, we obtain a gray level "contour map" of the given picture, provided that the gray level variation on the picture is sufficiently smooth.

Even when there are only two types of regions, it may be advantageous to use two thresholds in order to reduce the noisiness of the thresholded picture. Specifically, let us choose $t_1 < t_2$ such that all background points have levels below t_2, while all object points have levels above t_1, and some have levels above t_2. We first threshold at t_2 to extract these "core" object points; and we also accept points whose levels exceed t_1 as object points, provided they lie close to the core points. If we used t_1 alone, many background points might be called object points, while if we used t_2 alone, the objects would be incomplete. This method can be regarded as a simple example of region growing; such methods will be discussed in greater detail in Section 10.4.

d. Smoothing and thresholding

The gray level subpopulations corresponding to the different types of regions in a picture will often overlap. Under these circumstances, segmenting the picture into regions by thresholding becomes difficult, since wherever we put the threshold, it is impossible to cleanly separate the overlapping subpopulations.

We can usually alleviate this overlap problem by smoothing the picture before thresholding it. For example, we can simply locally average the picture, i.e., replace the gray level at each point by an average of neighboring gray levels. Within a given type of region, averaging dampens local gray level fluctuations, and hence reduces the variability of the gray levels, while preserving the mean gray level, as discussed in Section 6.4.2. Of course, averaging also blurs the borders of the regions; but thresholding will still extract the regions more or less correctly, though it will smooth out irregularities in their borders.

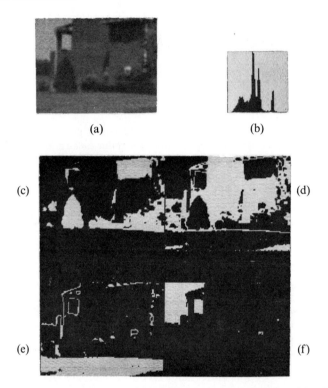

Fig. 7 Histogram peak sharpening by local averaging. (a) Result of 3 × 3 local averaging of the house picture in Fig. 5a. (b) Gray level histogram of (a); note that, as compared with Fig. 5b, the peaks are smoother, and the tall peak has been separated into two distinct peaks. (c)–(f) Result of applying three thresholds to (a); analogous to Figs. 5c–5e, using ranges [5, 19], [20, 30], [31, 40], and [43, 51]. Note that the grass and brick have been separated.

Figure 7 illustrates how local averaging can sharpen the peaks on a picture's histogram and make them easier to separate. (The unsmoothed picture and its histogram were shown in Fig. 5.)

An extreme example of histogram improvement by local averaging is shown in Fig. 8. Here the original picture contains only two gray levels, but they occur in different proportions in the two regions, so that the regions have different means. When we smooth this picture, the gray level subpopulations in the two regions shrink toward these means, and eventually become separable by thresholding.

It has already been pointed out that this local averaging method will not preserve the details of region shapes, since it will tend to smooth out irregularities at the borders between the regions. Note also that if there are three (or more) types of regions, say S_1, S_2, S_3, where the average gray level for

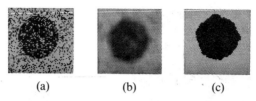

(a)   (b)   (c)

Fig. 8   Extreme example of region extraction by averaging and thresholding. (a) Original picture: densely dotted region on a sparsely dotted background. Since all dots have the same gray level, the densely dotted region cannot be extracted by thresholding. (b) Results of smoothing (a); the gray level at point (i, j) of (b) is the average of the gray levels in a 7×7 square centered at (i, j) in (a). (c) Results of thresholding (b) at 32. The densely dotted region has now been extracted.

S_2 is intermediate between those for S_1 and S_3, then averaging cannot be used to extract S_2 if S_1 and S_3 are adjacent to one another, since a strip along the border between S_1 and S_3 will have the same average as S_2.

e.   Variable thresholding

In many cases, no single threshold gives good segmentation results over an entire picture. Suppose, for example, that the picture shows dark objects on a light background, but that it was made under conditions of uneven illumination. The objects will still contrast with the background throughout, but both background and objects may be much lighter on one side of the picture than on the other. Thus a threshold that nicely separates objects from background on one side may accept too much of the background as belonging to the objects, or reject too much of the objects as belonging to the background, on the other side. The same remarks apply when a scene contains shadows, or when the picture was obtained by a sensor whose sensitivity or light gathering power varies from point to point (as in optical "vignetting').

If the uneven illumination is described by some known function of position in the picture, one could attempt to correct for it using gray level correction techniques (Section 6.2.1), after which a single threshold should work for the whole picture. If this information is not available, one can divide the picture into blocks and apply threshold selection techniques to each block. If a block contains both objects and background, its histogram should be bimodal, and the valley bottom should yield a *local threshold*, suitable for separating objects from background in that part of the picture. If a block contains objects only, or background only, it will not have a bimodal histogram, and no such threshold will be found for it; but a threshold can still be assigned to it by interpolation from the local thresholds that were found for nearby bimodal blocks. (Some smoothing of the resulting thresholds may be necessary, since if a threshold changes abruptly from one block to the next, artifacts may result.) An example of local histogramming and threshold interpolation is shown in Fig. 9.

Another situation in which local thresholding may be helpful is where the objects to be extracted are very small and sparse (bubble tracks, stars, etc.), so that the picture consists almost entirely of background, and the objects produce no detectable peak on its histogram. This problem is alleviated by taking histograms of small blocks of the picture; for blocks that contain objects, these histograms should be bimodal, so that thresholds can be selected for these blocks and interpolated to the rest of the picture.

Alternatively, if there exists a threshold t such that every object contains points with levels higher than t, but the background contains no such points, we can threshold the picture at t and then examine the neighborhoods of the above threshold points, with the aim of finding a local threshold that will separate object from background in each of these neighborhoods. (This is evidently more general than the two-threshold method described in (c).) This method is appropriate if the objects are relatively small and do not occur too close together. The neighborhoods used should be large enough to ensure that they contain both object and background points, so that the histograms of the gray levels in the neighborhoods will be bimodal.

Rather than looking for blocks having bimodal histograms, and using them to define local thresholds, we can choose blocks that have low variability, and use their average gray levels as "cluster centers," since any such block presumably lies within a single type of region. We can then classify the pixels on the basis of their distances from these centers. This approach may be appropriate in cases where there are several types of regions.

f. Local maxima

One sometimes wants to find *local maxima* in a picture, i.e., to extract points which have higher value, with respect to some local property, than the nearby points. Typically, one would also require these points to have values above some low threshold t; but once t is exceeded, all relative maxima are accepted, no matter what their absolute sizes. Local maximum finding can thus be regarded as an extreme case of local thresholding. The following are some cases in which local maxima are of interest:

(a) When a matching operation has been applied to a picture, we may want to find the best matches, even though they are far from perfect matches, as long as the degree of match exceeds some low threshold. If, however, two or more high matches occur too close together, we would normally want to keep only the highest of them, representing the position of best match. This is because if the shape or pattern that is being matched occurs at position (x, y), say, we do not want to detect it also at nearby positions where it would

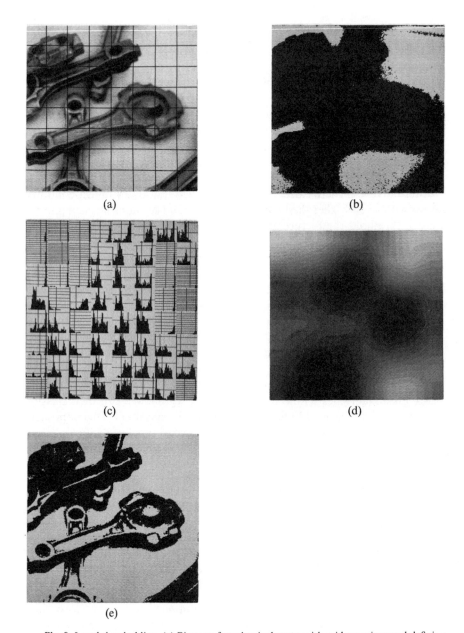

(a)

(b)

(c)

(d)

(e)

Fig. 9 Local thresholding. (a) Picture of mechanical parts, with grid superimposed defining blocks. (b) Result of applying a single threshold to (a), obtained by fitting a mixture of two Gaussians to the histogram of (a). (c) Histograms for the individual blocks in (a); the lines provide an indication of the relative scale of these histograms. A threshold was obtained for each block by fitting a mixture of two Gaussians to its histogram. (d) Display of the array of thresholds obtained by interpolation from these block thresholds. (e) Result of applying these thresholds to every point of (a).

have to overlap its occurrence at (x, y). For examples of matching operation outputs, which show local peaks at the positions of best match, see the figures in Section 9.4.

(b)   When edges (or lines, etc.) are detected in a picture, a given edge may be detected in more than one position (see Section 10.2); if we kept all of these, the resulting edge would be thick. We can keep a detected edge thin by rejecting nonmaxima in the direction across the edge, i.e., we ignore an edge detection value if there exists a higher value nearby in that direction. (On the other hand, we should not suppress an edge value if there is a higher value nearby along the edge, since we do not want the points along the edge to compete with one another.) Examples of nonmaximum suppression in edge detection will be given in Section 10.2.2.

Still other types of local maxima are of interest in the analysis of shapes, e.g., points of an object at which the distance to the background is a local maximum constitute the "skeleton" of the object; see Section 11.1.2a.

10.1.3   Spectral and Spatial Classification

In Section 10.1.2 we discussed how to segment a picture by classifying its pixels on the basis of their gray levels. If we are given a color picture, or a picture obtained by a multispectral scanner, we can classify the pixels on the basis of their colors or spectral signatures; this generally yields a much more refined classification, since it is based on several property values rather than a single value. Even in the case of a black-and-white picture, it is possible to measure a set of properties (e.g., local properties) for each pixel, and use the values of these properties as a basis for classification. Methods of pixel classification based on spectral or spatial ($=$ local) properties are discussed in this section.

a.   Spectral classification

As we saw in Section 10.1.2, homogeneous regions in a black-and-white picture often give rise to peaks on the picture's gray level histogram, i.e., to heavily populated regions in "gray level space." Analogously, homogeneous regions in a color picture may give rise to clusters (i.e., heavily populated regions) in color space. For example, suppose that colors are represented by triples of numbers representing the strengths of their red, green, and blue components (or their components in some other color coordinate system); see Section 3.3. Thus the color of each pixel defines a point in (red, green, blue) space, and if the picture contains a large region of pixels all having approximately the same color, say (r_0, g_0, b_0), we should expect to find a

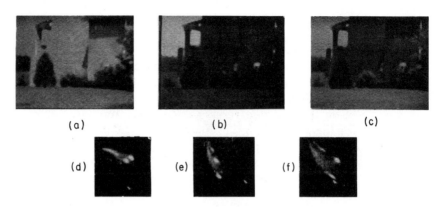

Fig. 10   Color space. (a)–(c) Red, green, and blue components of the house picture. (d)–(f) Projections of the (red, green, blue) feature space for this picture, projected onto the (red, green), (green, blue), and (blue, red) planes.

cluster of points in the vicinity of (r_0, g_0, b_0). Analogous remarks apply to pictures obtained by multispectral scanners; the "color coordinates" are now the intensities in each of the spectral bands.

Figures 10a–10c show the red, green, and blue components of a color picture of a house (the same one shown in Figs. 5 and 7). Figures 10d–10f display the (red, green, blue) space for this picture, projected onto the (red, green), (green, blue), and (blue, red) planes; high densities of points are displayed as light. (The densities have been logarithmically scaled to make variations more visible.) In other words, these displays show scatter plots of the (red, green), (green, blue) and (red, blue) values. Substantial clustering is apparent.

We can obtain a very good segmentation of the house picture by extracting the pixels whose color coordinates lie in specified ranges. For example, suppose that we use the pairs of ranges defined in Fig. 11 for the red and blue components. If we display the pixels that lie in each range as black, and all other pixels as white, we obtain the set of segmentations shown in Figs. 11a–11h. By combining sets of these ranges, we can further improve the segmentation, as shown in Figs. 11i and 11j. Combinations of ranges are needed because some of the clusters are elongated in a diagonal direction, and cannot be segmented using upright rectangles in (red, blue) space. Note that parallel, diagonal clusters will overlap when projected on either axis, and so cannot be separated by thresholding with respect to either coordinate alone.

Automatic detection of clusters in color space (or any other multidimensional feature space) is much more complicated than histogram peak detection, since clusters can have complex shapes and can interact in many ways

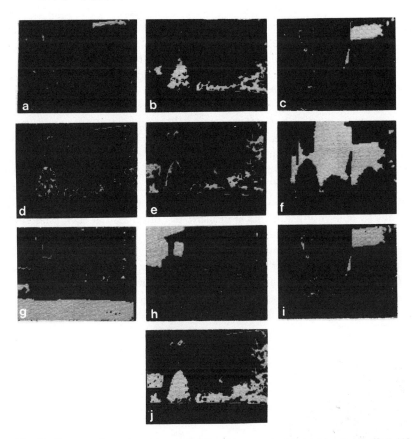

Fig. 11 Segmentation using color space. The ranges are {red level, blue level}: {[15, 20], [7, 12]}, {[13, 19], [13, 23]}, {[21, 28], [11, 17]}, {[20, 22], [18, 23]}, {[18, 26], [24, 33]}, {[37, 44], [22, 31]}, {[27, 39], [37, 44]}, {[37, 41], [46, 49]}, for (a)–(h), respectively. Pixels lying in each range are displayed as white, all others as black. (i)–(j) Unions of ranges {a, c} and {b, d, e}. These, together with ranges f, g, and h, give a fairly good segmentation into shadow, shrubbery, brick, grass, and sky.

(e.g., one cluster can surround another). If the clusters are relatively "globular" and do not overlap greatly, the simple method of extracting them described in Section 10.1.1c can be used.

b. Local property thresholding

There are many ways of segmenting a picture by thresholding the values of local properties (other than gray level) measured at each point. For example, as we shall see in Sections 10.2–10.3, local features such as edges, lines, spots, etc., are detected in a picture by applying local operators that yield high values at points where such features are present and low values elsewhere; one can then detect the features by thresholding these values. Similarly, points where the picture matches a given template can be detected by applying the template everywhere and thresholding the match values; on template matching see Section 9.4.

In the examples just given, the result of the segmentation is not a set of homogeneous regions, but rather a relatively sparse set of "special" points (e.g., points that lie on edges). We shall now discuss how to use local property thresholding to extract regions from a picture. For example, suppose that the picture is composed of "busy" regions and "smooth" regions, such as the bushes and the grass in the bottom part of the house picture (Figs. 5, 7, and 10). It should be possible, in principle, to segment the picture into these regions by computing some local measure of "busyness" at each point, and thresholding the values of this measure.

A difficulty with this simple idea is that the values of local properties such as busyness tend to be highly variable. Thus the peaks on a busyness histogram corresponding to smooth and busy regions will overlap, since even the busy regions will contain many points at which the busyness is low. This is illustrated in Figs. 12a and 12b, which show the busyness values for the

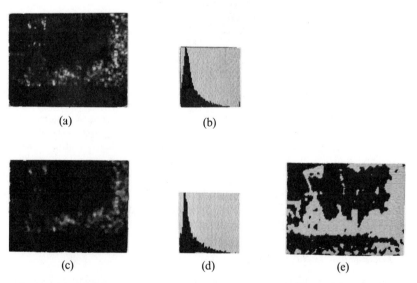

(a)   (b)

(c)   (d)   (e)

Fig. 12 Segmentation using "busyness." (a) Busyness values for the house picture, displayed as gray levels (high busyness = light). (b) Histogram of these values. (c) Result of 3 × 3 local averaging of (a). (d) Histogram of (c). (e) Points of (c) having values ≥ 12, displayed as white. The shrubbery has been extracted, but so have many other parts of the picture.

house picture, and the histogram of these values. [Busyness was measured by $\min(v_x, v_y)$, where in the neighborhood

$$A \quad B \quad C$$
$$D \quad E \quad F$$
$$G \quad H \quad I$$

v_x and v_y are defined as

$$v_x = |A - B| + |B - C| + |D - E| + |E - F| + |G - H| + |H - I|$$
$$v_y = |A - D| + |D - G| + |B - E| + |E - H| + |C - F| + |F - I|$$

The min was used in order to reduce the values obtained at edges, since the v measure should have low values in the direction along an edge; on the use of a max of absolute differences for edge detection see Section 10.2.] The busy regions give rise to a histogram peak that extends away from zero, but that still overlaps with the peak near zero corresponding to the smooth regions.

Better results can be obtained if we smooth the busyness values by locally averaging them over a neighborhood of each point. As we saw in Section 10.1.2d, averaging reduces the variability of the values, and so makes the histogram peaks easier to separate. Figure 12c shows the busyness values for the house picture after averaging, and Fig. 12d shows their histogram; a slight improvement is detectable. Figure 12e shows the segmentation obtained by thresholding the averaged busyness values.

An artificial example of segmentation by thresholding averaged local property values is shown in Fig. 13. In part (a) the central region is very busy and the surrounding region perfectly smooth. Part (b) displays the values of a simple difference operator applied to (a); these values are all zero in the surround, but they are not all high in the center. However, if we locally average (b), we eliminate the low values in the center, so that the picture can be segmented by thresholding, as shown in part (c).

(a)   (b)   (c)

Fig. 13 Segmentation by thresholding of averaged local property values. (a) Salt-and-pepper noise (0's and 63's) on a solid gray background; the central region cannot be extracted by averaging and thresholding, since its average gray level is the same as that of the background. (b) Result of applying a difference operator to (a). (c) Result of 11×11 local averaging of (b), followed by thresholding at 37.

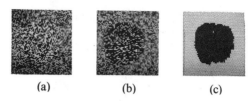

(a)   (b)   (c)

Fig. 14 Same as Fig. 13, except that the background consists of uniformly distributed random gray levels.

Another artificial example is shown in Fig. 14. Here both the central and surrounding regions are busy, but the former has higher contrast; thus if we apply the difference operator, we obtain a higher range of values in the center than in the surround. These ranges overlap substantially, but by local averaging we can separate them well enough to permit segmentation by thresholding. It should be mentioned that in Figs. 13a and 14a the average gray levels of the center and surround are the same, so that these pictures cannot be segmented by locally averaging and thresholding their gray levels; some local property other than gray level must be used.

The general technique just described—computing a local property at every point, then smoothing by local averaging, and finally thresholding—can be used to segment many types of pictures composed of regions that differ in visual texture. The use of mean values of local properties to discriminate textures will be discussed further in Section 12.1.5c.

An advantage of using local properties, rather than properties such as gray level and color, to classify pixels is that the latter remain the same no matter how the pixels are rearranged. For example, a picture composed of salt-and-pepper noise has the same gray level histogram as a picture whose top half is black and bottom half white. Properties such as busyness, on the other hand, do depend strongly on the arrangement of the pixels, so that their histograms provide more reliable information about the contents of the picture.

c. Spatial classification

Thresholding the values of a single local property provides only a limited segmentation capability. A more powerful approach is to classify the pixels on the basis of several local property values, possibly including gray level. Several examples of this approach are discussed in the following paragraphs.

One possibility is to use as properties the gray levels of the pixel and of some of its neighbors, or perhaps average gray levels computed over sets of neighbors. In a smooth region, these gray levels will all be closely similar, whereas in a busy region they will differ; thus pixels belonging to smooth and busy regions should give rise to points in different parts of this property space. In practice, however, it is difficult to obtain good segmentations into smooth and busy regions using sets of gray levels as properties. To see why this is so, consider a smooth region and a busy region each of which has average gray level z, and suppose that we use as properties of a pixel its own gray level and the average gray level of its eight neighbors. In (gray level, average gray level) space, the smooth region gives rise to a tight cluster of points centered at (z, z), while the busy region gives rise to a looser, elongated cluster (the gray level values are more variable than the average gray level values), also centered around (z, z). Since these clusters are concentric, they are impossible to separate.

Using several (average) gray levels as properties is not particularly useful for discriminating between smooth and busy regions, but it does have advantages where we want to discriminate between regions that differ in gray level. To see this, consider a picture composed of smooth light and dark regions, say having gray levels near z_1 and z_2, respectively. In (gray level, average gray level) space, the interior pixels of the light regions give rise to a cluster centered at (z_1, z_1), while the dark region interiors yield a cluster centered at (z_2, z_2); thus both of these clusters are on the "diagonal" of the scatter plot. On the other hand, pixels near the borders between light and dark regions give rise to off-diagonal points, since for such pixels, the gray level and average gray level differ significantly. Suppose, then, that we construct a histogram of the gray levels of those pixels whose (gray level, average gray level) values lie on or near the diagonal. These pixels should all lie in the interiors of the light or dark regions; thus the peaks on this new histogram should be more cleanly separated than those on the original histogram, since the pixels near the borders, which may have gray levels intermediate between z_1 and z_2, have been eliminated. Conversely, suppose that we consider only the gray levels of those pixels whose (gray level, average gray level) values lie far from the diagonal. These pixels should lie on the region borders; thus their average gray level should be a good threshold for discriminating between the regions. This approach to obtaining more useful histograms for segmentation is illustrated in Fig. 15.

Similar remarks to those in the last paragraph apply if we use gray level and its rate of change (e.g., the value of a suitable difference operator; see Section 10.2) as properties. In (gray level, absolute difference) space, the interior pixels of the light regions give rise to a cluster centered near $(z_1, 0)$, since the difference values are low in the region interiors, and similarly the dark interiors yield a cluster near $(z_2, 0)$. On the other hand, the border pixels yield difference values that differ significantly from zero. Thus in a histogram of the gray levels of those pixels that have near-zero difference values, the peaks should be more clearly separated than in the original

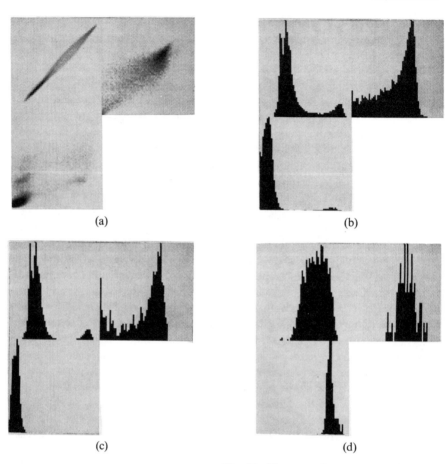

(a)

(b)

(c)

(d)

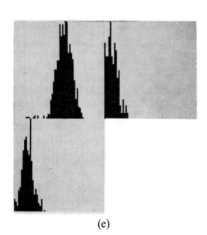

(e)

Fig. *15* Histogram improvement using (gray level, local average gray level) space. (a) Scatter plots of gray level (*y*-axis) versus 3×3 average gray level (*x*-axis) for the pictures in Fig. 1; high values are displayed as dark. (b) Histograms of the original pictures (same as Fig. 2; repeated here in compact format for comparison with the improved histograms). (c) Histograms of the gray levels of the points on the diagonals ($x = y$) of the scatter plots. The small peak in the handwriting picture histogram has been obliterated, but the cloud picture's histogram now has a definite valley. (d)–(e) Histograms of the gray levels of the 1% of the points farthest away from the diagonal on each side (i.e., with gray level much higher than local average gray level or vice versa). These histograms are unimodal; for each picture, the average of the peak values, or of the means, of the two histograms yields a good threshold.

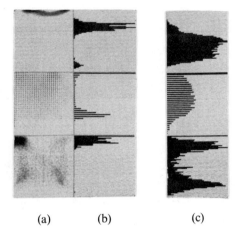

(a)   (b)   (c)

Fig. 16   Histogram improvement using (gray level, absolute difference) space. (a) Scatter plots of gray level (x-axis) versus difference (negative y-axis) for the pictures in Fig. 1. [The plots and histograms for this version of the cloud picture have only even-valued gray levels.] (b) Histograms of the gray levels of the points on the top rows of the scatter plots, i.e., the points having zero difference value. As in Fig. 15, the small peak in the handwriting histogram has been nearly obliterated, but the cloud histogram now has a valley. (c) Histograms of the gray levels of the 10 % of the points having highest difference values. The chromosome and cloud histograms are now unimodal, and their peaks or means yield good thresholds; the handwriting histogram is still bimodal, but the peaks are approximately equal, and the mean yields a good threshold.

picture's histogram. Conversely, the mean gray level of those pixels that have high difference values should be a good threshold. These remarks are illustrated in Fig. 16.

As a final example of the use of several local properties to classify pixels, the (average gray level, average busyness) scatter plot for the house picture is displayed in Fig. 17a; the smoothed gray level and busyness histograms for this picture were shown in Figs. 7 and 12. We see that there are two clusters that have essentially the same gray level range but different busyness ranges, while there are several clusters that have the same (low) busyness range but different gray level ranges; thus neither gray level alone nor busyness alone can be used to extract all the clusters. The pixels belonging to the five most prominent clusters in this space are displayed in Fig. 17b; these results are quite similar to, though not as clean as, the results shown in Fig. 11 using (red, blue) space.

Rather than dealing with a set of (spectral or spatial) properties simultaneously, in a multidimensional property space, we can use them one at a time to segment the picture recursively. Specifically, we choose a feature whose histogram displays an isolated peak, and extract the regions whose

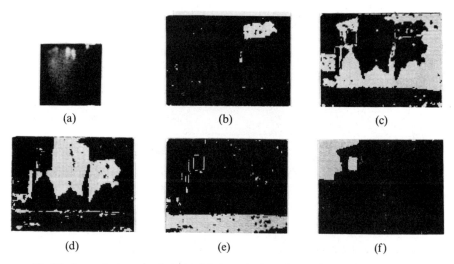

Fig. 17 Segmentation using (gray level, busyness) space. (a) Scatter plot of gray level (*x*-axis) versus busyness (negative *y*-axis) for the house picture. Note the two clusters at the upper left that overlap in gray level but differ in busyness; these correspond to the shadow and shrubbery. (b)–(f) Points of the picture that belong to each of the five clusters in (a). These given a fairly good segmentation into shadow, shrubbery, brick, grass, and sky; compare Fig. 11.

feature values belong to that peak. Deletion of these regions may improve the histograms of the remaining features, so that isolated peaks are more likely to occur on them, and we can repeat the process.

d. *Probability transforms*

For any local property z, let $p(z_0)$ be the probability that the value of z is z_0. We can estimate these probabilities for a given picture by histogramming the values of z; the fraction of times value z_0 occurs is an estimate of $p(z_0)$. Let (i, j) be a pixel for which the value of z is z_0; then $p(z_0)$ is a measure of the "typicality" of (i, j) with respect to the property z. Evidently, if (x, y) belongs to a histogram peak, its typicality is high, while if it belongs to a valley, it has low typicality.

More generally, let z and w be any two local properties, and let $p(z_0 | w_0)$ be the conditional probability that z has value z_0, given that w has value w_0. These probabilities can be estimated from the scatter plot of (z, w) values for the given picture. For any pixel (i, j) having $z = z_0$ and $w = w_0$, $p(z_0 | w_0)$ is a measure of the "conditional typicality" of (i, j) with respect to the pair of properties (z, w).[§] "Joint typicality" can be defined analogously.

[§] Of course, one can use a function of the (conditional) probability, rather than the probability itself; an example might be the entropy, i.e., the log of the probability.

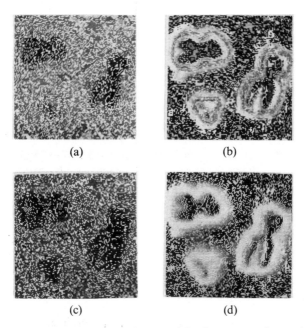

(a)   (b)

(c)   (d)

Fig. 18 Conditional typicality values for part of the chromosome picture: (a) 8-neighbor Laplacian magnitude, (b) Roberts gradient magnitude, (c) 4-neighbor average gray level, (d) right neighbor's gray level, all conditioned on the given gray level.

As an illustration of this concept, Fig. 18 shows conditional typicality values, suitably scaled, based on four pairs of properties, for part of the chromosome picture in Fig. 1a.

Typicality values can be used as properties for classifying pixels. Note that they are not local properties, since their values depend on statistics obtained from the entire picture. On their use in texture analysis see Haralick [26].

Conditional probabilities can also be used to define clusters of z values. For any z_0, the estimated probability vector $p(w|z_0)$ can be regarded as a vector-valued property of z_0, and we can attempt to detect clusters with respect to this property (i.e., to find sets of z_0's for which the patterns of joint occurrences of w are all similar). Note that here we are classifying the z values themselves, rather than the pixels. For the details see Rosenfeld *et al.* [67].

Alternatively, suppose that we have found a set of nonoverlapping clusters C_i that represent different types of regions. We now want to extend these clusters so as to classify the remaining parts of the picture. For each C_i, we construct a histogram of the gray levels (or other feature values) of the neighbors of the pixels in C_i, but excluding neighbors that themselves lie in C_i. Peaks in this conditional histogram represent clusters that co-occur with C_i. If a cluster co-occurs with C_i but with no other C_j, we can associate it with C_i for segmentation purposes. For the details see Aggarwal [2].

10.2 EDGE DETECTION

This and the next section deal with local operations that can be used to detect various types of local features, such as edges and curves, in a picture. As pointed out in Section 10.1, this is a special case of pixel classification based on local property values; we decide whether or not the feature is present by thresholding the value produced by the local operation. The main purpose of these two sections is to describe various operations that are useful for local feature detection, and to discuss their capabilities and limitations.

Local features usually involve abrupt changes in gray level, which may take several geometrical forms:

(1)   An *edge*: the gray level is relatively consistent in each of two adjacent, extensive regions, and changes abruptly as the border between the regions is crossed. An ideal edge has a steplike cross section, as shown in Fig. 19a; a more realistic example, incorporating blur and noise effects, is Fig. 19b.

(2)   A *line* or *curve*: the gray level is relatively constant except along a thin strip. In cross section, this yields a sharp spike, as shown in in Fig. 19c. It should be mentioned that lines often occur in association with edges (highlights on edges of blocks; membranes separating parts of a cell; roads running between fields bearing different crop types); an idealized version of the resulting combined cross section is shown in Fig. 19d.

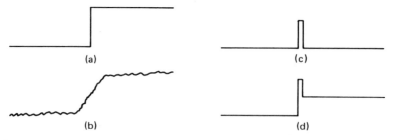

Fig. 19   Idealized edge cross sections: how the gray level changes as we move across the edge. (a) Perfect step edge. (b) Noisy, blurred step edge. (c) Perfect "spike" line. (d) Line combined with step edge.

(3)   A *spot*: the gray level is relatively constant except at one location. This too looks like a spike in cross section, but this is now true for cross sections in all directions.

Section 10.2.1 discusses difference operators that respond to changes in gray level or average gray level; these include digital versions of the gradient and Laplacian operators. Section 10.2.2 deals with edge detection by mask matching or surface fitting. Section 10.2.3 treats criteria for edge detection, and also discusses the detection of changes in color or in local property values. Section 10.3 will treat operations that respond selectively to lines, curves, spots, or features having given shapes.

Edges and curves are "local" features in the sense that they are detectable by examining local neighborhoods; but they may also be very long. Methods of linking local edge or curve responses into global edges or curves will be discussed in Section 10.3.4.

Ideally, one can segment a picture into regions by detecting the edges of the regions; the regions are then the connected components of the nonedge points. This approach may be desirable if some of the regions are small, or there are regions having many different gray level ranges, so that the regions are not readily extractable by pixel clustering and classification. Note, however, that due to blur or noise, the edges will often have gaps, so that they will not completely surround the regions. On the problem of filling in a region from an incomplete border see Section 11.2.2e.

10.2.1   Difference Operators

Difference operators, which yield high values at places where the gray level is changing rapidly, were introduced in Section 6.3. In this section we discuss the responses of such operators to edges.

a.   The gradient

As indicated in Section 6.3, if $\partial f/\partial x$ and $\partial f/\partial y$ are the rates of change of a function f in any two perpendicular directions, then the rate of change in any direction θ (measured from the x-axis) is a linear combination of these:

$$\frac{\partial f}{\partial x'} = \frac{\partial f}{\partial x} \cos \theta + \frac{\partial f}{\partial y} \sin \theta$$

Moreover, the direction in which this rate of change has the greatest magnitude is $\tan^{-1}((\partial f/\partial y)/(\partial f/\partial x))$, and this magnitude is $\sqrt{(\partial f/\partial x)^2 + (\partial f/\partial y)^2}$. The vector having this magnitude and direction is called the *gradient* of f.

To illustrate these concepts, let $f(x, y) = a(x \cos \varphi + y \sin \varphi) + b$; then we have $\partial f/\partial x = a \cos \varphi$ and $\partial f/\partial y = a \sin \varphi$. Thus the rate of change at any point (x, y) in direction $\tan^{-1}((\partial f/\partial y)/(\partial f/\partial x)) = \tan^{-1}(\sin \varphi/\cos \varphi) = \varphi$ is

$$\frac{\partial f}{\partial x} \cos \varphi + \frac{\partial f}{\partial y} \sin \varphi = a \cos^2 \varphi + a \sin^2 \varphi = a$$

while that in the perpendicular direction $\varphi + (\pi/2)$ is

$$\frac{\partial f}{\partial x} \cos\left(\varphi + \frac{\pi}{2}\right) + \frac{\partial f}{\partial y} \sin\left(\varphi + \frac{\pi}{2}\right) = a(-\cos \varphi \sin \varphi + \sin \varphi \cos \varphi) = 0$$

If we used a directional derivative as a measure of edge strength, its response would vary with the orientation of the edge. To avoid this, we can simply use the magnitude of the gradient, since this automatically gives the rate of change in the direction of greatest steepness.

For a digital picture, analogously, we could use first differences instead of first derivatives, e.g.,

$$(\Delta_x f)(x, y) \equiv f(x, y) - f(x - 1, y)$$

$$(\Delta_y f)(x, y) \equiv f(x, y) - f(x, y - 1)$$

Note that these are digital convolution operators which convolve f with the patterns

$$[-1 \quad 1] \quad \text{and} \quad \begin{bmatrix} 1 \\ -1 \end{bmatrix}$$

respectively.§ We could then combine $\Delta_x f$ and $\Delta_y f$ by taking, e.g., the square root of the sum of the squares. However, it does not seem correct to combine $\Delta_x f$ and $\Delta_y f$ at the same position (x, y), since the differences that these operators measure are not symmetrically located with respect to (x, y); Δ_x uses a pair of pixels centered at $(x - \frac{1}{2}, y)$ while Δ_y uses a pair centered at $(x, y - \frac{1}{2})$.

We can avoid this objection by using difference operators other than Δ_x and Δ_y. The following are two simple possibilities:

(1) $$(\Delta_{2x} f)(x, y) \equiv f(x + 1, y) - f(x - 1, y)$$

$$(\Delta_{2y} f)(x, y) \equiv f(x, y + 1) - f(x, y - 1)$$

---

§ More precisely, they cross-correlate f with these patterns; they convolve f with the 180° rotations of these patterns. We will ignore this distinction in what follows.

These operators measure the horizontal and vertical changes in f "across (x, y)." Note that their values do not depend on $f(x, y)$ itself. They are the convolutions of f with

$$[-1 \quad 0 \quad 1] \quad \text{and} \quad \begin{bmatrix} 1 \\ 0 \\ -1 \end{bmatrix}$$

respectively; they measure "central differences."

(2)
$$(\Delta_+ f)(x, y) = f(x + 1, y + 1) - f(x, y)$$
$$(\Delta_- f)(x, y) = f(x, y + 1) - f(x + 1, y)$$

These operators measure the 45° and 135° diagonal changes in f, using pairs of points that symmetrically surround $(x + \frac{1}{2}, y + \frac{1}{2})$. (Recall that in computing the gradient, derivatives in any pair of perpendicular directions can be used.) They convolve f with

$$\begin{bmatrix} 0 & 1 \\ -1 & 0 \end{bmatrix} \quad \text{and} \quad \begin{bmatrix} 1 & 0 \\ 0 & -1 \end{bmatrix}$$

The digital gradient operator based on Δ_+ and Δ_- is known as the *Roberts* operator.

Note that in (1), differences are taken across two-unit distances, while in (2), distances of $\sqrt{2}$ are involved. In order to obtain rates of change per unit distance, Δ_{2x} and Δ_{2y} should be divided by 2, while Δ_+ and Δ_- should be divided by $\sqrt{2}$. Note also that (1) responds to a perfect step edge in two positions, i.e., when (x, y) is adjacent to the edge on either side; thus it gives two-point thick responses to edges.

How should the two difference operators (Δ_{2x} and Δ_{2y}, or Δ_+ and Δ_-) be combined into a digital "gradient magnitude"? In analogy with the non-digital case, we could take their RMS (i.e., the square root of the sum of their squares). However, as we shall now show, we obtain better orientation invariance if we combine them in a simpler way. Let E_H, E_V, E_{D+}, and E_{D-} be perfect digital step edges of height h running in directions 0°, 90°, 45°, and 135°, respectively, i.e.,

hhhh	hh00
hhhh,	hh00,
0000	hh00
0000	hh00

hh00		00hh
hh00		00hh
hh00 ,	and	00hh
hh00		00hh

The magnitudes of the responses of Δ_{2x}, Δ_{2y}, Δ_+ and Δ_- at these edges are as follows:

	E_H	E_V	E_{D+}	E_{D-}
$|\Delta_{2x}|$	0	h	h	h
$|\Delta_{2y}|$	h	0	h	h
$|\Delta_+|$	h	h	0	h
$|\Delta_-|$	h	h	h	0

Thus the responses of $\sqrt{\Delta_{2x}^2 + \Delta_{2y}^2}$ to these edges are h, h, $h\sqrt{2}$, and $h\sqrt{2}$, a bias of $\sqrt{2}$ in favor of the diagonal edges; and the responses of $\sqrt{\Delta_+{}^2 + \Delta_-{}^2}$ are $h\sqrt{2}$, $h\sqrt{2}$, h, and h, giving a bias against the diagonal edges. These biases are eliminated if we use the maximum of the absolute values, rather than the RMS, as our gradient magnitude; this makes the responses all equal to h. Note that this has two further advantages:

(1) It avoids the computational cost of the square and square root operations.

(2) It yields values in the same range as the original gray scale, which is convenient for display purposes. (The RMS, on the other hand, can have values as high as $\sqrt{2}$ times the maximum gray level, if the two Δ's are each equal to that maximum.)

Exercise 10.5.   Another possible way of combining the two Δ's is to use the sum of their absolute values. Verify, however, that this introduces an even worse bias for or against diagonal edges, and that it also has an even larger range of possible values. ∎

Exercise 10.6.   Prove that for all nonnegative a, b we have

$$\max(a, b) \leqslant \sqrt{a^2 + b^2} \leqslant a + b$$

and

$$(a + b)/\sqrt{2} \leqslant \sqrt{a^2 + b^2} \leqslant \max(a, b)\sqrt{2}$$

Prove also that if $a \geqslant b$, we have

$$\sqrt{a^2 + b^2} \leqslant a + \frac{b}{2} \leqslant \frac{\sqrt{5}}{2}\sqrt{a^2 + b^2} \quad ∎$$

The values of $\max(|\Delta_+|, |\Delta_-|)$ and of $\max(|\Delta_{2x}|, |\Delta_{2y}|)$ for a set of pictures are shown in Fig. 20.

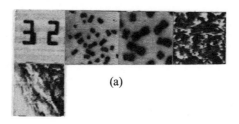

(a)

(b)   (c)

Fig. 20   Difference magnitudes for a set of pictures. (a) Originals (characters; chromosomes at two scales; two types of cloud cover); (b) $\max(|\Delta_+|, |\Delta_-|)$; (c) $\max(|\Delta_{2x}|, |\Delta_{2y}|)$.

It should be pointed out that if we want to associate edge values with the "cracks" between pixels, rather than with the pixels themselves, it becomes perfectly acceptable to use Δ_x and Δ_y as edge detectors, without combining them.

b.   The Laplacian

As we saw in Section 6.3, the Laplacian $\partial^2 f/\partial x^2 + \partial^2 f/\partial y^2$ is also an orientation-invariant derivative operator. Its analog for digital pictures is given by

$$(\nabla^2 f)(x, y) \equiv [f(x + 1, y) + f(x - 1, y) + f(x, y + 1)$$
$$+ f(x, y - 1)] - 4f(x, y)$$

which is the digital convolution of f with

$$\begin{bmatrix} & 1 & \\ 1 & -4 & 1 \\ & 1 & \end{bmatrix}$$

We recall that $\nabla^2 f$ is proportional to $f - \bar{f}$, where $\bar{f}(x, y) \equiv \frac{1}{5}[f(x, y) + f(x + 1, y) + f(x - 1, y) + f(x, y + 1) + f(x, y - 1)]$, so that $\bar{f}$ is the result of locally averaging f.

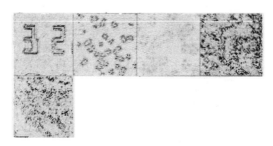

Fig. 21 Laplacian magnitudes $|\nabla^2 f|$ for the set of pictures in Fig. 20a. The original values are all quite small; for display purposes, they have *not* been divided by 4.

The digital Laplacian, since it is a second-difference operator, has zero response to linear ramps, but it responds to the "shoulders" at the top and bottom of a ramp, where there is a change in the rate of change of gray level. Thus it responds on each side of an edge, once with positive sign, once with negative. If we want only positive responses, we can use the absolute value $|\nabla^2 f|$, or the positive value $(\nabla^2 f)^+$ ($\equiv \nabla^2 f$ when $\nabla^2 f \geqslant 0$; $\equiv 0$ when $\nabla^2 f < 0$). Note that these responses can have values as high as four times the maximum gray level; if we want to insure that the gray level range is preserved, we should divide them by 4. The $|\nabla^2 f|$ responses, for the same set of pictures as in Fig. 20, are shown in Fig. 21; for visibility they have not been divided by 4.

The digital Laplacian does respond to edges, but it responds even more strongly to corners, lines, line ends, and isolated points. In fact, consider the pattern

$$
\begin{array}{cccccc}
1 & 1 & 1 & 1 & 1 & 1 \\
1 & \underline{1} & 1 & & & \\
1 & 1 & & & & \\
1 & & & & & 1 \\
\end{array}
$$

where the blanks are 0's. The $|\nabla^2 f|$ responses to this pattern are

$$
\begin{array}{cccccccc}
 & 1 & 1 & 1 & 1 & 1 & 1 & \\
1 & 2 & 1 & 1 & 2 & 2 & 3 & 1 \\
1 & 1 & \underline{0} & 2 & 2 & 1 & 1 & \\
1 & 1 & 2 & 2 & & & 1 & \\
1 & 3 & 2 & & & 1 & 4 & 1 \\
 & 1 & & & & & 1 & \\
\end{array}
$$

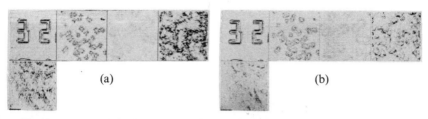

Fig. 22 (a) Max of absolute second differences, and (b) absolute difference between mean and median, for the pictures in Fig. 20a.

where the 0 corresponds to the underlined interior point, and the 4 to the isolated point. Thus in a noisy picture, the noise will produce higher Laplacian values than the edges, unless it has much lower contrast. (Note also that the response to the diagonal edge is twice as strong as the responses to the horizontal and vertical edges, so that the digital Laplacian is not actually orientation-invariant.) The digital gradient responses to the pattern, on the other hand, never exceed 1, so that the gradient responds to edges as strongly as it does to noise. Thus the gradient would ordinarily be a better edge detector than the Laplacian.

If we use the maximum, rather than the sum, of the (absolute) second differences in the x and y directions, we obtain a "pseudo-Laplacian" operator that still does not respond to ramps, but is somewhat less sensitive to noise than the Laplacian. Another such operator is the (absolute) difference between the mean and the median, computed over the same neighborhood of the given pixel. Outputs of these operators, for the same set of pictures, are shown in Fig. 22.

As we saw in Section 6.3, when we apply the Laplacian to a picture, low spatial frequencies are weakened, while high ones remain relatively intact. Thus high-pass spatial frequency filtering (suppressing low frequencies, and retaining high ones) should have effects similar to those of applying the Laplacian. These effects are illustrated, for the same set of pictures, in Fig. 23. Again, the results do not seem to be as useful as those obtained with gradient operators.

c. Differences of averages

We can reduce the effects of noise on the responses of a difference operator by smoothing the picture before applying the operator. In particular, we can locally average before differencing—or, equivalently, we can use an operator that computes differences of local averages. Such operators are discussed in the following paragraphs.

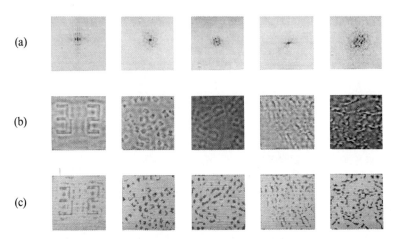

Fig. 23   Edge detection by high-pass spatial frequency filtering, applied to the pictures of Fig. 20a. (a) Power spectra (64 × 64). (b) Reconstructed pictures with spatial frequency range 1–10 (in all directions from the origin) suppressed. (c) Results of setting all but the darkest 15% of the points in (b) to zero.

Suppose that we take averages over 2 × 2 neighborhoods, e.g.,

$$\bar{f}_4(x, y) \equiv \tfrac{1}{4}[f(x, y) + f(x + 1, y) + f(x, y + 1) + f(x + 1, y + 1)].$$

In defining a difference operator based on such averages, we should not use the averages at adjacent pixels, since the neighborhoods for such pixels overlap, and the differencing will cancel out the common values and weaken the response. For example, we have

$$
\begin{aligned}
\bar{f}_4(x, y) - \bar{f}_4(x - 1, y) &= \tfrac{1}{4}[f(x, y) + f(x + 1, y) + f(x, y + 1) \\
&\quad + f(x + 1, y + 1) - f(x - 1, y) - f(x, y) \\
&\quad - f(x - 1, y + 1) - f(x, y + 1)] \\
&= \tfrac{1}{4}[f(x + 1, y) + f(x + 1, y + 1) - f(x - 1, y) \\
&\quad - f(x - 1, y + 1)]
\end{aligned}
$$

which is a weakened difference of two two-pixel averages. Rather, we should take differences of averages that come from adjacent, but non-overlapping, neighborhoods, e.g., $(\bar{\Delta}_{4x} f)(x, y) \equiv \bar{f}_4(x + 1, y) - \bar{f}_4(x - 1, y)$, which is the convolution of f with

$$\frac{1}{4}\begin{bmatrix} -1 & -1 & 1 & 1 \\ -1 & -1 & 1 & 1 \end{bmatrix}$$

An operator based on differences of averages will respond "blurrily" to an edge in several positions; for example, the responses of $|\bar{\Delta}_{4x}|$ to the pattern

$$\cdots$$
$$\cdots \quad 0 \quad 0 \quad 0 \quad 1 \quad 1 \quad 1 \quad \cdots$$
$$\cdots \quad 0 \quad 0 \quad 0 \quad 1 \quad 1 \quad 1 \quad \cdots$$
$$\cdots$$

are

$$\cdots$$
$$\cdots \quad 0 \quad \tfrac{1}{2} \quad 1 \quad \tfrac{1}{2} \quad 0 \quad 0 \quad \cdots$$
$$\cdots \quad 0 \quad \tfrac{1}{2} \quad 1 \quad \tfrac{1}{2} \quad 0 \quad 0 \quad \cdots$$
$$\cdots$$

These responses can be sharpened by suppressing nonmaxima in the direction across the edge, i.e., setting a response to zero if there is a stronger response sufficiently close to it in that direction, on either side. [Here "sufficiently close" means "closer than the size of the averaging neighborhood"; e.g., if we had used 3×3 averages, the responses to the edge would have been $\cdots \quad 0 \quad \tfrac{1}{3} \quad \tfrac{2}{3} \quad 1 \quad \tfrac{2}{3} \quad \tfrac{1}{3} \quad 0 \quad \cdots$, so that we would want to suppress nonmaxima out to distance 2.][§] Nonmaxima should not be suppressed in the direction along the edge, since the edge would then compete with itself. An operator based on differences of $n \times n$ averages followed by nonmaximum suppression should not be used to detect the edges of regions that are narrower than n, to avoid competition between the opposite edges of a single region. Of course, we would not ordinarily use a large operator to detect edges of small regions, since its responses would be weak; e.g., the response of $|\bar{\Delta}_{4x}|$ to an isolated "1" surrounded by 0's is only $\tfrac{1}{4}$ (in eight positions!).

We can reduce the blurredness of the responses to edges, while still retaining some smoothing power, by averaging only in the direction along the edge. For example, we can use the operator $\bar{\Delta}_{2x}$ defined by

$$(\bar{\Delta}_{2x} f)(x, y) \equiv \tfrac{1}{2}[f(x, y) + f(x, y + 1) - f(x - 1, y) - f(x - 1, y + 1)]$$

[§] Suppression out to a smaller distance would usually also be effective, since the responses vary smoothly, so that a nonmaximal response should have a stronger one near it on one side.

which is the convolution of f with

$$\frac{1}{2}\begin{bmatrix} -1 & 1 \\ -1 & 1 \end{bmatrix}$$

This responds sharply to a vertical step edge. (However, as we shall see below, it does not respond sharply to diagonal edges.)

To obtain a "corase digital gradient" that is insensitive to orientation, we can combine two operators that take differences in perpendicular directions. For example, we can combine $\bar{\Delta}_{4x}$ with $\bar{\Delta}_{4y}$, defined as the convolution of f with

$$\frac{1}{4}\begin{bmatrix} 1 & 1 \\ 1 & 1 \\ -1 & -1 \\ -1 & -1 \end{bmatrix}$$

[Note that $\bar{\Delta}_{4x}$ and $\bar{\Delta}_{4y}$ are both symmetric around $(x + \frac{1}{2}, y + \frac{1}{2})$.] Similarly, we can combine $\bar{\Delta}_{2x}$ with $\bar{\Delta}_{2y}$, which convolves f with

$$\frac{1}{2}\begin{bmatrix} 1 & 1 \\ -1 & -1 \end{bmatrix}$$

If we want values that are symmetric around (x, y), we can use $\bar{\Delta}_{3x}$ and $\bar{\Delta}_{3y}$, defined as the convolutions of f with

$$\frac{1}{3}\begin{bmatrix} -1 & 0 & 1 \\ -1 & 0 & 1 \\ -1 & 0 & 1 \end{bmatrix} \quad \text{and} \quad \frac{1}{3}\begin{bmatrix} 1 & 1 & 1 \\ 0 & 0 & 0 \\ -1 & -1 & -1 \end{bmatrix}$$

respectively.

Some care is required in deciding just how to combine the two perpendicular operators, since unlike the operators considered in Section 10.2.1a, the ones based on averages have weaker, blurred responses to edges that are not optimally oriented. For example, the responses of $\bar{\Delta}_{2x}$, $\bar{\Delta}_{3x}$, and $\bar{\Delta}_{4x}$ to a 45° edge of height h are as follows:

$$\bar{\Delta}_{2x}: \quad \cdots \quad 0 \quad 0 \quad h/2 \quad h/2 \quad 0 \quad 0 \quad \cdots$$

$$\bar{\Delta}_{3x}: \quad \cdots \quad 0 \quad 0 \quad h/3 \quad 2h/3 \quad 2h/3 \quad h/3 \quad 0 \quad 0 \quad \cdots$$

$$\bar{\Delta}_{4x}: \quad \cdots \quad 0 \quad 0 \quad h/4 \quad 3h/4 \quad 3h/4 \quad h/4 \quad 0 \quad 0 \quad \cdots$$

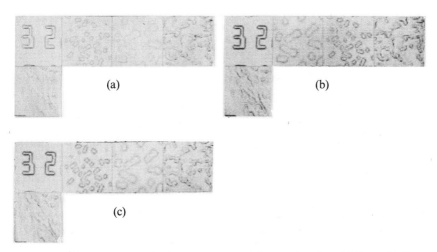

Fig. 24   Difference magnitudes for the same set of pictures, based on (a) $\bar{\Delta}_{2x,y}$, (b) $\bar{\Delta}_{3x,y}$, (c) $\bar{\Delta}_{4x,y}$, using the max.

This suggests that one might want to use the sum of the absolute values, or the RMS, of the two operators, rather than the maximum of their absolute values as in Section 10.2.1.[§]

Digital gradient values based on $\bar{\Delta}_{2x,y}$, $\bar{\Delta}_{3x,y}$, and $\bar{\Delta}_{4x,y}$ using the max are shown in Fig. 24 for the same pictures used in Figs. 20–23. (The $\bar{\Delta}_3$ operator is sometimes called the *Prewitt operator*.) An example involving a much greater amount of averaging is shown in Fig. 25. Here the input (a) is the same as in Fig. 8a; as we see in part (b), an ordinary digital gradient is of little use in detecting the edge of the central region, since the noise has the same contrast as that edge. Part (c) shows gradient values based on differences of averages taken over 8×8 neighborhoods, combined using the

---

[§] Haralick (personal communication) has pointed out that the RMS response magnitude of the $\bar{\Delta}_2$ operator is always $1/\sqrt{2}$ times that of the Roberts operator. Indeed, for the 2×2 neighborhood

$$AB$$
$$CD$$

we have

$$
\begin{aligned}
(A + B - C - D)^2 + (A + C - B - D)^2 &= 2(A^2 + B^2 + C^2 + D^2 + AB - AC \\
&\quad - AD - BC - BD + CD + AC - AB \\
&\quad - AD - CB - CD + BD) \\
&= 2(A^2 + B^2 + C^2 + D^2 - 2AD - 2BC) \\
&= 2((A - D)^2 + (B - C)^2)
\end{aligned}
$$

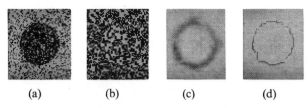

(a)   (b)   (c)   (d)

Fig. 25 "Gradient" based on differences of averages. (a) Input picture (same as Fig. 8a). (b) Roberts gradient of (a); this is of little use in detecting the edge of the densely dotted region. (c) Gradient of (a) based on averages over 8×8 square neighborhoods (see text). (d) Results of suppressing nonmaxima in (c); we keep only horizontal maxima of the vertical differences, and vertical maxima of the horizontal differences.

maximum of the absolute values. In part (d), nonmaximum suppression has been applied to the differences of averages before combining them.

The operators considered so far are all based on unweighted averages, but weighted averages can also be used. An example is the *Sobel operator*, whose x and y components are the convolutions of f with

$$\frac{1}{4}\begin{bmatrix} -1 & 0 & 1 \\ -2 & 0 & 2 \\ -1 & 0 & 1 \end{bmatrix} \quad \text{and} \quad \frac{1}{4}\begin{bmatrix} 1 & 2 & 1 \\ 0 & 0 & 0 \\ -1 & -2 & -1 \end{bmatrix}$$

This operator gives greater weight to points lying closer to (x, y); its response to diagonal edges is therefore not weakened as much as that of the Prewitt operator (*Exercise*: verify this!).§ Conversely, to reduce the weakening of the response to a blurred edge, one might want to give greater weight to points that lie farther from (x, y). MacLeod and Argyle have proposed operators that use "Gaussian" patterns of weights; in the former operator, the points close to (x, y) are weakly weighted, and in the latter case, strongly weighted. The effect of stronger weighting near (x, y) is also obtained if we multiply together a set of operators of different sizes; here a point near (x, y) receives greater weight because it is involved in a greater number of the operators.

Marr [39] has developed a refined approach to edge detection and description by comparing the outputs of operators of several sizes. He classifies detected features as edges, extended edges, shading edges, or lines, and estimates their position, orientation, contrast, and fuzziness. He then groups edges whose types and descriptions (roughly) match, to obtain a "primal sketch" of the given scene.

§ Note that the weights in the Sobel operator do not fall off linearly with distance. To obtain an "isotropic" weighting, we should replace the 2's by $\sqrt{2}$'s [21].

10.2.2 Edge Matching and Fitting

In Section 10.2.1 we considered gradient operators which estimate the edge magnitude and direction at a point using difference operators in two perpendicular directions. In this section we discuss edge operators that involve matching local patterns, or fitting surfaces, to the picture at the given point.

a. Mask matching

Since an ideal edge is a steplike pattern, one approach to detecting edges is to match such patterns, in various orientations, with the given picture. We can then take the orientation that gives the best match at a given point as the edge orientation at that point, and the magnitude of this best match as a measure of the edge strength.

In choosing edge patterns or "masks" to match with a picture, it is customary to use masks that represent second differences of step edges (see Section 9.4.2 on matching based on second differences). Such masks turn out to be much like the difference operators of Section 10.2.1. For example, the second differences of the step edge that has cross section $\cdots aaabbb \cdots$ are $\cdots 00(b - a)(a - b) \cdots$; thus in a 2×2 neighborhood, a second-difference mask for a vertical step edge would have values proportional to

$$\begin{array}{cc} -1 & 1 \\ -1 & 1 \end{array}$$

which is just the $\bar{\Delta}_{2x}$ operator of Section 10.2.1c. Similarly, if we consider a blurred step edge having cross section $\cdots aaa[(a + b)/2]bbb \cdots$, the second differences become $\cdots 00[(b - a)/2]0[(a - b)/2]00 \ldots$, so that in a 3×3 neighborhood, a second-difference mask for a vertical edge of this type would have values proportional to

$$\begin{array}{ccc} -1 & 0 & 1 \\ -1 & 0 & 1 \\ -1 & 0 & 1 \end{array}$$

which is the Prewitt operator $\bar{\Delta}_{3x}$. When we are comparing matches with masks in various orientations, we can ignore the constant factor corresponding to the edge contrast, and we can also ignore the denominator in the normalized cross-correlation (see Section 9.4.1), since it is the same for all of the masks; we thus need only cross-correlate (or convolve) the masks themselves with the picture (we will see that the set of masks is closed under $180°$ rotation).

In summary: to detect edges by mask matching, we convolve a set of difference-operator-like masks, in various orientations, with the picture. The

mask giving the highest value at a given point determines the edge orientation at that point, and that value determines the edge strength; indeed, in
our example, the value is just the contrast of the edge.

Virtually any of the difference operators in Section 10.2.1 can be used to
define sets of masks for edge detection by mask matching. For example, we
can use the masks

$$1, -1, \qquad -1, 1, \qquad \begin{matrix} 1 \\ -1 \end{matrix}, \qquad \begin{matrix} -1 \\ 1 \end{matrix}$$

and, if desired, also the masks

$$\begin{matrix} 0 & 1 \\ -1 & 0 \end{matrix}, \quad \begin{matrix} 0 & -1 \\ 1 & 0 \end{matrix}, \quad \begin{matrix} 1 & 0 \\ 0 & -1 \end{matrix}, \quad \text{and} \quad \begin{matrix} -1 & 0 \\ 0 & 1 \end{matrix}$$

corresponding to "forward" and "backward" first differences in four or
eight directions; in other words, we can compute $f(x, y) - f(u, v)$, where (u, v)
ranges over the four (or eight) neighbors of (x, y), and use the maximum of
these values as an edge measure. Note that these sets of operators are symmetric with respect to (x, y), which removes the objection to first-difference
operators raised in Section 10.2.1. As another example, we can use the
masks

$$-1 \; 0 \; 1, \qquad \begin{matrix} 1 \\ 0 \\ -1 \end{matrix} \qquad \begin{matrix} 1 \\ 0 \\ -1 \end{matrix} \qquad \text{and} \qquad \begin{matrix} 1 \\ 0 \\ -1 \end{matrix}$$

corresponding to central differences in four directions, and take the maximum resulting value as an edge measure. The outputs of these operators,
for the same set of pictures as in Figs. 20–24, are shown in Fig. 26.

The difference-of-averages operators of Section 10.2.1c can also be used
to define edge masks. The following are some commonly used examples
(we have omitted the constant factor in each case):

(1)

1	1	1		1	1	0		1	0	−1		0	−1	−1
0	0	0		1	0	−1		1	0	−1		1	0	−1
−1	−1	−1		0	−1	−1		1	0	−1		1	1	0

−1 −1 −1	−1 −1 0	−1 0 1	0 1 1
0 0 0	−1 0 1	−1 0 1	−1 0 1
1 1 1	0 1 1	−1 0 1	−1 −1 0

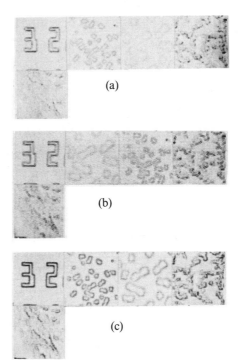

(a)

(b)

Fig. 26 Difference magnitudes for the five pictures, based on (a) max of absolute differences between each point and its four neighbors; (b) same, but using eight neighbors; (c) max of four central differences centered at each point.

(c)

(These are a generalization of the components of the Prewitt operator; the constant factor is $\frac{1}{3}$.)[§]

(2)

1	2	1		2	1	0		1	0	−1		0	−1	−2
0	0	0		1	0	−1		2	0	−2		1	0	−1
−1	−2	−1		0	−1	−2		1	0	−1		2	1	0

−1	−2	−1		−2	−1	0		−1	0	1		0	1	2
0	0	0		−1	0	1		−2	0	2		−1	0	1
1	2	1		0	1	2		−1	0	1		−2	−1	0

[§] Of course, we need not actually use all eight of these operators, since the last four are the negatives of the first four. Similar remarks apply to some of the other sets of masks described in this section.

(These generalize the Sobel operator; the constant factor is $\frac{1}{4}$.)

(3)

1	1	1		1	1	1		1	1	-1		1	-1	-1
1	-2	1		1	-2	-1		1	-2	-1		1	-2	-1
-1	-1	-1		1	-1	-1		1	1	-1		1	1	1

-1	-1	-1		-1	-1	1		-1	1	1		1	1	1
1	-2	1		-1	-2	1		-1	-2	1		-1	-2	1
1	1	1		1	1	1		-1	1	1		-1	-1	1

These masks, proposed by Prewitt, give positive weight to each five consecutive neighbors of the center point, and negative weight to the remaining three neighbors; the center point itself is given weight -2 in order to make the sum of all the weights zero (so that the response in a region of constant gray level is zero), and the constant factor is $\frac{1}{5}$. Note that these masks resemble corners rather than straight step edges.

(4)

5	5	5		5	5	-3		5	-3	-3		-3	-3	-3
-3	0	-3		5	0	-3		5	0	-3		5	0	-3
-3	-3	-3		-3	-3	-3		5	-3	-3		5	5	-3

-3	-3	-3		-3	-3	-3		-3	-3	5		-3	5	5
-3	0	-3		-3	0	5		-3	0	5		-3	0	5
5	5	5		-3	5	5		-3	-3	5		-3	-3	-3

These masks, proposed by Kirsch, give negative weights to each five consecutive neighbors and positive weights to the other three, with the weights chosen so that they sum to zero; the center point can thus be given weight zero, and the constant factor is $\frac{1}{15}$.

The edge masks defined above respond to patterns other than edges; for example, they respond to (off-center) points or lines. One can define "mask" operators that respond more selectively by incorporating logical conditions into their definitions; for example, in using the mask

$$
\begin{array}{rrr}
1 & 1 & 1 \\
0 & 0 & 0 \\
-1 & -1 & -1
\end{array}
$$

we could require that each of the top three gray levels be greater than the corresponding bottom one (or vice versa), and if not, we could set the output to zero. Analogous modifications can be made in most of the operators defined in this section. The use of such nonlinear "mask" operators to selectively detect lines will be discussed in detail in Section 10.3.

b. Step fitting

Another approach to edge detection is based on fitting an ideal step edge to the given picture $f(x, y)$ at the given point P, which for simplicity we take to be the origin. A step edge of slope θ through P, having values a and b on its two sides, is defined by the function

$$
s(x, y) = \begin{cases} a & \text{if} \quad x \sin \theta \geq y \cos \theta \\ b & \text{otherwise} \end{cases}
$$

We want to find a, b, and θ that minimize some measure of the distance between f and s. These values then provide estimates of the contrast $|b - a|$ and slope θ of the edge at P, if one exists. (The intercept of the edge can also be taken as a parameter, if we do not want to require that it pass exactly through the origin. In this case we need not carry out the edge detection process at every point P; it suffices to use a set of neighborhoods that cover the picture, and find a best fitting step edge that passes through each of these neighborhoods, not necessarily through its center.) Suppose that the origin P is at the point where four pixels A, B, C, D meet, i.e.,

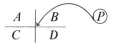

and we want to minimize the sum of the squared errors between $s(x, y)$ and $f(x, y)$ in a circular neighborhood centered at P and contained in the union of these four pixels. (We could use a larger neighborhood, but we restrict it to overlapping only four pixels in order to keep the computations simple.) If θ is in the first quadrant, the sum of squared errors is proportional to

$$
e^2 \equiv (b - A)^2 + (a - D)^2 + \frac{2\theta}{\pi}(a - B)^2 + \left(1 - \frac{2\theta}{\pi}\right)(b - B)^2
$$

$$
+ \frac{2\theta}{\pi}(b - C)^2 + \left(1 - \frac{2\theta}{\pi}\right)(a - C)^2 \tag{8}
$$

Similar expressions can be written for the cases where θ is in the other three quadrants. One could attempt to minimize e^2 by taking partial derivatives of (8) with respect to a, b, and θ and equating them to zero; but since (8) is linear in θ, this cannot yield a true minimum.

An alternative method of finding a best-fitting step edge is to expand both $f(x, y)$ and $s(x, y)$ in terms of a set of orthogonal basis functions, and use the sum of the squared differences between corresponding coefficients as an error measure. To keep the computations simple, we truncate the expansion to a few terms; of course, this implies that the error measure is only approximate. Hueckel [28] was the first to take this approach, using a set of eight basis functions defined on a disk; he also solved the problem of determining a best-fitting edge/line (see Fig. 19d) [29], where the edge/line or edge is not required to pass through the center of the disk. Nevatia [52] used a subset of Hueckel's basis; O'Gorman [53] used a set of two-dimensional Walsh functions defined on a square; Meró and Vassy [42] used only two basis functions, defined by diagonally subdividing a square, to determine edge orientation; and Hummel [30] used a set of optimal basis functions derived from the Karhunen–Loève expansion of the local image values. The details of these results will not be given here; but to indicate how this approach works, we will work out a simple example using only three basis functions defined on a disk, namely

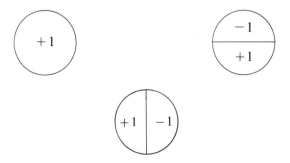

Let the coefficients of

$$f(x, y) \equiv \begin{array}{cc} A & B \\ C & D \end{array}$$

with respect to these basis functions be f_0, f_1, and f_2, respectively; then readily we have

$$f_0 = \frac{A + B + C + D}{4} \equiv \frac{S}{4}, \qquad f_1 = \frac{-A - B + C + D}{4},$$

$$f_2 = \frac{A - B + C - D}{4} \tag{9}$$

Similarly, let the coefficients of $s(x, y)$ be s_0, s_1, and s_2; then readily

$$s_0 = \frac{a + b}{2}$$

$$s_1 = \frac{\theta}{2\pi}(b - a) + \frac{\pi - \theta}{2\pi}(a - b) = \frac{2\theta - \pi}{2\pi}(b - a) \qquad \text{if} \quad 0 \leqslant \theta \leqslant \pi$$

$$= \frac{\theta - \pi}{2\pi}(a - b) + \frac{2\pi - \theta}{2\pi}(b - a) = \frac{3\pi - 2\theta}{2\pi}(b - a) \quad \text{if} \quad \pi \leqslant \theta \leqslant 2\pi$$

$$s_2 = \frac{\theta + \dfrac{\pi}{2}}{2\pi}(b - a) + \frac{\dfrac{\pi}{2} - \theta}{2\pi}(a - b) = \frac{\theta}{\pi}(b - a) \qquad \text{if} \quad -\frac{\pi}{2} \leqslant \theta \leqslant \frac{\pi}{2}$$

$$= \frac{\theta - \dfrac{\pi}{2}}{2\pi}(a - b) + \frac{\dfrac{3\pi}{2} - \theta}{2\pi}(b - a) = \frac{\pi - \theta}{\pi}(b - a) \qquad \text{if} \quad \frac{\pi}{2} \leqslant \theta \leqslant \frac{3\pi}{2}$$

$$\tag{10}$$

We now want to minimize $E^2 \equiv (f_0 - s_0)^2 + (f_1 - s_1)^2 + (f_2 - s_2)^2$; because of the way s_1 and s_2 are defined, this must be done separately for θ in each quadrant. Actually, since a and b are interchangeable, by symmetry it suffices to treat the first and second quadrants. In fact, in (10), if we replace θ by $\theta + \pi$ and interchange a and b in $[(3\pi - 2\theta)/2\pi](b - a)$, we obtain $[(2\theta - \pi)/2\pi](b - a)$; and similarly $[(\pi - \theta)/\pi](b - a)$ yields $(\theta/\pi)(b - a)$.

In the first quadrant we have

$$E^2 = \left(\frac{S}{4} - \frac{a + b}{2} \right)^2 + \left(\frac{-A - B + C + D}{4} - \frac{2\theta - \pi}{2\pi}(b - a) \right)^2$$

$$+ \left(\frac{A - B + C - D}{4} - \frac{\theta}{\pi}(b - a) \right)^2 \tag{11}$$

Taking partial derivatives of (11) with respect to a and b and setting them equal to zero, we obtain

$$-\frac{1}{2}\left[\frac{S}{4} - \frac{a + b}{2} \right] + \frac{2\theta - \pi}{2\pi}\left[\frac{-A - B + C + D}{4} - \frac{2\theta - \pi}{2\pi}(b - a) \right]$$

$$+ \frac{\theta}{\pi}\left[\frac{A - B + C - D}{4} - \frac{\theta}{\pi}(b - a) \right] = 0$$

$$-\frac{1}{2}\left[\frac{S}{4} - \frac{a + b}{2} \right] - \frac{2\theta - \pi}{2\pi}\left[\frac{-A - B + C + D}{4} - \frac{2\theta - \pi}{2\pi}(b - a) \right]$$

$$-\frac{\theta}{\pi}\left[\frac{A - B + C - D}{4} - \frac{\theta}{\pi}(b - a) \right] = 0$$

$$\tag{12}$$

Adding gives immediately $S/4 = (a + b)/2$, or $a + b = S/2$. Taking the partial derivative of (11) with respect to θ and equating it to zero gives

$$\left[\frac{-A - B + C + D}{4} - \frac{2\theta - \pi}{2\pi}(b - a)\right]$$

$$+ \left[\frac{A - B + C - D}{4} - \frac{\theta}{\pi}(b - a)\right] = 0 \qquad (13)$$

or

$$\frac{C - B}{2} = \frac{4\theta - \pi}{2\pi}(b - a)$$

so that

$$\frac{(4\theta - \pi)(b - a)}{\pi} = C - B \qquad (14)$$

Also, substituting $S/4 = (a + b)/2$ and (13) in either equation of (12) gives

$$\frac{-A - B + C + D}{4} - \frac{2\theta - \pi}{2\pi}(b - a) = 0 \qquad (15)$$

so that we also have

$$\frac{A - B + C - D}{4} - \frac{\theta}{\pi}(b - a) = 0 \qquad (16)$$

If θ is not 0 or $\pi/2$, and $b \neq a$, we can divide (15) by (16) (or vice versa) to obtain

$$\frac{A - B + C - D}{-A - B + C + D} = \frac{\theta}{\theta - \pi/2} \qquad (17)$$

from which we readily have

$$\theta(2A - 2D) = \frac{\pi}{2}(A - B + C - D)$$

or (if $A \neq D$)

$$\theta = \frac{\pi}{4} \cdot \frac{A - B + C - D}{A - D} = \frac{\pi}{4}\left[1 - \frac{B - C}{A - D}\right] \qquad (18)$$

Combining this with (16) gives $b - a = A - D$, and combining this with $b + a = S/2$ gives

$$a = \frac{(-A + B + C + 3D)}{4} = \frac{S}{4} - \frac{(A - D)}{2}$$

$$b = \frac{(3A + B + C - D)}{4} = \frac{S}{4} + \frac{(A - D)}{2}$$

$$(19)$$

Note that by (15), (16), and the fact that $a + b = S/2$, we actually have $E^2 = 0$ for this solution. It is easily verified that if we assume $\theta = 0$ or $\theta = \pi/2$ in (11), and set the partial derivatives with respect to a and b equal to zero, we obtain special cases of this solution. In fact, for $\theta = 0$, we find that $A + C = B + D, b = (A + B)/2$, and $a = (C + D)/2$; while for $\theta = \pi/2$ we get $A + B = C + D, b = (A + C)/2$, and $a = (B + D)/2$.

In the second quadrant, analogously, we get

$$a = \frac{(A + 3B - C + D)}{4} = \frac{S}{4} - \frac{(C - B)}{2}$$

$$b = \frac{(A - B + 3C + D)}{4} = \frac{S}{4} + \frac{(C - B)}{2}$$

$$(20)$$

and

$$\theta = \frac{\pi}{4}\left[3 + \frac{A - D}{B - C}\right] \qquad (21)$$

It can be verified that the first and second quadrant solutions agree if $\theta = \pi/2$. Moreover, note that for (18) to actually lie in the first quadrant we must have

$$-1 \leqslant \frac{B - C}{A - D} \leqslant 1$$

which is evidently equivalent to $|B - C| \leqslant |A - D|$. Similarly, for (21) to actually lie in the second quadrant we must have

$$-1 \leqslant \frac{A - D}{B - C} \leqslant 1$$

which is evidently equivalent to $|A - D| \leqslant |B - C|$. Thus by comparing the magnitudes and signs of $A - D$ and $B - C$ we can choose the appropriate best-fitting step edge for the given neighborhood

$$A \quad B$$

$$C \quad D$$

Note that if $A - D = B - C$ we have $\theta = 0$ in (18) and $\theta = \pi$ in (21); moreover, in this case (19) and (20) also agree, with a and b interchanged. Similarly, if $A - D = C - B$, we have $\theta = \pi/2$ in both (18) and (21), and here (19) and (20) agree too.

In summary, the "best-fitting" step edge to

$$A \quad B$$
$$C \quad D$$

is found as follows:

If $|B - C| \leqslant |A - D|$, then $\theta = \dfrac{\pi}{4}\left[1 - \dfrac{B - C}{A - D}\right]$, and a, b are given by (19)

If $|B - C| \geqslant |A - D|$, then $\theta = \dfrac{\pi}{4}\left[3 + \dfrac{A - D}{B - C}\right]$, and a, b are given by (20)

The magnitude $|a - b|$ of the edge is $|A - D|$ in the first case, and $|B - C|$ in the second case; in other words, the magnitude is $\max(|A - D|, |B - C|)$. Note that this is just the magnitude of the Roberts operator of Section 10.2.1a. For example, if

$$A \quad B \qquad 1 \quad 2$$
$$ = $$
$$C \quad D \qquad 3 \quad 4$$

we have $|B - C| \leqslant |A - D|$, and (18) gives $\theta = (\pi/4)[1 - (-1/-3)] = \pi/6$, while (19) gives $a = 4$, $b = 1$—a reasonable result.

Exercise 10.7.   Let f_1 and f_2 be defined on a unit square by

$$f_1(x, y) = 1 \qquad \text{for} \quad x + y \geqslant 1; \qquad = -1 \qquad \text{for} \quad x + y < 1$$
$$f_2(x, y) = 1 \qquad \text{for} \quad x \leqslant y; \qquad = -1 \qquad \text{for} \quad x > y$$

Let g be a step edge of slope θ that passes through the square. Prove [42] that $\iint gf_2 / \iint gf_1 = \tan(\theta - \pi/4)$.   ∎

c.   Ramp and surface fitting

Another approach to edge detection by surface fitting is to fit a simple function to the gray levels in a neighborhood of a point, and use the gradient of that function as an estimate of the digital gradient at that point.

Let $f(x, y)$, $f(x + 1, y)$, $f(x, y + 1)$, and $f(x + 1, y + 1)$ be the gray levels at four neighboring points of a picture. Suppose that we want to fit a plane $z = ax + by + c$ to these four values. This cannot be done exactly, but we can find a plane for which some measure of the error in fit, e.g., the sum of the squared differences,

$$(ax + by + c - f(x, y))^2 + (a(x + 1) + by + c - f(x + 1, y))^2$$
$$+ (ax + b(y + 1) + c - f(x, y + 1))^2$$
$$+ (a(x + 1) + b(y + 1) + c - f(x + 1, y + 1))^2$$

is a minimum. To find this best-fitting plane, we differentiate the error expression with respect to each of a, b, and c, set the results equal to zero, and solve these equations for a, b, and c:

$$x(ax + by + c - f(x, y)) + (x + 1)(a(x + 1) + by + c - f(x + 1, y))$$
$$+ x(ax + b(y + 1) + c - f(x, y + 1))$$
$$+ (x + 1)(a(x + 1) + b(y + 1) + c - f(x + 1, y + 1)) = 0$$

$$y(ax + by + c - f(x, y)) + y(a(x + 1) + by + c - f(x + 1, y))$$
$$+ (y + 1)(ax + b(y + 1) + c - f(x, y + 1))$$
$$+ (y + 1)(a(x + 1) + b(y + 1) + c - f(x + 1, y + 1)) = 0$$

$$(ax + by + c - f(x, y)) + (a(x + 1) + by + c - f(x + 1, y))$$
$$+ (ax + b(y + 1) + c - f(x, y + 1))$$
$$+ (a(x + 1) + b(y + 1) + c - f(x + 1, y + 1)) = 0$$

It is easily verified that the solution is

$$a = \frac{f(x + 1, y) + f(x + 1, y + 1)}{2} - \frac{f(x, y) + f(x, y + 1)}{2}$$

$$b = \frac{f(x, y + 1) + f(x + 1, y + 1)}{2} - \frac{f(x, y) + f(x + 1, y)}{2}$$

$$c = \tfrac{1}{4}(3f(x, y) + f(x + 1, y) + f(x, y + 1) - f(x + 1, y + 1)) - xa - yb$$

The x and y partial derivatives of the plane $z = ax + by + c$ are just a and b. Thus for the best-fitting plane, these derivatives are the same as the operators $\bar{\Delta}_{2x}$ and $\bar{\Delta}_{2y}$ defined in Section 10.2.1c.

More generally, let us fit a degree-m polynomial $g(x, y)$ to the gray levels in an $n \times n$ neighborhood, where the number of coefficients $(m + 1)(m + 2)/2$ of g is less than n^2; then we can regard the gradient of g at the center of the neighborhood as an approximation to the gradient of the picture there. This gradient can be expressed in terms of the gray levels at points in the neighborhood, and this expression then defines a digital gradient operator. For example, if we least squares fit a second degree polynomial g_2 to a 3×3 neighborhood (i.e., $m = 2$, $n = 3$), it turns out that the x and y derivatives of g_2 are proportional to the components $\bar{\Delta}_{3x}$ and $\bar{\Delta}_{3y}$ of the Prewitt operator defined in Section 10.2.1c.

Exercise 10.8.   Verify this. Show also that $\bar{\Delta}_{3x}$ and $\bar{\Delta}_{3y}$ are proportional to the x- and y-coordinates of the centroid of the 3×3 neighborhood, where we take the center of the neighborhood as the origin; and analogously for $\bar{\Delta}_{2x}$ and $\bar{\Delta}_{2y}$ in a 2×2 neighborhood. ∎

Exercise 10.9.   What is the Laplacian of g_2 at the center of the neighborhood? What happens if we use a 4×4 neighborhood? ∎

10.2.3   Edge Detection and Generalizations

a.   Edge detection criteria

Up to now we have discussed operators that respond to edges, but we have not addressed the problem of *detecting* edges by thresholding the responses of such an operator. This problem can be formulated in decision-theoretic terms using the concepts introduced in Section 10.1. For example, suppose that there are two types of regions having probabilities $p(1)$ and $p(2)$, and that the probabilities of points of the two regions being border points are $p(12)$ and $p(21)$. Suppose further that we know the probability densities of edge operator responses Δ in the region interiors and on their borders, say $p(\Delta|1), p(\Delta|2), p(\Delta|12)$, and $p(\Delta|21)$. Then we can compute the probabilities $p(1|\Delta), p(2|\Delta), p(12|\Delta)$, and $p(21|\Delta)$ that a given edge response arises from an interior or border point. If the latter is more likely, we have detected an edge.

In certain simple cases, it is easy to estimate the probabilities $p(\Delta|1, 2, 12,$ and $21)$. For example, suppose that $\Delta = \Delta_x$, the horizontal first difference operator, and that the gray levels of adjacent points interior to a region are independent and have the same probability density $q(z|r)$, where $r = 1$ or 2. If the possible gray levels are $0, 1, \ldots, k$, then the possible values of Δ_x are $-k, \ldots, 0, \ldots, k$, and the probability that Δ_x has value h is

$$p(h|r) = \sum_z q(z|r)q(z + h|r).$$

Note that this is just the discrete autocorrelation of the gray level probability density q in region r. Similarly, the probability that Δ_x has value h at a border point is $p(h|rs) = \sum_z q(z|r)q(z + h|s)$, where $rs = 12$ or 21; this is just the cross correlation of $q(z|r)$ and $q(z|s)$. Analogous calculations can be made for more complex difference operators. Unfortunately, the gray levels of adjacent points are not usually independent, so that the analysis becomes much more complicated in realistic cases.

One can also detect edges based on a set of measurements made at a given point, or on measurements made at a set of points. An example of the latter approach, based on normal probability densities and independence of adjacent points, was treated by Griffith, to whose paper [23] the reader is referred for the details. The following simpler example was given by Yakimov-sky [84]. Suppose that we want to decide whether a given piece S of a picture is interior to a homogeneous region or overlaps two such regions. Suppose that each region has normally distributed gray levels with mean μ_i and standard deviation σ_i, and that the points in the region are mutually independent. If S contains points having gray levels $z_1, \ldots, z_m$, and these all come from the same region, then the maximum likelihood estimates of the mean and variance for that region are

$$\mu \equiv \frac{1}{m} \sum_{i=1}^{m} z_i \qquad \text{and} \qquad \sigma^2 = \frac{1}{m} \sum_{i=1}^{m} (z_i - \mu)^2$$

Given a point from the normal distribution having mean μ and variance σ^2, the probability that the point has gray level z is

$$\exp[-(z - \mu)^2/2\sigma^2]/\sqrt{2\pi}\,\sigma$$

Thus for the independent gray levels $z_1, \ldots, z_m$, the joint probability is

$$\exp\left[-\sum_{i=1}^{m}(z_i - \mu)^2/2\sigma^2\right]\bigg/(\sqrt{2\pi}\,\sigma)^m = \exp[-m/2]/(\sqrt{2\pi}\,\sigma)^m$$

On the other hand, suppose the first m' of the points come from one region and the remaining $m - m' = m''$ from another. Then the means and variances of the regions are estimated by

$$\mu_1 = \frac{1}{m'} \sum_{i=1}^{m'} z_i, \qquad \sigma_1^2 = \frac{1}{m'} \sum_{i=1}^{m'} (z_i - \mu_1)^2$$

$$\mu_2 = \frac{1}{m''} \sum_{i=m'+1}^{m} z_i, \qquad \sigma_2^2 = \frac{1}{m''} \sum_{i=m'+1}^{m} (z_i - \mu_2)^2$$

so that the joint probability is

$$\exp(-m'/2)/(\sqrt{2\pi}\,\sigma_1)^{m'} \cdot \exp(-m''/2)/(\sqrt{2\pi}\,\sigma_2)^{m''} = \exp(-m/2)/(\sqrt{2\pi})^m \sigma_1^{m'} \sigma_2^{m''}.$$

We would thus decide in favor of a single region or two regions according to whether the probability ratio

$$\sigma_1^{m'} \sigma_2^{m''}/\sigma^m$$

is greater or less than 1.

Exercise 10.10. Show that $\mu, \mu_1, \mu_2, \sigma, \sigma_1$, and σ_2 are related by (compare Exercise 10.1)

$$\mu = [m'\mu_1 + m''\mu_2]/m$$

$$\sigma^2 = [m'\sigma_1^2 + m''\sigma_2^2 + m'(\mu - \mu_1)^2 + m''(\mu - \mu_2)^2]/m \quad \blacksquare$$

To decide whether or not an edge passes through a given point of the picture, let S be a neighborhood of that point. Bisect S in various ways (e.g., horizontally, vertically, diagonally, etc.), and for each bisection, compute the probability ratio. If it is sufficiently small for some bisection, decide that an edge is present in that orientation. Note that if the bisection is perpendicular to the edge, the ratio will be close to 1.

Exercise 10.11. Given $p(1)$, $p(2)$, $q(z|1)$, and $q(z|2)$, and given a row of gray levels $f(i, j), \ldots, f(i + m, j)$, how do we find the most probable position of the edge between the two regions along this row? ∎

To detect edges between large regions without locating them precisely, we can apply a coarse difference operator in a set of regularly spaced positions; e.g., we can divide the picture into a grid of squares and take differences between pairs of adjacent squares. Even if the edge passes through the middle of a square, it will still give rise to a half-strength response, and so should be detectable. Finer difference operators can then be applied to locate the edge more accurately. This approach of using low-resolution operators to guide the application of higher-resolution operators is sometimes called *planning* [32]; it has many applications in picture processing and analysis.

b. Spectral and spatial edge detection

In the following paragraphs we discuss difference operators that make use of color (or spectral) components, or of local property values, rather than gray levels. Such operators can be used to detect color edges or texture edges of various types.

Difference operators can be applied to each component of a color picture, and the results can be combined in various ways, e.g., we can take their RMS, or the sum or maximum of their absolute values. For example, let f_r, f_g, and f_b be the red, green, and blue components. Then the RMS of $\Delta_x f_r$, $\Delta_x f_g$, and $\Delta_x f_b$ at (x, y) is

$$[(f_r(x, y) - f_r(x - 1, y))^2 + (f_g(x, y) - f_g(x - 1, y))^2$$
$$+ (f_b(x, y) - f_b(x - 1, y))^2]^{1/2}$$

which is just the Euclidean distance between the color vectors $(f_r(x, y), f_g(x, y), f_b(x, y))$ and $(f_r(x - 1, y), f_g(x - 1, y), f_b(x - 1, y))$. If we use the sum or maximum of the absolute values, rather than the RMS, we are introducing a bias in favor of or against edges that involve differences in more than one color component; but the amount of this bias is bounded (compare Exercise 10.6). Color coordinate systems other than (red, green, blue) can

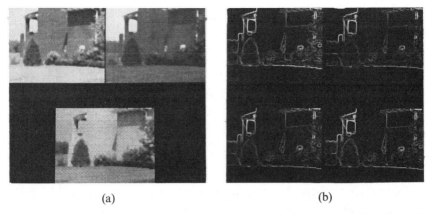

(a)   (b)

Fig. 27   (a) Red, green, and blue components of the house picture (same as Figs. 10a–10c). (b) Results of applying the Roberts operator to each component, and combining the results using max, mean, median, and RMS.

also be used. Figure 27 shows the three color components for a picture of a house, and the edge values obtained by applying the Roberts operator to each component and combining the results using both maximum and RMS, as well as mean and median. Of course, other operators can also be used, including operators based on differences of color averages.

If a picture contains two adjacent regions in which some local property has different average values, we can detect the edge between these regions by computing the local property at each point and then taking differences of averages. We have already seen an example of this in the case where the regions differ in average gray level (Fig. 25). Two other examples are shown in Figs. 28 and 29. Here the input pictures are the same as Figs. 13a and 14a;

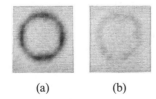

(a)   (b)

Fig. 28   "Gradient" based on difference of average local property values. (a) Differences of 8×8 local averages computed on the Roberts values of Fig. 13b. (b) Result of suppressing nonmaxima.

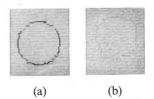

(a)   (b)

Fig. 29   Same as Fig. 28, but using Fig. 14b as input.

the regions do not differ in average gray level, but they differ in "busyness"; and this can be converted into a difference in average gray level by applying the Roberts operator at each point, as shown in Figs. 13b and 14b. Differences of 8 × 8 averages (as in Fig. 25), computed on the Roberts values, are shown in Figs. 28 and 29, both with and without nonmaximum suppression.

Other statistical measures can be used, instead of differences of averages, to detect edges between statistically different regions. A simple example is the *Fisher distance*, defined by $|\mu_1 - \mu_2|/\sqrt{\sigma_1{}^2 + \sigma_2{}^2}$, where μ_1, μ_2 are the averages (of gray level or other local property value) in the two neighborhoods, and σ_1, σ_2 are the corresponding standard deviations. This measure gives high weight to the difference of averages when the regions are smooth, and lower weights when they are busy or noisy, which is intuitively reasonable. More generally, if $p_1(z)$ and $p_2(z)$ are the histograms (of gray level or local property value) for the two neighborhoods, we can measure the difference between them by, e.g., $\sum_z |p_1(z) - p_2(z)|$. Such measures should be based on relatively large neighborhoods, since otherwise the measured statistics will not be reliable. They will thus respond to edges over a wide range of positions, and so should be used in conjunction with nonmaximum suppression. In general, any measure of differences between textures can be used to define difference operators for texture edge detection; see Section 12.1.5 on textural properties.

10.3   FEATURE DETECTION

This section deals with the detection of local features such as curves and spots in a picture. Of course, difference operators will respond to such features; but they will also respond to edges. In Section 10.3.1 we discuss operators that respond to lines, but not to edges (or spots); and in Section 10.3.2 we describe how to generalize these operators to respond to curves (possibly thick or dotted), and how to define analogous operators that respond to spots or other types of local features. Section 10.3.3 discusses the detection of curves that have given shapes (e.g., straight lines, circles, etc.) with the aid of coordinate transformations that map sets of points which lie on such curves into clusters of feature values.

In Section 10.4 we will discuss sequential methods of extracting curves from pictures by tracking or curve following. As a preliminary to this, Section 10.3.4 discusses methods of locally linking edge points into curves.

10.3.1 Line Detection

In this and the next section we consider, in some detail, methods of detecting simple patterns such as lines (or curves), spots, corners, etc., in a picture. We first discuss a linear matched filtering approach, and compare it with nonlinear filtering.

Let us first consider how to detect thin vertical straight line segments that are darker (i.e., have higher gray level) than their background. Since it is the shape (thin, vertical) that is important here, rather than the specific gray levels of the line and the background, it is reasonable to use a filter function based on a derivative of the line pattern, as indicated at the end of Section 9.4.2. In fact, it is customary to use a filter based on the Laplacian of an ideal line. In a digital picture, a vertical line of 1's on a background of 0's

$$\vdots$$
$$1$$
$$1 \qquad \text{(where all points not marked are 0's)}$$
$$1$$
$$\vdots$$

has digital Laplacian proportional to

$$
\begin{array}{ccc}
\vdots & \vdots & \vdots \\
-\tfrac{1}{2} & 1 & -\tfrac{1}{2} \\
-\tfrac{1}{2} & 1 & -\tfrac{1}{2} \\
-\tfrac{1}{2} & 1 & -\tfrac{1}{2} \\
\vdots & \vdots & \vdots
\end{array}
$$

The 3×3 operator h_V which convolves the picture with

$$
\begin{array}{ccc}
-\tfrac{1}{2} & 1 & -\tfrac{1}{2} \\
-\tfrac{1}{2} & 1 & -\tfrac{1}{2} \\
-\tfrac{1}{2} & 1 & -\tfrac{1}{2}
\end{array}
$$

is commonly used as a detector for thin vertical lines. Note that this operator gives zero output on a constant or ramp portion of the picture, as well as on a horizontal line (e.g., $\cdots 111 \cdots$ on a background of 0's). On the other hand,

h_V has a high positive output for a dark vertical line on a light background; for example, if the line is

$$\vdots \quad \vdots \quad \vdots \quad \vdots \quad \vdots$$

$$\cdots \quad b \quad b \quad a \quad b \quad b \quad \cdots$$

$$\cdots \quad b \quad b \quad a \quad b \quad b \quad \cdots \qquad \text{(where } a > b\text{)}$$

$$\cdots \quad b \quad b \quad a \quad b \quad b \quad \cdots$$

$$\vdots \quad \vdots \quad \vdots \quad \vdots \quad \vdots$$

the output of h_V is

$$\vdots \qquad \vdots \qquad\qquad \vdots \qquad\qquad \vdots \qquad \vdots$$

$$\cdots \quad 0 \quad 3(b-a)/2 \quad 3(a-b) \quad 3(b-a)/2 \quad 0 \quad \cdots$$

$$\cdots \quad 0 \quad 3(b-a)/2 \quad 3(a-b) \quad 3(b-a)/2 \quad 0 \quad \cdots$$

$$\cdots \quad 0 \quad 3(b-a)/2 \quad 3(a-b) \quad 3(b-a)/2 \quad 0 \quad \cdots$$

$$\vdots \qquad \vdots \qquad\qquad \vdots \qquad\qquad \vdots \qquad \vdots$$

which has the high positive value $3(a-b)$ at points of the line, flanked by negative values $3(b-a)/2$ just alongside the line.

In spite of its plausible behavior in the cases just considered, the operator h_V has serious shortcomings; it responds to many patterns that are not linelike, and it may respond to them more strongly than it does to lines. For example, at a vertical step edge of the form

$$\vdots \quad \vdots \quad \vdots \quad \vdots$$

$$\cdots \quad a \quad a \quad b \quad b \quad \cdots$$

$$\cdots \quad a \quad a \quad b \quad b \quad \cdots$$

$$\cdots \quad a \quad a \quad b \quad b \quad \cdots$$

$$\vdots \quad \vdots \quad \vdots \quad \vdots$$

the output of h_V is

$$\vdots \qquad \vdots \qquad\qquad \vdots \qquad \vdots$$

$$\cdots \quad 0 \quad 3(a-b)/2 \quad 3(b-a)/2 \quad 0 \quad \cdots$$

$$\cdots \quad 0 \quad 3(a-b)/2 \quad 3(b-a)/2 \quad 0 \quad \cdots$$

$$\cdots \quad 0 \quad 3(a-b)/2 \quad 3(b-a)/2 \quad 0 \quad \cdots$$

$$\vdots \qquad \vdots \qquad\qquad \vdots \qquad \vdots$$

Thus its peak positive output, $3(a - b)/2$, is only half the peak output from a line of the same contrast; but the peak output from a step edge of high contrast can be greater than the peak output from a line of lower contrast. Similarly, at an isolated point, say

$$
\begin{array}{ccccc}
\vdots & \vdots & \vdots & \vdots & \vdots \\
\cdots \; b & b & b & b & b \; \cdots \\
\cdots \; b & b & b & b & b \; \cdots \\
\cdots \; b & b & a & b & b \; \cdots \\
\cdots \; b & b & b & b & b \; \cdots \\
\cdots \; b & b & b & b & b \; \cdots \\
\vdots & \vdots & \vdots & \vdots & \vdots
\end{array}
$$

the output of h_V is

$$
\begin{array}{ccccc}
\vdots & \vdots & \vdots & \vdots & \vdots \\
\cdots \; 0 & 0 & 0 & 0 & 0 \; \cdots \\
\cdots \; 0 & (b-a)/2 & a-b & (b-a)/2 & 0 \; \cdots \\
\cdots \; 0 & (b-a)/2 & a-b & (b-a)/2 & 0 \; \cdots \\
\cdots \; 0 & (b-a)/2 & a-b & (b-a)/2 & 0 \; \cdots \\
\cdots \; 0 & 0 & 0 & 0 & 0 \; \cdots \\
\vdots & \vdots & \vdots & \vdots & \vdots
\end{array}
$$

which again is weaker than that from a line of the same contrast, but may be stronger than that from a line of lower contrast.

Exercise 10.12.   If a picture is two-valued (with values a and b, say), verify that h_V takes on its maximum possible positive value $3(a - b)$ at points that lie on vertical lines, and nowhere else. Thus in the two-valued case, h_V detects vertical lines more strongly than anything else. ∎

Exercise 10.13.   It has been suggested that if we want an operator that responds to vertical lines but not to isolated points, we can subtract the output of the horizontal operator h_H which convolves the picture with

$$
\begin{array}{ccc}
-\tfrac{1}{2} & -\tfrac{1}{2} & -\tfrac{1}{2} \\
1 & 1 & 1 \\
-\tfrac{1}{2} & -\tfrac{1}{2} & -\tfrac{1}{2}
\end{array}
$$

pointwise from the output of h_V. This seems plausible at first glance, since h_H has output zero at a vertical line, so that subtracting it has no effect; while both h_H and h_V have output $a - b$ at an isolated point, so that they cancel when we subtract. What happens, however, at the neighbors of an isolated point? ∎

These remarks suggest that if we want an operator which responds to thin vertical lines, we must use something more elaborate than our simple linear operator h_V. We shall now define a nonlinear operator h_V' that has the desired properties. At the point (x, y) of the picture g, the output of h_V' is defined to be the same as that of h_V, i.e.,

$$g(x, y + 1) + g(x, y) + g(x, y - 1) - \tfrac{1}{2}(g(x - 1, y + 1)$$
$$+ g(x + 1, y + 1) + g(x - 1, y) + g(x + 1, y)$$
$$+ g(x - 1, y - 1) + g(x + 1, y - 1))$$

whenever

$$g(x, y + 1) > g(x - 1, y + 1), \qquad g(x, y + 1) > g(x + 1, y + 1)$$
$$g(x, y) > g(x - 1, y), \qquad\qquad g(x, y) > g(x + 1, y)$$
$$g(x, y - 1) > g(x - 1, y - 1), \qquad g(x, y - 1) > g(x + 1, y - 1)$$

and zero otherwise. In other words, $h_V' = h_V$ at (x, y) provided that point (x, y) in g is darker than each of (x, y)'s horizontal neighbors, and the same is also true of (x, y)'s two vertical neighbors. Clearly these conditions are exactly what we mean by saying that a thin dark vertical line passes through (x, y). If the conditions are not met, the output of h_V' is defined to be zero. Note that h_V' no longer has negative values at points adjacent to a dark vertical line.

Exercise 10.14.   Compute the output of h_V' in the vicinity of a step edge and of an isolated point. ∎

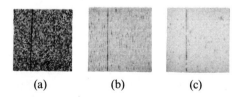

(a)   (b)   (c)

Fig. 30   Vertical line detection by linear and nonlinear operators. (a) Input picture. The noise has normally distributed gray levels with mean 32 and standard deviation 9; the line has constant gray level 50. (b) Result of applying operator h_V (see text). Note that the surviving noise is streaky. (c) Result of applying operator h_V'. The breaks in the line are due to adjacent noise points which cause the operator's detection criteria to be violated.

Exercise 10.15. Let g consist of a thin black line (value 1) on a background of salt-and-pepper noise (value 1 has probability p, 0 has probability $1 - p$). Compute the probability densities of the values of h_V and h_V': (a) at a point of the black line; (b) at a point of the background. ∎

Examples of the output produced by h_V and h_V' are shown in Fig. 30.

It should be pointed out that although h_V' does not respond to isolated points it does respond to certain spotlike patterns. For example, it responds (with value 3) at the center point of the pattern

$$
\begin{array}{ccccccc}
 & & & 1 & & & \\
 & & 1 & 2 & 1 & & \\
 & 1 & 2 & 3 & 2 & 1 & \\
 & & 1 & 2 & 1 & & \\
 & & & 1 & & &
\end{array}
$$

Of course, it is not difficult to modify h_V' so that it no longer responds to this pattern; for example, we can ignore responses at points where the nonlinear horizontal operator h_H' also responds (compare Exercise 10.13). Examples such as this, however, illustrate the nontriviality of designing operators that respond only to a specified class of patterns.

Exercise 10.16. A possible alternative to h_V' is the "semilinear" operator, defined to have the same output as h_V when

$$g(x, y + 1) + g(x, y) + g(x, y - 1) > g(x - 1, y + 1)$$
$$+ g(x - 1, y) + g(x - 1, y - 1)$$

and

$$g(x, y + 1) + g(x, y) + g(x, y - 1) > g(x + 1, y + 1)$$
$$+ g(x + 1, y) + g(x + 1, y - 1)$$

and zero otherwise. Show that this operator responds to isolated points, but not to step edges. Still another alternative is the operator that has the same output as h_V when

$$g(x, y + 1) > [g(x - 1, y + 1) + g(x + 1, y + 1)]/2$$
$$g(x, y) > [g(x - 1, y) + g(x + 1, y)]/2$$
$$g(x, y - 1) > [g(x - 1, y - 1) + g(x + 1, y - 1)]/2$$

and zero otherwise. Show that this operator responds to step edges, but not to isolated points. ∎

10.3.2 Curve and Spot Detection

In Section 10.3.1 we described a simple nonlinear operator h_V' designed to detect thin, vertical, dark lines on a light background. We now discuss how to modify this design to handle broken lines, thick lines, lines having other slopes, and lines that contrast with their background in other ways, as well as curves and various types of local features.

To detect light lines on a dark background, we need only reverse the inequalities, and change the signs, in the definition of h_V'. In other words, at each point (x, y) of the picture g, we require that $g(x, y)$ be less than the values of its two horizontal neighbors, and the same at each of (x, y)'s vertical neighbors; and if these conditions are fulfilled, we give h_V' the value

$$\tfrac{1}{2}[g(x - 1, y + 1) + g(x + 1, y + 1) + g(x - 1, y)$$
$$+ g(x + 1, y) + g(x - 1, y - 1) + g(x + 1, y - 1)]$$
$$- [g(x, y + 1) + g(x, y) + g(x, y - 1)]$$

at (x, y).

A more complicated definition is needed if we want to detect, say, a gray line whose background is white on one side and black on the other side. (Such a line might result from digitization of a perfect step edge). Here we must require that $g(x, y)$ be greater than $g(x - 1, y)$ and less than $g(x + 1, y)$, and similarly at the points $(x, y \pm 1)$. The directions of the inequalities should be consistent at all three points (x, y) and $(x, y \pm 1)$; otherwise, we would detect the columns of a checkerboard as "lines," which would normally not be desirable.

Still further modification is needed if we want to detect a line that differs in "texture" from its background, e.g., a line whose points are alternately black and white, on a background of solid gray. We could attempt to detect this line by the fact that it contrasts with the background at each point (even though the sign of the contrast changes from point to point), but this approach would also detect checkerboard columns as "lines." Instead, we can detect the line by the fact that it is "busy" while its background is smooth. For example, we can apply a vertical absolute difference operator to the picture; this will have high value at every point of the line, but zero value elsewhere in the picture. Thus we now have a new picture in which the line is darker than its background, and can be detected using h_V'. The method used here is analogous to our method of detecting "texture edges" by first converting them to average gray level edges, as described in Section 10.2.3b.

We next describe how to extend our methods to thick lines ("streaks"), say having thickness $2t$. Suppose that we average the given picture g, using averaging neighborhoods of radius t; let the average for the neighborhood centered at (x, y) be $g^{(t)}(x, y)$. We can define a thick analog $h_V'^{(t)}$ of our h_V' operator by simply replacing the individual gray levels used in the original definition by adjacent, nonoverlapping averages. For example, we require that $g^{(t)}(x, y)$ be greater than both of $g^{(t)}(x \pm 2t, y)$, and similarly at the two vertically adjacent, nonoverlapping positions $(x, y \pm 2t)$. When these conditions are fulfilled, the values of $h_V'^{(t)}$ is defined analogously to that of h_V', but using $g^{(t)}(u, v)$'s instead of $g(u, v)$'s. Other types of streak contrast are handled similarly. (When we are dealing with thick streaks rather than thick lines, a wide variety of types of "texture contrast" becomes possible; see Section 10.2.3b for examples.) Note that $h_V'^{(t)}$ will detect a streak of width $2t$ most strongly when it is centered on the streak, but it will also detect the streak as we move it sideways to nearby positions, just as was the case with the "coarse" edge detection operators in Section 10.2.1c. If we want to locate the streak exactly, we should suppress nonmaxima in the direction across the streak, just as in Section 10.2.1c.

A streak detector based on averages over neighborhoods of radius t will detect streaks of width $2t$ strongly; it will also respond to streaks of other widths (between 0 and $6t$), but the more the size differs from $2t$, the weaker its response will be. Streak detectors are most commonly implemented only for small values of t, e.g., using 2×2 averages.

Exercise 10.17. Show that the maximum response of our streak detector to a streak of width w is proportional to $2tw$ for $0 \leqslant w \leqslant 2t$; to $t(6t - w)$ for $2t \leqslant w \leqslant 6t$; and is zero otherwise. ∎

Exercise 10.18. Design an algorithm for detecting streaks of a given width by finding pairs of step edges (one a step up, the other a step down) that have a given separation. (On the use of this approach to handle a range of widths see Cook and Rosenfeld [14].) ∎

We can also extend our approach to broken, e.g., "dotted" lines (or streaks). To this end, we redefine h_V' so that it depends on more than just the three vertically consecutive points (x, y) and $(x, y \pm 1)$. For example, we can examine the five points (x, y), $(x, y \pm 1)$, and $(x, y \pm 2)$, and require that at least three of these points be darker than their two horizontal neighbors. If this condition is satisfied, we give our new h_V' the value

$$\sum_{k=-2}^{2} g(x, y + k) - \frac{1}{2} \sum_{k=-2}^{2} [g(x - 1, y + k) + g(x + 1, y + k)]$$

and value 0 otherwise. The generalization to handle dotted streaks, and other types of contrast, are analogous.

Finally, we consider operators that detect lines in directions other than the vertical. As long as our operators depend only on small neighborhoods, the number of directions that need be considered is not large. A straight line that passes through the center of a 3×3 neighborhood must intersect that neighborhood in one of the following 12 patterns (the center of the neighborhood has been underlined):

$$
\begin{array}{ccccccc}
a & a & a & a & a & a & \quad a \\
\underline{a} & \underline{a} & \underline{a} & \underline{a} & \underline{a} & \quad\underline{a} & \quad\underline{a} \\
a & \ a & a & a & a & \ \ a & a
\end{array}
$$

$$
\begin{array}{ccccc}
a\underline{a}a & a\underline{a} & a & \quad a & \underline{a}a \\
 & \ a & \underline{a}a & a\underline{a} & a
\end{array}
$$

(see Section 11.3.3b on the characterization of digital straight lines). To obtain a complete set of thick-line detection operators, we would need an analog of $h_{\rm v}'$ for each of these patterns. For the five near-vertical patterns, the h' operator would be defined by comparing the three "a" points with their horizontal neighbors; for the five near-horizontal patterns, we would compare the points with their vertical neighbors; and for the two diagonal patterns, we could do either type of comparison.

A similar set of operators could be used for streak detection, with the a's representing adjacent nonoverlapping averages rather than individual points. For dotted line detection, a much larger set would be needed, since a straight line can intersect (say) a 5×5 neighborhood in many ways. (*Exercise*: how many?) If linear operators (h rather than h') are used, it is not necessary to use all of the possible operators, since all lines will still be detected, some more strongly than others.

We can use sets of local line detection operators not only to detect straight lines having arbitrary slopes, but also to detect arbitrary smooth curves. (We assume here that a smooth curve never turns as much as 90° within a 3×3 neighborhood; thus we need not consider patterns such as

$$
\begin{array}{cc}
a & a \\
\underline{a}, & \underline{a}a \\
a &
\end{array}
$$

and their rotations.) Some examples of curve detection, curved streak detection, and dotted curve detection by this method are shown in Figs. 31 and 32.

Methods similar to those used in this and the preceding section can be used to detect types of patterns other than lines and smooth curves. For example, to detect isolated dots in g that contrast with their surroundings, we can use the operator d whose value is $|\nabla^2 g(x, y)|$ when $g(x, y)$ is greater than all four of its neighbors and zero otherwise. (We would not want to

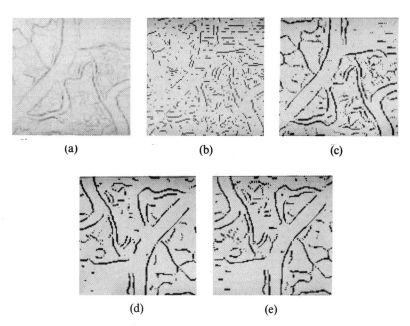

Fig. 31 Curve and curved streak detection. (a) Input picture: Output of an edge detection operation applied to part of a LANDSAT picture. (b) Result of thin curve detection operation (see text), applied to (a). (c) Result of streak detection operation, based on averages over 2 × 2 neighborhoods (see text), applied to (a). (d)–(e) Results of repeating the same streak detection operation twice more, i.e., applying it again to (c), then again to (d). The noise is now almost all gone.

use ∇^2 itself as a dot detector, since as pointed out in Section 10.2.1b, it responds to edges and lines as well as dots.) We can generalize d to detect coarse spots by using averages over adjacent, nonoverlapping neighborhoods instead of single points; and we can handle spots that differ texturally from their background using the methods of Section 10.2.3b.

Exercise 10.19. Design operators that detect (a) ends of curves; (b) sharp angles on curves[§]; (c) branchings or crossings of curves. You need only do this for one orientation. ▌

10.3.3 Detecting Curves of Given Shapes: Hough Transforms

Another approach to line or curve detection involves applying a coordinate transformation to the picture such that all the points belonging to a curve of a

---

[§] Another approach to detecting angles on digital curves will be described in Section 11.3.3c.

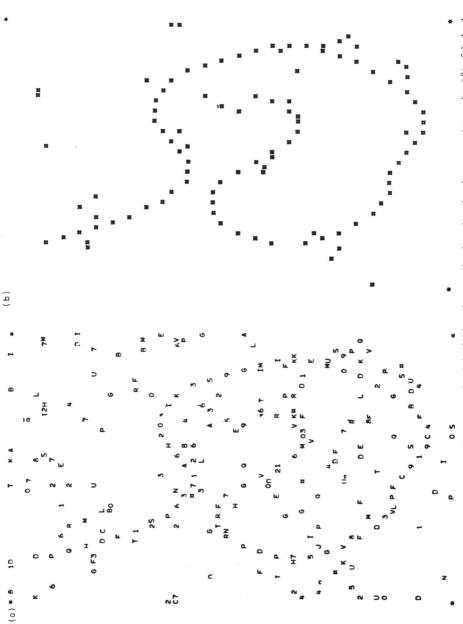

Fig. 32 Dotted curve detection. (a) Input picture: gray levels (1–31) have been randomly assigned to the curve points and to 6 % of the background points. (b) Result of dotted curve detection operation (see text), applied to (a).

given type map into a single location in the transformed space. This approach maps global features (sets of points belonging to a given curve) into local features, which may be easier to detect. In the following paragraphs we illustrate this idea with several examples.

As a very simple example, suppose that we want to detect straight edges or lines passing through a given point P. If we take P as the origin, any such straight line has an equation of the form $y = mx$. Let $P_1, \ldots, P_n$ be a set of detected points (e.g., points at which there is an above-threshold edge or line response), say having coordinates (x_i, y_i) relative to P, $1 \leqslant i \leqslant n$. Let us construct a one-dimensional array in which the value at position w is the number of P_i's such that $y_i/x_i = w$. (More generally, we can let the value of w be the sum of the edge or line responses at all such P_i's.) Then if many of the P_i's lie on a straight line, say of slope m, we may expect a peak at $w = m$ in this array. Note that this method does not depend on the connectedness of the collinear points; it detects straight lines even if they are dotted. Note also that if there are many P_i's that do not belong to straight lines through P, they will give rise to a high background noise level in the w array.

As another simple example, suppose that we want to detect lines having a known slope θ. If we sum the gray levels along all lines parallel to θ, we obtain a one-dimensional projection of the given picture. If the picture contains many points that lie on the line of slope θ and intercept b, the projection should contain a peak at the position corresponding to b. This type of projection can be obtained by scanning the picture through a slit oriented at θ, or scanning it with a "flying line" of slope θ. The use of projections to measure properties of a picture will be discussed further in Section 12.1.4a.

The simple methods just described only detect lines that all pass through a given point or that all have a given slope. To detect all possible straight lines, we must use a two-dimensional transformed space, in the form of an array having coordinates (u, v). For example, for each detected point $P_i = (x_i, y_i)$, let us construct the line $v = -x_i u + y_i$ in the uv-plane; in other words, we add 1 (or an amount proportional to the strength of P_i) to each (u, v) that lies on this line. It is easily verified that if the P_i's are collinear, the corresponding lines all pass through the same point. (If the P_i's lie on a near-vertical line, this point recedes to infinity, so that the lines become parallel. We will ignore this case here.) Thus when we have constructed the lines corresponding to all the P_i's, we should obtain peaks at the points in (u, v) space that correspond to collinear sets of P_i's, and the positions of these points determine the equations of the corresponding lines.

Exercise 10.20. Verify that collinear points do give rise to concurrent lines; compute the coordinates of the common (u, v) point corresponding to a given line in the xy-plane. ∎

Straight-line detection becomes simpler if we know not only the positions, but also the slopes, of the detected edge or line points P_i. If the slope at $P_i = (x_i, y_i)$ is m_i, we know that the equation of the line must be $y = y_i + (x - x_i)m_i$; this corresponds to the point $(u, v) = (m_i, y_i - x_i m_i)$ in (slope, intercept) space. If we have many P_i's that lie on the same line, and map them all into their corresponding points in (u, v) space, we should obtain a large peak at or near $(m_i, y_i - x_i m_i)$. [Here again, we map the P_i's into (u, v) space by adding 1 at the appropriate position for each detected P_i, or by adding an amount proportional to the edge or line strength at P_i.] In practice, the peak will be smeared out, since the slope estimates m_i will not be very accurate; for example, if we obtain them from a set of 3×3 edge or line detection masks, they can only be accurate up to $45°$, and their accuracy will be similarly limited if we obtain them from a 3×3 gradient operator.

The (slope, intercept) parameter space used in the foregoing examples has the disadvantage that the slope becomes infinite for vertical lines, as already mentioned. To avoid this, we can use a different space, e.g., slope, distance from origin. It is easily verified that the line of slope θ_i through $P_i = (x_i, y_i)$ has closest distance d_i from the origin given by $d_i = r_i \sin(\theta_i - \varphi_i)$, where $r_i = \sqrt{x_i^2 + y_i^2}$ and $\tan \varphi_i = y_i/x_i$. The parameters θ_i and d_i uniquely define this line. Thus if we map the P_i's into (θ, d) space, collinear P_i's (having the proper slopes θ_i) will yield a peak at position (θ_i, d_i) in this space. However, this method too has the disadvantage that the estimate of θ_i at P_i is inaccurate, and as a result, the estimate of d_i is even less accurate. This is illustrated in Fig. 33.

Analogous methods can be used to detect other types of curves. Consider a two-parameter family of curves $F(x, y, a, b) = 0$; then each $P_i = (x_i, y_i)$ gives rise to a curve $C_i: 0 = F(x_i, y_i, a, b)$ in (a, b) space, and if we have many P_i's that all lie on the same curve, say $F(x, y, a_0, b_0) = 0$, all of the corresponding C_i's should intersect in the point (a_0, b_0). If we are given not only the coordinates of P_i but also the slope of $F(x, y, a, b) = 0$ at P_i, then in general we can map this information into a point, or a finite set of points (a_i, b_i) in (a, b) space, and if we have many (point, slope) values for the same curve, this should yield a peak at or near the corresponding point in parameter space.

To further illustrate these ideas, consider the problem of detecting arcs of circles that pass through a given point, which we take to be the origin. Any such circle has an equation of the form $(x - a)^2 + (y - b)^2 = a^2 + b^2$. Given $P_i = (x_i, y_i)$, we want to map it into the curve C_i in (a, b) space defined by $(x_i - a)^2 + (y_i - b)^2 = a^2 + b^2$; note that C_i is the straight line $2x_i a + 2y_i b = x_i^2 + y_i^2$. If we have many P_i's that all lie on a circle with the same center (a_0, b_0), the corresponding lines C_i will all pass through the point (a_0, b_0).

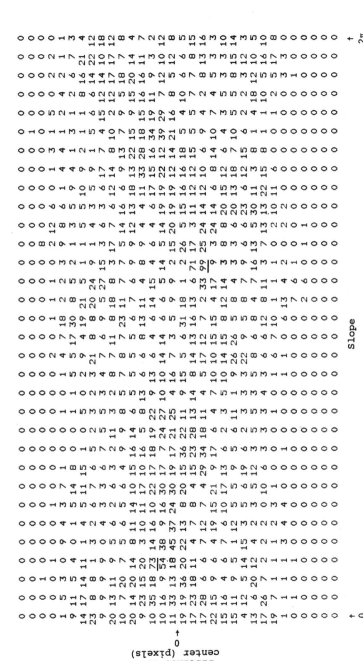

Slope

Distance from
center (pixels)

Fig. 33 Straight-line detection using a "Hough transform." The transform is derived from the edge detector output of Fig. 31a; the origin is taken to be at the center of the picture. For each point, the edge magnitude (RMS of x and y differences) is added to the transform in position (θ, d), where θ is slope (arc tangent of the ratio of the x and y differences) and d is perpendicular distance from the origin. The magnitude sums have been normalized to the range 0–99, and θ and d have been quantized to 32 values each. The peaks (underlined) correspond to the major edges in diagonal directions. The approximately horizontal and vertical edges are harder to detect.

Suppose next that we know not only the coordinates (x_i, y_i) but also the slope m_i at each point P_i. Now the slope y' of the circle $(x - a)^2 + (y - b)^2 = a^2 + b^2$ at (x_i, y_i) is determined by

$$2(x - a) + 2(y - b)y' = 0$$

so that $(x_i - a) + m_i(y_i - b) = 0$. Thus P_i maps into the point (a, b) of C_i that satisfies

$$2x_i a + 2y_i b = x_i{}^2 + y_i{}^2, \qquad a + m_i b = x_i + m_i y_i$$

Readily, this is the point

$$a = x_i - \frac{m_i(x_i{}^2 + y_i{}^2)}{2(m_i x_i - y_i)}, \qquad b = y_i + \frac{x_i{}^2 + y_i{}^2}{2(m_i x_i - y_i)}$$

If we have many P_i's belonging to the same circle $(x - a_0{}^2) + (y - b_0{}^2) = a_0{}^2 + b_0{}^2$, this method should yield a peak at (a_0, b_0).

Exercise 10.21. Suppose that we are interested only in circles that are tangent to a given line at a given point. Choose a coordinate system so that the line is the x-axis and the point is the origin; thus a circle of radius r has equation $x^2 + (y \pm r)^2 = r^2$, or $(x^2 + y^2)/y = \pm 2r$. Show how to detect points lying on such a circle using a one-dimensional parameter space. ∎

As another example, suppose that we want to find circles of a given radius r that have arbitrary centers, so that their equations are of the form $(x - a)^2 + (y - b)^2 = r^2$. In this case, each $P_i = (x_i, y_i)$ maps into the circle $(a - x_i)^2 + (b - y_i{}^2) = r^2$ in (a, b) space. If we have many P_i's that lie on the circle with center (a_0, b_0), all of the corresponding circles will pass through the point (a_0, b_0). If we know the slope m_i at P_i, an alternative possibility is to map (x_i, y_i, m_i) into the point at distance r from (x_i, y_i) in direction perpendicular to m_i (note that we are not using a parameter space now); this point should be at or near the center of the circle, so that if we have many P_i's that lie on the same circle, we should obtain a peak around this point.

Hough transforms that detect families of ellipses can also be defined, but we will not give the details here. It is usually desirable to use the Hough space iteratively, i.e., each time a peak is found in the space, we remove all picture points that contributed to it; this decreases the background noise level in the space, and makes other peaks easier to find.

10.3.4 Edge and Curve Linking

If a picture consists of regions, or objects on a background, that have simple, well-defined borders, and we apply an edge detection operator to it, we expect to obtain edge points most of which lie on smooth curves corresponding to the region borders. In practice, of course, we will also obtain many "noise" edge points that do not lie on borders, and conversely there will be border points at which we do not detect edges. This section briefly discusses methods of improving the set of edge responses by deleting noise responses and filling gaps.

Since the desired edge responses should lie on smooth curves, one possible approach to obtaining cleaner responses is to apply curve detection operators to the edge output. In particular, the nonlinear curve detectors described in Section 10.3.2 should eliminate much of the noise, while preserving the edge points that do lie on curves provided the gaps in the curves are not too wide. Repeated application of these operators should further strengthen the noise cleaning effect. This approach was illustrated in Fig. 31. At the same time, the dotted curve detectors of Section 10.3.2 can be easily modified to fill gaps. If the desired borders or curves are straight, techniques based on the Hough transform can be used to eliminate noise responses and fill gaps.

Another possible approach is to examine the neighbors of each edge point in (or close to) the direction along the edge, i.e., approximately perpendicular to the gradient direction. If other edge points are found in this direction on both sides, and if their slopes do not differ too greatly from the slope of the given edge point, it is linked to them; if not, it can be deleted. This approach can be used even if definite edge detection decisions have not been made; one can strengthen edge responses if directionally consistent responses are found at neighboring points, and weaken them otherwise, and this process can be iterated. Of course, the same method can be applied to cleaning up curve detector responses. Iterative reinforcement methods of this type will be discussed further in Section 10.5, where examples of their effects will be given.

Edge responses can also be improved by making use of both edge detection and thresholding (or, more generally, edge detection and spectral or spatial pixel classification). When a picture is thresholded, a set of borders between the above-threshold and below-threshold regions is defined. Noise edge responses can be largely eliminated by discarding all responses that do not lie on (or near) these borders. At the same time, gaps in edges can be filled by linking the edge points along the borders, using border following techniques; this makes it possible to fill large gaps without the need for extensive search. An example of this approach is shown in Fig. 34. More generally, one can use a variable threshold or a set of thresholds, and link edge points that are neighbors along the borders defined over a range of thresholds. Conversely, edge detection can be used in support of object extraction by thresholding; given an above-threshold object, we can check whether its border consists largely of edge points, and reject it as a "noise object" if not, as illustrated in Fig. 35.

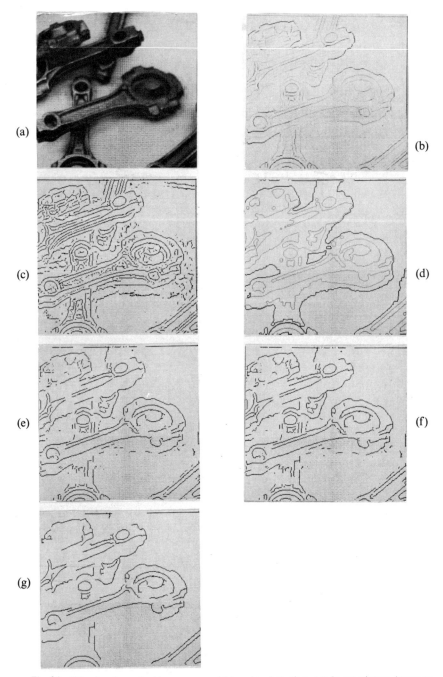

Fig. 34 Edge/border coincidence as an aid in edge detection. (a) Input picture (same as Fig. 9a). (b) Edge values for (a). (c) Points having nonzero edge values. (d) Border points for (a) using two variable thresholds, obtained by fitting either two or three Gaussians to each local histogram. (e) Edge points coinciding with these border points. (f) Edge points of (e) linked by filling gaps of length ≤ 2 along the borders. (g) Connected components of points in (f) having more than 30 points.

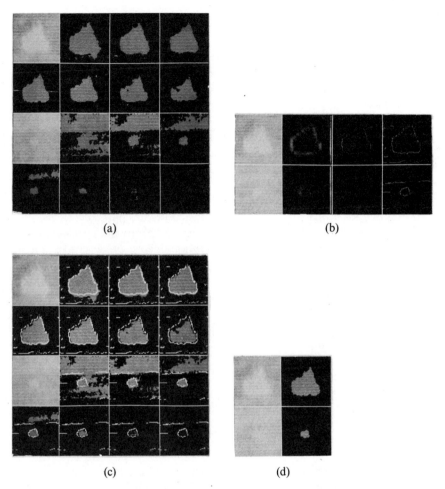

Fig. 35 Edge/border coincidence as an aid in threshold selection. (a) Input pictures (objects in an infrared image), and results of applying several different thresholds to them. (b) Edges in input pictures: differences of averages, maxima, thresholded maxima. (c) Result of overlaying edge maxima on each thresholded picture. (d) Above-threshold objects whose borders best coincide with the edge maxima.

When large pieces of edges or curves have been extracted from a picture, they can be further linked into global curves. The criteria for linking two curves into a larger curve might depend on their strengths (i.e., the average strengths of the original responses), lengths, and mutual alignment or "good continuation." Of course, if global curves of specific shapes are desired, this can also be taken into account in defining the linking criteria.

10.4 SEQUENTIAL SEGMENTATION

In most of the segmentation methods described up to now, the processing that was done at each point of the picture did not depend on results already obtained at other points. Thus these methods can be regarded as operating on the picture "in parallel," i.e., at all points simultaneously; and they could be implemented very efficiently on a suitable parallel computer. This section deals with segmentation methods in which we do take advantage, in processing a point, of the results at previously processed points. In these inherently sequential methods, the processing that is performed at a point, and the criteria for accepting it as part of an object, can depend on information obtained from earlier processing of other points, and in particular, on the natures and locations of the points already accepted as parts of objects.

Sequential segmentation methods have a potential advantage over parallel methods, with respect to their computational cost on a conventional, sequential computer. In the parallel approach, the same computations must be performed at every point of the picture, since our only basis for accepting or rejecting a given point is the result of its own local computation. If we want our segmentation process to be reliable, these computations may have to be relatively complex. When using the sequential approach, on the other hand, we can often use simple, inexpensive computations to *detect* possible object points. Once some such points have been detected, more complex computations can be used to *extend* or "*track*" the object(s). The latter computations need not be performed at every picture point, but only at points that extend objects that have already been detected. The detection computations may have to be performed at every point, but they can be relatively cheap, since they need only detect *some* points of every object while rejecting nonobject points.

Sequential methods are so flexible, and can be defined in so many different ways, that we will not attempt to discuss specific techniques in detail as we did in the previous sections. Rather, we will describe general classes of methods, and try to suggest the many variations that are possible.

10.4.1 Edge and Curve Tracking

a. Raster tracking

We begin our discussion of sequential methods with a very simple example. Suppose that the objects to be extracted from the given picture are thin, dark, continuous curves whose slopes never differ greatly from 90°. In this case, we can extract the objects by tracking them from row to row of the picture, as we scan the picture row by row in the manner of a TV raster.

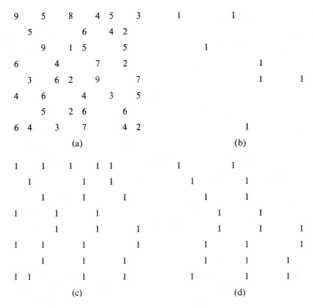

Fig. 36 Simple example of raster tracking. (a) Input picture (non-0's). (b) Result of thresholding (a) at 7. (c) Result of thresholding (a) at 4. (d) Result of tracking with $d = 7$, $t = 4$ (see text).

Specifically, in each row we accept any point whose gray level exceeds some relatively high threshold d; this is our detection criterion. In addition, once a point (x, y) on the yth row has been accepted, we accept any neighbor of (x, y) on the $(y - 1)$st row, i.e., we accept any of the points $(x - 1, y - 1)$, $(x, y - 1)$, and $(x + 1, y - 1)$, provided that the accepted points have gray level above some lower threshold t; this is our tracking criterion.

This trivial example of detection and tracking is illustrated in Fig. 36. Note that if we used the d threshold alone, or the t threshold alone, the curves would not be extracted correctly; but when we combine the two thresholds, in conjunction with row-by-row tracking, we are able to extract the curves. To achieve the same results by a parallel method, we would have to use some sort of line detection operation (see Sections 10.3.1–10.3.2) at each point; this would be relatively costly, since in line detection we would have to compare *every* point to its neighbors on the rows above and below it, whereas here we did so only for points already being tracked.

In the following paragraphs we indicate some of the many possible ways that our simple example can be modified and extended. We still consider only raster tracking; extensions to thin curves having arbitrary slopes, and to solid objects, will be discussed later in this section.

The detection and tracking acceptance criteria can involve local property values other than gray level. For example, they can be based on local contrast, as measured, for example, by the value of some derivative operator such as the gradient. Note that the gradient defines not only a degree of contrast, but also a direction (of highest rate of change of gray level at the point); in tracking, one could look for successor points along the perpendicular direction, which should be the direction "along" the curve being tracked. The tracking criterion can also depend on a comparison between the current point (x, y) and its candidate neighbors, so as to disciminate against candidates whose gray levels, etc., do not "resemble" that of (x, y), if desired.

In some cases, we may want to accept all candidate points that satisfy the tracking criterion. In other cases, however, it may be preferable only to accept the best one(s). For example, it may turn out that accepting all of them would make a curve thick, which contradicts our *a priori* knowledge that the curves are thin.

The tracking criterion should, in general, apply to some region below each currently accepted point (x, y), rather than just to the point's immediate neighbors on the row below it. Moreover, the criterion can depend on the candidate point's position in this region. For example, it can be more stringent for points that are farther from (x, y), so as to discriminate against large gaps in the curves, or it can be more stringent for points that are less directly below (x, y), so as to discriminate against obliquity.

Similarly, the tracking criterion should depend not just on a single currently accepted point (x, y), but on a set A of already accepted points. This makes it possible, for example, to fit a line or curve to the points in A, and make the criterion more stringent for candidate points that are far from this curve. In this way, we can discriminate against curves that make sharp turns. If desired, we can give greater weight to the more recently accepted points of A than to the "older" points. ("Tracking" based on sets of already accepted points will be further discussed later under the heading of region growing.)

The preceding remarks can be summarized in the following generalized raster tracking algorithm.

(1) On the first row (or line of the raster) accept all points that meet the detection criterion. Take each such point to be the initial point of a curve C_i that is to be tracked.

(2) On any current row other than the first row:

(a) For each curve C_i currently being tracked, apply the appropriate tracking criterion to the points in its acceptance region; adjoin the resulting accepted points to C_i. We recall that this criterion may depend on the distances and directions of these points from the end of C_i or from some curve that extends C_i, as well as on the gray levels, contrasts, etc., of the points. If no new points are accepted into C_i, tracking of C_i has terminated. Note that a curve C_i may branch into two or more curves, in which case we must track them all; or two or more curves may merge into a single curve, in which case we need only track that one from there on.

(b)   In addition, apply the detection criterion to points that do not belong to any acceptance region; if any points meet this criterion, take them to be initial points of new curves C_i.

(3)   When the bottom row is reached, the tracking process is complete.

This algorithm is analogous to an algorithm given in Section 11.3.1 for tracking connected components; but the present algorithm is much less specific, because of the wide variety of detection and tracking criteria that could be used.

b.   Omnidirectional tracking

Up to now we have restricted ourselves to tracking methods based on a single raster scan. This has the disadvantage that the results depend on the orientation of the raster, and the direction in which it is scanned. For example, if a strong curve gradually becomes weaker (lighter, lower contrast, etc.) as we move from row to row, our tracking scheme may be able to follow it, since the acceptance criteria for a curve that is already being tracked are relatively permissive; but if we were scanning in the opposite direction, so that the curve starts out weak and gets stronger, we might not detect it for quite a while, since our detection criteria are relatively stringent. We can overcome this problem by scanning the picture in both directions, carrying out our tracking procedure for each of the scans independently, and combining the results, but this doubles the computational cost, and the results may not "fit" well.

Further, raster tracking has the disadvantage that it breaks down for curves that are very oblique to the raster lines. For example, if we scan row by row, we cannot track curves that are nearly horizontal, since the crossings of successive rows by such a curve are many columns apart. One way to overcome this problem is to use two perpendicular rasters, e.g., the rows and the columns, so that any curve meets at least one of the rasters at an angle between 45 and 135°. If we carry out detection and tracking independently for the two rasters, and combine the results, we should be able to track every part of the curve using at least one of the rasters. Not only does this approach double the computational cost, but it may also cause us to miss parts of the curve, since when it becomes too oblique to be tracked by one raster, it will be picked up by the other raster only if it meets the more stringent detection criteria.

These directionality problems can be largely avoided if we use omni-directional curve-following, rather than raster tracking. Given the current point (x, y) on a curve that is being tracked, a curve-following algorithm examines a neighborhood of (x, y), and picks a candidate for the next point. If the curves being tracked can branch or cross, it may be necessary to pick more than one next point; in that case, all but one of the chosen points are stored for later investigation, and the tracking proceeds with the one re-maining point as next point.

The candidates for next point are evaluated on the basis of their satisfying a tracking criterion for acceptance (dark, high contrast, etc.). They need not be immediate neighbors of (x, y), since we want to be able to tolerate small gaps in curves; but the tracking criterion should discriminate against points far from (x, y), since a connected curve would normally be preferable to one with gaps. They need not lie on the side of (x, y) directly opposite the already accepted points, since our curves need not be straight lines; but the criterion should discriminate against points that deviate too much from this direction, since a smooth curve is normally preferable to one that makes sharp turns. All of this is, of course, analogous to the raster tracking case, except that all the neighbors of the current point, not just those on some "next row," are candidates for being the next point.

At any stage, the next point must not be a point that has already been accepted at a previous stage. If there are no candidates that have not already been accepted, the tracking terminates. We can now go back to any points that were stored for later investigation, and resume tracking these, to find other curve branches. In particular, when we first detect a curve point, we track that curve in one direction until it terminates, and then go back to the originally detected point to track the curve in the other direction, if possible. When no points remain to be investigated, we can resume searching the picture for other points that meet the detection criterion; these points must be on curves that do not cross the curves that we have already tracked.

We can summarize this approach to curve-following in the form of a generalized algorithm:

(1)   Scan the picture systematically, looking for points that meet the detection criterion. When such a point is found, it becomes the "current point."

(2)   Examine the neighborhood of the current point and apply an appropriate tracking criterion. As before, this criterion may depend on the gray level, contrast, etc., of the candidate point, as well as on its distance and direction from the current point, or from some curve that extends the curve branch currently being tracked.

(a)   If no points as yet unaccepted meet this criterion, tracking of that branch has terminated. In this case, we take the next point on list L [see (c)] as our new current point, and resume tracking. If list L is empty, we go back to step (1).

(b)   If the unaccepted points meeting the criterion all appear to lie on a single curve, we accept the closest of them as belonging to the curve branch being tracked, take it as the new current point, and go back to step (2).

(c)   If these points appear to lie on more than one curve, we may conclude that the curve being tracked has branched, or has crossed another curve. In this case we put all but one of the closest ones on list L for later investigation. The remaining closest point is accepted as the new current point, as in case (b), and we go back to step (2).

(3)   When the systematic scan is finished, the algorithm has terminated.

This algorithm is analogous to an algorithm given in Section 11.2.2 for following the borders of connected components; but the present algorithm is much less specific, because of the wide variety of acceptance criteria that could be used.

Edge tracking (i.e., tracking the high-gradient points around the border of a region) is analogous to curve tracking. The acceptance criteria are based on the gradient magnitude; information about the gray levels of the regions whose edges are being tracked can also be used, if available. The gradient direction can provide useful information; in fact, smooth borders can sometimes be tracked by simply moving perpendicularly to the gradient direction from point to point. Ordinarily, edge tracking would have to be omnidirectional, since region boundaries are closed curves. In principle, tracked edges should never branch, and should never terminate except by returning to the starting point or running off the edge of the picture.

Omnidirectional tracking, as we have seen, is much more flexible than raster tracking, but it has the disadvantage that it requires access to the points of the picture in a nonprespecified order, as determined by the shapes of the edges or curves being tracked. Raster tracking, on the other hand, accesses the picture row by row, which is very convenient if the picture must be accessed from peripheral storage devices.

c.   Search techniques in tracking

The tracking techniques described up to now make their acceptance decisions in a specified sequence (row by row of the raster, point by point along the curve or around the object boundary), and never reverse a decision once made. An important extension of these techniques is to allow backup, i.e., if an acceptance decision seems to be leading to a series of poor subsequent acceptances, one can go back and alter the decision.

5		5	5
5		5	5
7	4	7	4
3	5	3	5
1	6	1	6
1	6	1	6
	6		6
(a)		(b)	(c)

Fig. 37 Tracking with backup. (a) Input picture (values not shown are 0's). (b) Result of picking the darkest candidate point on each row. (c) Result of picking the second-best choice on the third row.

To illustrate this idea, suppose that we are using raster tracking to find the best possible thin, dark curve that crosses the picture from top to bottom. If we pick the darkest candidate point at each successive row, we may get trapped in a blind alley. As shown in Fig. 37, when backup is allowed, we can avoid this.

In order to formulate the backup concept more precisely, suppose that we are given some method of evaluating the curves (or object borders, etc.) that are being tracked, e.g., using criteria of contrast, smoothness, etc. For any curve C, let $\varphi(C)$ be the value of C. For example, $\varphi(C)$ might be the average contrast of C, minus its average curvature. (On the measurement of curvature for digital curves, see Section 11.3.3.) Given any point (x, y), let $\hat{\varphi}(x, y)$ be an estimate of the value of the best curve in the picture that passes through (x, y). For example, we might take $\hat{\varphi}(x, y)$ to be $\max(\varphi(C))$ for all curves C so far found between the starting point and (x, y).

At any stage of the tracking process, we have a list L of already accepted points from which the tracking might continue [compare (b) above]. Initially, L consists of just the starting point, We compute $\hat{\varphi}$ for all the points on L, and pick the point (x, y) for which $\hat{\varphi}$ is highest. We now examine the neighbors of (x, y), find those that are candidates for acceptance, and put them on the list L; this completes one stage of tracking. The point (x, y) need no longer be on L, since its neighborhood has already been examined. We should remember (x, y), however, if we want to avoid reaccepting points that have already been accepted.

If we use this procedure, the point picked for examination of its neighbors at a given stage is not necessarily a neighbor of the point picked at the previous stage. Rather, at each stage, the tracking process backs up, if necessary, to the highest-value point whose neighbors have not yet been examined. Indeed, if we use the $\hat{\varphi}$ just described, this procedure will pick the highest contrast, straightest curve C yet found at the given stage, and extend C by accepting the appropriate neighbors of its end point. One could also make the value of a curve depend on its length, so that short curves have higher values than long ones. In practice, this may result in an excessive amount of backup. It may be desirable, in fact, to limit the distance over which backup is allowed. For example, once the point (x, y) has been picked, we can discard from L all points that are more than some given distance away from (x, y).

As a simple example of how this tracking procedure operates, consider Fig. 37. We take the "5" in the top row as the starting point; let its coordinates be $(0, 6)$. Initially, this is the only point available for examination. Its sole candidate neighbor is $(0, 5)$, and we have $\hat{\varphi}(0, 5) = 5 - 0 = 5$ (the average gray level is 5, the curvature is 0). This is now the only available point, but it has two candidate neighbors, $(0, 4)$ and $(1, 4)$. Now

$$\hat{\varphi}(0, 4) = \frac{17}{3} - 0 = \frac{17}{3}$$

while

$$\hat{\varphi}(1, 4) = \frac{14}{3} - \frac{1}{3} \cdot \frac{\pi}{4}$$

since there is a 45° turn on the best curve leading to $(1, 4)$; hence we pick $(0, 4)$ as the point to investigate next. Its neighbor $(0, 3)$ has $\hat{\varphi}(0, 3) = 5$, and this in turn has neighbor $(0, 2)$ with $\hat{\varphi}(0, 2) = \frac{21}{5}$. Since $\hat{\varphi}(0, 2) < \hat{\varphi}(1, 4)$, the tracking process now backs up to $(1, 4)$ and examines its neighbors, thus discovering the stronger branch of the curve.

The $\hat{\varphi}$ evaluation function just described is only one of many devices that could be used to control the order in which the tracking process examines points. In general, the tracking problem can be formulated as one of searching for an optimum curve between the top and bottom rows of the picture (say), where optimality might be defined in terms of maximizing a function such as φ. In principle, mathematical programming techniques can be used to find optimum curves, although this approach tends to be computationally costly. Alternatively, heuristic search techniques can be used; if it is properly defined, such a technique is guaranteed eventually to find an optimum path. Ideally, a heuristic technique would not only evaluate a given point on the basis of how it can be reached from the starting point, but would also attempt to look ahead and estimate how the goal (a curve that crosses the picture) can be reached from the given point.

Search methods are an improvement over classical tracking techniques, since they allow look-ahead and backup, rather than making irrevocable decisions at each step. Still more flexibility is achieved if the evaluation functions are modified during the course of the search, based on whatever information has been obtained about the nature of the curves that are actually present in the picture.

10.4.2 Region Growing

We now consider the problem of sequentially extracting solid objects or regions, rather than edges or thin curves, from a picture. We will usually assume in this section that the goal is to extract objects of a specific type, rather than to completely partition the picture. Partitioning will be discussed in Section 10.4.3.

As indicated in the preceding section, the criteria used for edge and curve extraction typically include contrast (or edge strength) and shape (e.g., straightness, smoothness, etc.). A broader range of criteria can be used in extracting objects or regions; they include region homogeneity (in gray level, texture, etc.) and contrast with the background, strength of the region's edges, size, shape simplicity, conformity to a desired texture or shape, and so on. Each of these criteria can be measured in many different ways: in subsection (a) below we describe several ways of measuring homogeneity.

Subsection (b) discusses row-by-row "region tracking," and Subsections (c)–(d) treat omnidirectional "region aggregation" and "region merging"; these correspond to raster and omnidirectional curve tracking. In Subsection (e) we discuss some methods of grouping regions into objects; here the objects are not necessarily homogeneous, but are defined by relationships among the regions. Finally, Subsection (f) discusses the use of scene models in guiding the process of region growing or grouping.

a.   Region homogeneity

Many of the region growing and picture partitioning techniques discussed in this and the next section make use of criteria based, at least in part, on region homogeneity. The following are a few of the ways in which the homogeneity of a region S can be measured:

(1)   *Region/subregion similarity.* Subdivide $S = S_0$ in some arbitrary way into a few large pieces $S_1, S_2 \ldots$ (e.g., if it is square, divide it into quadrants). Compute a set of gray level statistics for S and for each of the S_i's; these might include first-order statistics such as the mean and variance of gray level; second-order statistics such as those used for texture description (see Section 12.1.5); statistics of various local properties (e.g., mean gradient magnitude); and so on. This gives us a feature vector for each S_i. If these feature vectors are all sufficiently close to one another, we say that S is homogeneous.

(2) *Region uniformity.* Let $V(S)$ be a measure of the gray level variability of S; examples are the variance or standard deviation, the gray level range or interquartile range, etc. Then we can use, e.g., $1/(1 + V(S))$ as a measure of S's uniformity; this is 1 when S has constant gray level, and goes to zero as V becomes large.

(2′) *Approximability by a constant.* Another way of looking at region uniformity is that it measures how well the gray levels on S can be approximated by a constant. In particular, let us consider two simple measures of the approximation error: maximum absolute error and mean-squared error. Let S have m points, and let the gray levels at these points be $z_1, \ldots, z_m$.

Max error. Let a and b be the min and max of the z's, respectively. Then when we approximate S by a constant c, the maximum absolute error is $\max[|a - c|, |b - c|]$. Readily, this error is minimized by taking $c = (b + a)/2$; the max error is then $(b - a)/2$.

Mean-squared error. In this case the error when S is approximated by c is $e \equiv (1/m) \sum_{i=1}^{m} (z_i - c)^2$. If we differentiate e with respect to c and equate to zero, we obtain $\sum_{i=1}^{m} (z_i - c) = 0$, or $c = (1/m) \sum_{i=1}^{m} z_i$, so that $c \equiv \mu$ is the mean gray level of S. For this choice of c, e is $(1/m) \sum_{i=1}^{m} (z_1 - \mu)^2 \equiv \sigma^2$, the gray level variance of S. Thus the mean-squared approximation error is minimized by taking $c = \mu$, and this minimum error is σ^2.

(3) *Approximability by a function.* More generally, we can measure the uniformity of S by how well its gray levels can be approximated by some standard function (e.g., a plane or some polynomial surface).

b. *Region tracking*

One approach to sequential object extraction is to track runs of object points from row to row of the picture; this is analogous to raster curve tracking (Section 10.4.1a). The general idea is as follows: On the first row, each run of detected points is taken as the start of an object. On subsequent rows, we accept points that are adjacent to already accepted runs, if they satisfy a tracking acceptance criterion; and we also accept other points if they satisfy the more stringent detection criterion. The accepted points on a new row constitute a new set of runs, and the process repeats. Runs may merge, split, or terminate, as in the thin curve case. The tracking acceptance criterion will depend, in general, on the natures and positions of the candidate points relative to the already accepted runs. On the use of run tracking for shape description see Section 11.3.2c.

Like curve tracking, object or region tracking can be regarded as a search or optimization problem. With each run we can associate a figure of merit based on its homogeneity, distinctness from its surround, resemblance— in gray level or texture—to the corresponding run on the previous row (or to the part of the object so far extracted), smoothness with which it continues that run, degree to which it conforms to the desired object size or shape, etc. A straightforward approach is to choose the runs of highest merit on each row, but it may turn out afterwards that other choices would have led to an overall higher merit, so that backtracking (e.g., readjustment of the run end points) may be desirable. For further discussion of search and optimization techniques in tracking, see Section 10.4.1c.

As in the case of raster curve tracking, the main disadvantage of this approach is its direction dependence; the acceptance criteria depend on the object points that have already been encountered. Moreover, we must assume that this dependence makes a significant difference as regards which points get accepted; if it did not, we could have used a parallel segmentation technique.

c. Region aggregation

A much more flexible approach to sequential extraction of objects is to "grow" the objects in all directions; this is analogous to omnidirectional curve tracking (Section 10.4.1b). The concept is as follows: We start with a point that meets a detection criterion; the choice of starting points will be discussed further below. Let O be a currently accepted piece of an object. We examine all its neighboring points, and incorporate any of them that meet a growth acceptance criterion. As previously, this criterion can depend both on the natures and positions of the candidate points. For example, acceptance of a point (x, y) may depend on the closeness of (x, y)'s gray level to the average gray level of O; and it may also depend on (x, y)'s position relative to O, and on the size and shape of O itself, since we may want to bias the growing process to favor certain types of object sizes and shapes. When new points are accepted, they are adjoined to O, and the process is repeated with the resulting new O. The growth terminates when no acceptable neighbors exist.

As a very simple illustration of object extraction by region aggregation, suppose that a picture contains two regions, A and B, consisting primarily of gray levels 0 and 2, respectively, but that both regions also contain occasional noise points of gray level 1. Let us accept a point into a growing region if their gray levels differ by 1 or less. If we use this criterion, the noise points in each region will become part of that region, but the two regions will remain separate. (1's on the border between A and B may be accepted into either A or B, depending on which of the regions is grown first.) Note that we could obtain a similar result by averaging the picture and then thresholding it at 1; but this would result in some smoothing of the boundary between A and B.

When tracking or growing arbitrary objects rather than thin curves, the acceptance criteria can be based on texture rather than on gray level or contrast. We can test candidate "cells" (small groups of points), rather than single points, for inclusion in an object, and accept them if their textures resemble that of the object sufficiently closely. For some methods of comparing textures see Section 10.2.3b (regarding texture edge detection). We can also use edge detection to set up barriers to growth, i.e., we never accept a neighbor into an object if there is an edge between it and the object.

To initiate the growth process, one should use a point or cell that is very typical of a given type of object or region. Such points can be identified by first measuring a set of property values at every point, and finding clusters in these values, as in Section 10.1; a point whose values are near the center of a cluster is likely to be a typical representative of its class of points. (If ideal property values for classes are known *a priori*, closeness to the ideal values can be used as a starting criterion.) Once the growth has been initiated, the classification of successive points or cells as to whether or not they belong to the given cluster may depend on their property values relative to those of the growing object (i.e., the part of the cluster already accepted), as well as on their relative spatial positions. Thus region growing can be regarded as a sequential clustering or classification process.

The results of region aggregation will almost always depend on the choices of starting points and the order in which points or cells are added to the growing objects. Here again, optimization or heuristic search techniques can be used to control the process: one can compute a merit for each possible choice, and pick one having the highest merit. Another possibility is to grow the region "in parallel," i.e., to simultaneously adjoin every acceptable adjacent cell to it, repeatedly. Note, however, that this makes it much harder to control the properties (homogeneity, etc.) of the growing region.

d. Region merging

Another approach to region growing is to first construct a set of highly uniform regions, e.g., connected components of constant gray level, or, more generally, regions extracted by some preliminary segmentation process. Growth then takes place by starting with one of these regions and merging neighboring regions with it, one at a time. Here again, the choice of which neighbor to merge should depend both on the similarity of the regions (in gray level, color, texture, etc.) and on the size and shape of the resulting merged region. The merit of a given merge might reflect, e.g., a tradeoff between the increase in gray level variability and the decrease in shape complexity. For example, we might want to eliminate small "noise" regions, even though they do not resemble the neighboring larger region(s), since this

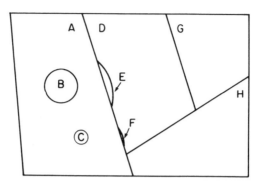

Fig. 38   Sketch illustrating region merging possibilities (see text).

will result in shape simplification, and will also eliminate "regions" that actually represent pieces of the borders between pairs of larger regions.

As a simple illustration of how region merging might work, consider the set of regions sketched in Fig. 38. Let the gray levels of these regions be as follows:

Region:	A	B	C	D	E	F	G	H
Gray level:	0	2	2	5	4	3	3	4

In this example, we might allow C to merge with A, even though their gray levels differ by 2, because C is small, but not B, because it is larger. E could merge with D, since their gray levels differ by only 1, and so might F even though it differs by 2, since it appears to be a piece of border zone between A and D. H could merge with either G or D, depending on which of these merges is considered first.

The following are some examples of quantitative size and shape criteria that can serve as a basis for merging. As we shall see in Section 11.3.4a, one way to measure shape complexity is in terms of p^2/A, where p^2 is perimeter (e.g., the number of border points in the given region) and A is area (the total number of points). If q is the length of common border of the two regions, i.e., the number of pairs of adjacent points that belong to the two regions, and they have areas A_1, A_2 and perimeters p_1, p_2 respectively, then $q \leqslant \min(p_1, p_2)$, and the p^2/A measure for the merged region is $(p_1 + p_2 - 2q)^2/(A_1 + A_2)$. This is evidently greatest when $q = 0$, i.e., when the regions have no common border; and it is least when one of the regions surrounds the other, since q is then as large as possible. Thus if we use the p^2/A of the merged region to measure the cost of a merge, we will be more likely to merge two regions the larger their common border is relative to their perimeters. A more refined possibility is to take the strength of the common border into account, i.e., to count the number q_t of common-border point pairs that differ in gray level by less than some threshold t; the merge merit computation can then take into account the size of q_t relative to $p_1 + p_2$ or to $\min(p_1, p_2)$. For additional remarks about possible merging criteria, see Sections 10.4.3c and 10.4.3e, where we also show how merging can be performed by processing the graph representing the region properties and adjacencies.

e. Region linking

Up to now we have generally assumed that the goal of region growing is a homogeneous region or object. At a higher level, we can look for sets of regions that are parts of a given (nonhomogeneous) object or pattern. In the following paragraphs we give several examples of extracting such related sets of regions.

(1)   Suppose that the pattern of regions is specified by a model, i.e., we know what types of regions should compose it, and how these regions should be arranged. In this case, finding the pattern is essentially a matching task. Such tasks were discussed in Section 9.4.3.

(2)   In some cases one can formulate reasonable rules for associating regions together that are useful in a variety of situations, and not only for patterns defined by specific models. The Gestalt laws of organization (similarity, proximity, good continuation, closure, etc.), discussed in Section 3.4, are rules of this sort. For example, one would normally use "good continuation," based on agreement in slope and alignment, to link pieces of curves into global smooth curves, as discussed in Section 10.3.4.

(3)   An interesting example of region-linking rules is provided by the work of Guzman [25] on the interpretation of straight-line drawings as representations of three-dimensional polyhedra. Consider, for example, the three types of branch points shown in Fig. 39a. The " Y " type, in which the three angles are all less than 180°, looks like a trihedral angle viewed from above its vertex; thus the three regions that meet at a Y could be the faces of such an angle, i.e., could belong to a single polyhedron. On the other hand, an "A" branch point, which has one angle greater than 180°, looks like a trihedral angle viewed from the side; here the two regions bounded by the angles that are less than 180° could be faces of the trihedral, thus belonging to a single polyhedron, whereas the third region appears not to belong to the same polyhedron. Finally, a " T " branch point, where one angle is exactly 180°, looks like a place where one polyhedron disappears behind another, and there is no good evidence that any of the three regions belong to the same polyhedron. Given a line drawing, suppose that we "link" regions that meet at branch points in accordance with the rules just suggested: all three regions at Y are linked, the two regions at an A whose angles are less than 180° are

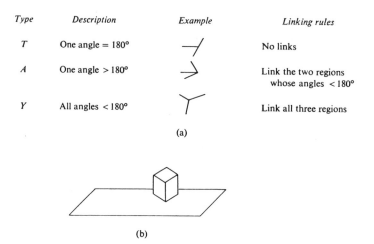

Fig. 39 Use of branch point properties to link regions into "bodies." (a) Three types of branch points and their linking rules. (b) If these rules are applied to the simple drawing shown here, the faces of the cube become doubly linked to one another, but no other links are formed.

linked, and nothing is linked at a *T*. For the drawing shown in Fig. 39b, these rules create two links between each pair of faces of the cube, and no links between any other regions in the drawing. Thus, using these rules, we would have grounds for merging the three cube faces into a single polyhedron. These ideas are discussed further in the Appendix to Chapter 12.

f. Model-guided region growing

The region growing schemes described in this section have been primarily "general-purpose," i.e., designed to extract homogeneous objects or closely related sets of regions from a picture. They begin with an initial "cell" (a point, a small block of points, or a uniform region), chosen to be highly typical of a class or cluster. At successive stages, merge merits are computed for neighboring pairs of cells or region fragments, and a best choice is made; the merits may depend on such properties as homogeneity, edge strength, shape simplicity, etc. The process stops when no acceptable merges remain to be made.

Ideally, region growing should be based on a model for the object(s) being extracted from the picture, and the merges at each step should be compatible with this model. General-purpose merge criteria such as those mentioned in the preceding paragraph should be used only in the absence of a more specific model. In the following paragraphs we mention some simple ways in which models can be used to guide region growing.

Suppose that we have a model for the class of pixels expected to be present in the desired object(s), e.g., in terms of their spectral signatures. We can then begin region growth at a point that is highly representative of this class, as mentioned earlier; and we can control the growth so as to insure that the distribution of spectral features at the points of the growing object stays close to the desired distribution. If we also know something about the desired object sizes or shapes, we can in principle use this knowledge too to guide the growth; but this is generally harder, especially at the early stages of the growth process.

It is sometimes possible to make use of a "map" that represents the expected arrangement of parts in the given picture. For example, this might be feasible when dealing with pictures of terrain taken from a known position, with x rays of specific body parts, or with known patterns such as printed circuits. The choices of starting points for region growth, of merge criteria, and of barriers to growth can then be based on the properties, locations, sizes, and shapes of the growing regions relative to the regions on the map. The use of models to control picture partitioning will be discussed further in the next section.

10.4.3 Partitioning

In the previous section we discussed methods of extracting regions from a picture by repeated aggregation or merging. Many of these methods can also be used to partition a picture into regions. More generally, *picture partitions* can be constructed by starting with an initial partition and allowing both merging and splitting of regions. The initial partition may be trivial (e.g., the entire picture is the sole region; each pixel or $k \times k$ block of pixels is a "region"; the connected components of constant gray level are the regions; etc.), or it can be the result of a previous segmentation process.

The criteria for merging and splitting should depend, if possible, on how well the partition conforms to a model for the given class of pictures. In the absence of such a model, one can use general-purpose criteria based on such factors as region homogeneity, distinctiveness of adjacent regions, edge strength, size, shape simplicity, and so on, just as in the case of region growing. There are many ways of defining these factors; in Subsection (a) below we discuss some ways of measuring the goodness of a partition as regards region homogeneity.

Partition building can be based on merging alone, starting with single pixels, cells, or uniform regions; approaches of this type are discussed in Subsections (b)–(c). At the other extreme, we can start with the entire picture as a single region, and partition it by repeated splitting; such methods are treated in Subsection (d). It is preferable to allow both splitting and merging, as discussed in Subsection (c). The use of scene models in partitioning is considered in Subsection (f).

a.   Piecewise homogeneity

The simplest way of measuring the goodness of a partition with respect to homogeneity is to use the average homogeneity (see Section 10.4.2a) of the regions in the partition; note that by this criterion, the partition into individual pixels, or into components of constant gray level, is perfectly homogeneous, since each of its regions is. A less trivial possibility might be the following: For each region S, let $S_1, S_2, \ldots$ be its neighboring regions, and let H be a homogeneity measure; then $H(S)/\max_i H(S \cup S_i)$ measures how homogeneous S is as compared with what would happen if it were merged with its most similar neighboring region, and one could use the average of this quantity (over S) as a measure of the goodness of the given partition. Still another idea would be to use average homogeneity times average edge strength, where the latter is averaged over all the region borders.

If we define region homogeneity to mean that each region is closely approximable by a function of some standard form (e.g., by a constant), then a good partition corresponds to a good piecewise approximation of the picture by such functions. It is straightforward to determine best piecewise approximations when the partition is given, but we are concerned with the much harder case where the partition itself is allowed to vary (but, e.g., the number of regions is fixed; if even this is not fixed, we can always define zero-error piecewise approximations by taking the regions to be single pixels).

In one dimension, variable-partition minimum-error piecewise approximations by splines have been extensively studied, and there are several important theorems about properties of such approximations, using the maximum error or the mean-squared error as the quantity to be minimized. For an introduction to this subject see Pavlidis [56], especially Sections 2.7–2.11. In two dimensions, the study of such approximations is much more difficult, since a partition has many degrees of freedom.

Intuitively, in a good partition the region borders should generally coincide with strong edges. In one dimension, this observation has a rigorous formulation due to McClure (cited by Pavlidis [56], p. 165): In a least squares approximation to a function on a variable partition using splines of order k, as the number of regions becomes large, the subdivision points are distributed as the $2/(2k + 1)$ power of the $(k + 1)$st derivative of the function. For $k = 0$ (i.e., piecewise constant approximation), we have $k + 1 = 1$ and $2/(2k + 1) = 2$, so that the subdivision points tend to be distributed as the square of the first derivative, i.e., as the square of the gradient magnitude. Thus a reasonable way to partition a single row of a digital picture in order to obtain a good piecewise constant approximation would be at points where the absolute first differences are highest. Similarly, if we want a good piecewise linear approximation, we should partition at points where the second differences are high; and so on. Here again, the situation is much more complicated in two dimensions, but one can still attempt to partition a picture into regions whose boundaries have strong edge values (for good piecewise constant approximation) or line values (for good piecewise linear approximation).

Since there are so many possible criteria for picture partitioning, we will not give specific examples here. For an extensive treatment of this approach the reader is referred to the book by Pavlidis [56].

b. Tracking

Row-by-row tracking techniques can be used to partition a picture by merging runs on successive rows. We partition the first row, using any desired segmentation technique—perhaps a one-dimensional region growing or partition building technique, based on the ideas in this or the preceding section. We similarly partition each subsequent row, but now we require not only that the partition be as "good" as possible in its own right, but also that it be as compatible as possible with the partition(s) of the preceding row(s). The criteria for computing the merit of a partition are analogous to those for run selection, as discussed in Section 10.4.2b. As a simplification, we can partition each row independently, and then apply region merging techniques (see below) to the resulting set of runs.

In implementing the run merging process for row-by-row region growing or picture partitioning, we can represent the runs on each row by a list of endpoints, and associate with each run the properties that are relevant to the merge merit (average gray level, length, etc.). The merge decisions can then be made by processing these lists, and merges can be represented by linking their entries in the lists. When this process is complete, the linked sets of runs represent the connected regions produced by the merging (see Section 11.3.1b). The use of graphs in region merging will be discussed further in (c) below.

c. Merging

In picture partitioning by merging, one usually starts with an initial partition into single pixels, small blocks or "cells" of pixels, or highly uniform regions produced by a prior segmentation process (e.g., connected components of constant gray level). One then computes merge merits for pairs of adjacent regions (or cells, etc.), and merges a pair having highest merit.

This process is repeated until no acceptable merges remain. The procedure is exactly analogous to that in Section 10.4.2c, except that there the goal was to grow regions of a particular type, rather than to partition the entire picture into regions.

Partitioning by region merging can be carried out in parallel by proceeding as follows: We start with the initial set of uniform regions. At any given stage, we compute merge merits for all adjacent pairs of regions. Let $\{(S_i, S_i')|i = 1, 2, \ldots\}$ be a set of pairs that have mutually highest merge merit, i.e., for each i, S_i prefers S_i' over all other candidates, and vice versa. (If exact ties in merit value are ignored, the set of such mutually compatible pairs is uniquely defined.) It is then safe to merge all these pairs simultaneously. On the other hand, it would not be safe to simultaneously merge every region with its most compatible region, since if the compatibilities are not mutual, this could lead to contradictions; for example, S_1 might want to merge with S_2 and S_2 with S_3, which would result in all three merging, even if S_1 and S_3 were incompatible.

Region merging can be implemented by representing the regions and their adjacencies in the form of a graph, in which the nodes represent regions, and two nodes are joined by an arc if the corresponding two regions are adjacent. (We may suppose that the points of the initial region were distinctively labeled, and that each node knows the label of its corresponding region. Such region adjacency graphs will be discussed further in Section 11.3.1e.) We associate with each node the properties of its region that are relevant to the merge decision (average gray level, area, etc.), and similarly we associate with each arc the relevant properties of that pair of regions (length and strength of common border, etc.). The region merges can then be carried out by contracting the graph, recomputing the properties whenever two nodes are merged (areas add, average gray levels are averaged, etc.). During the merging process, we keep track of which of the original nodes have been merged into each of the current nodes. When the process is complete, we can thus easily construct a new labeled picture in which the points of the final regions have distinctive labels.

Figure 40 shows the region adjacency graph for the sketch in Fig. 38. If we associate the properties of area and average gray level with each node of this graph, we can determine from the graph which pairs of adjacent nodes can be merged, and can recompute the properties for the merged nodes, as just described.

d. Splitting

At the opposite extreme, we can start with the entire picture as a single region, and partition it by repeated splitting. Of course, it is impractical to

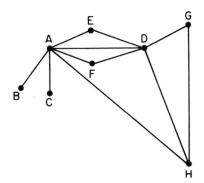

Fig. 40   Region adjacency graph for Fig. 38.

evaluate all possible ways of subdividing the picture (or a region); we therefore allow only certain simple types of subdivision. Examples:

(1)   *Bisection.*   If the entire picture is not homogeneous, we divide it into
quadrants; if a quadrant is not homogeneous, we divide it into subquadrants;
and so on.

(2)   *Triangulation.*   If the picture is not homogeneous, we divide it
into four triangular sectors that meet at an interior point (e.g., a point having
gray level farthest from the mean); if a triangle is not homogeneous, we
similarly divide it into four subtriangles; and so on.

These processes must stop, since the regions get smaller at each step, and
eventually we reach single pixels, which must be homogeneous.

If we use only splitting, and do not allow merging at all, the final partition
may contain adjacent pairs of regions that have identical characteristics.
For example, if there is a homogeneous region at the center of the picture,
this will happen when we use bisection, since the first subdivision into
quadrants will split this region into four parts. This situation can be rectified
if merging, as well as splitting, is permitted.

In one dimension (e.g., in partitioning a single row of a picture; see also
Section 11.3.3c on segmentation of borders and curves), it is possible to use
"natural" subdivision points, such as edges, rather than using a fixed subdivision scheme such as bisection. One can also try to shift the existing
segmentation points in such a way as to improve the goodness of the partition, rather than creating new segmentation points by splitting. This
should substantially reduce the amount of subsequent merging that is
likely to be needed. In two dimensions, however, it is not obvious how to
define natural subdivision lines or curves, and it is not practical to examine
a large set of possible dividing lines in order to determine the best ones, or to

evaluate the effects of shifting the borders between regions, since these have many degrees of freedom.

The effects of splitting by repeated bisection can be represented by a tree of degree 4 (a "quadtree") in which the root represents the entire picture and the leaves represent the (sub)quadrants that have not yet been split. When a (sub)quadrant is split, the corresponding leaf becomes the parent of four new leaves. The representation of pictures and regions by quadtrees will be discussed further in Section 11.1.2b.

e. Split-and-merge

We have seen that allowing both splitting and merging is preferable to either merging alone or splitting alone if we want to produce good partitions. The general idea is to start with a given initial partition; see the possibilities mentioned at the beginning of this section. We merge adjacent regions if the resulting new region is sufficiently homogeneous, and we split a region if it is not homogeneous enough. We can also try to adjust the boundaries between regions in such a way as to increase the regions' homogeneity; it may be easier to decide how to adjust an existing boundary than to choose a new one. Note that whenever we split a region, its parts become candidates for merging with the adjacent regions, since they may now resemble these regions.

Of course, the criteria for merging (and splitting) need not be based solely on inhomogeneity; they may also depend on size and shape (see, e.g., the examples in Section 10.4.2d). In particular, we may be willing to tolerate a lower degree of homogeneity if it results in a simpler shape; and we will often want to eliminate small "noise" regions, whether they are entirely surrounded by a single large region or are on the borders between such regions. The criteria may also depend on relative, rather than absolute, (in)homogeneity; for example, we merge S_1 with S_2 if $V(S_1 \cup S_2)$ is small compared with $V(S_1)$ and $V(S_2)$, where V is a measure of variability, and similarly we split S into S_1 and S_2 if $V(S)$ is large compared with $V(S_1)$ and $V(S_2)$. If homogeneity is measured by the closeness to which the region can be approximated by a function of a given type, then merge merit can also be defined in terms of the similarity between the two approximations.

Pavlidis ([56], especially Sections 5.2–5.4) describes a split-and-merge process that uses bisection for splitting; thus the splitting can be implemented using a quadtree representation [see (d) above], while the merging can make use of the region adjacency graph [(c) above]. We can start with a uniform subdivision of the picture into square blocks, rather than taking the entire picture as a single initial block; in this case the quadtree can be used for some of the merging steps (when all four children of a node are merged) as well as for splitting.

Starting with a nontrivial initial partition and using both splitting and merging may have computational cost advantages (see Pavlidis [56], Section 5.2). Of course, the partition obtained from a split-and-merge process is not necessarily optimal. However, it should be pointed out (Pavlidis [56], Section 7.8) that the number of regions in such a partition should not greatly exceed the number in an optimal partitioning using the same homogeneity criterion; in one dimension, in fact, if we use the proper sort of uniformity criterion, it must be within a factor of two of the optimum. (Proof: Let the split-and-merge segmentation points be $P_1, \ldots, P_m$, and the optimal segmentation points be $Q_1, \ldots, Q_n$. If three consecutive P's, say P_{i-1}, P_i, P_{i+1}, fell between two consecutive Q's, say Q_j and Q_{j+1}, we could merge segments $[P_{i-1}, P_i]$ and $[P_i, P_{i+1}]$, since subintervals of the uniform interval $[Q_j, Q_{j+1}]$ must be alike. Hence at most two P's can fall between each two consecutive Q's, which implies that $m \leq 2n$.) Thus split-and-merge appears to be the preferred approach to picture partitioning, in the absence of special knowledge about the desired partition.

f. Model-guided partitioning

Given any partition of a picture into regions, and given a model describing the regions that the picture is supposed to contain, we can in principle estimate the probability of the partition, based on the properties of the picture regions (gray level, texture, size, shape, position, etc.) and their relationships to one another. In practice, however, this will usually not be feasible, because of the large number of possible correspondences between picture regions and model regions that must be considered. A simplified approach, introduced by Feldman and Yakimovsky [20], will be outlined in the following paragraphs.

Suppose that we independently estimate the probability that each picture region S is any one of the model regions; and similarly we estimate the probability that each border between a pair of picture regions corresponds to any given border in the model. (This border probability depends on the individual region probabilities, as well as on intrinsic properties of the border.) If we treat all these estimates as independent, we can multiply them to obtain estimates of the probabilities of various global interpretations of the picture in terms of the model, and we can then pick the interpretation that has the highest probability. We can also evaluate potential region merges in terms of how they would increase the probability of the most likely interpretation. Of course, the estimates are not really independent, but this approach can still provide useful guidance to the region merging process.

Even when we treat the probabilities as independent, the task of finding a most probable global interpretation is still very tedious. It is generally necessary to use search techniques to find an interpretation that has high probability. Bounds on the probability of the most likely global interpretation can be found by, e.g., picking the most likely interpretation of each region and multiplying the probabilities of these; or deciding that the highest-probability interpretation is correct, and refining the probabilities of the other inter- pretations based on this information with the aid of conditional inter- pretation probabilities. This latter process can be iterated, so that a sequence of best choices is made, until an interpretation has been assigned to every region; of course, the results depend on the order of the choices, and many sequences may have to be tried. Further details of Feldman and Yakimovsky's approach will not be given here. A parallel method of iterative probability refinement ("relaxation") will be discussed in Section 10.5. A combination of relaxation and heuristic search is used for model-guided region growing by Tenenbaum and Barrow [7, 75].

10.5   ITERATIVE SEGMENTATION: "RELAXATION"

In the previous sections of this chapter we have discussed both parallel and sequential methods of segmentation. Parallel methods make the classifi- cation decision at each point independently of the decisions at other points; thus they could be applied at each point simultaneously, if a suitable parallel processing capability were available. (Note that the decision at a point may depend on the gray levels, or local property values, at other points, but it does not depend on having previously classified those points.) Most sequential methods, on the other hand, do make use of previous decisions, both in choosing the points to be classified next (e.g., neighbors of points already classified) and in defining the classification criteria to be used (e.g., in terms of resemblance to the points already classified). This makes sequential methods fundamentally more powerful than parallel methods, since they can "learn as they go" to define highly precise criteria for classification. On the other hand, sequential methods cannot be speeded up greatly even if parallel processing is available, and their results usually depend on the order in which points are examined.

This section describes an iterative approach to segmentation which makes fuzzy or probabilistic classification "decisions" at every point in parallel, at each iteration, and then adjusts these decisions at successive iterations based on the decisions made at the preceding iteration at neighboring points. This approach is sometimes called *relaxation*, because of its resemblance to a class of iterative numerical methods. The relaxation approach is order-independent, and can be greatly speeded up by parallel processing, since each iteration is parallel, and typically only a few iterations are necessary. On the other hand, it is more powerful than one-shot parallel methods, since its initial classifications are refined, at each iteration, based on the local context. It makes tentative, rather than firm, classifications at each stage, and repeatedly reconsiders them, unlike the other types of methods, which usually make decisions at each point only once (except in cases where sequential methods allow backtracking).

As an informal illustration of how relaxation operates, let us briefly consider the problem of detecting smooth curves in a picture. (This example will be treated in greater detail later in this section.) We begin by applying a set of directional local operators, such as those defined in Section 10.3, at each point of the picture. In a parallel approach to curve detection, we would determine the curve slope at a given point P by choosing the operator that gives the highest response at P; and we would decide whether or not a curve is present at P by thresholding this response. In the relaxation approach, we use the relative strengths of the directional responses at P to define initial "probabilities" that the curve through P (if any) has each of the possible slopes; and we use the absolute strengths of these responses, compared with the maximum possible response, to define an initial probability that no curve is present at P. We then adjust these initial probabilities at each point by examining the probabilities at neighboring points. Let p_i be the curve probability at P for slope θ_i; then the curve probability q_j for slope θ_j at a neighboring point Q contributes an increment to p_i whose strength depends on how smoothly slope θ_j at Q continues slope θ_i at P. At the same time, the no-curve probability q_0 at any neighbor Q contributes an increment to the no-curve probability p_0 at P. This incrementation process is carried out in parallel for all pairs of neighboring points and all pairs of probabilities. Typically, when this is done, the curve probabilities for the appropriate slopes become high for points that do lie on smooth curves, while the no-curve probabilities become higher for points that do not. Thus the decision as to whether or not a curve is present at each point, and the choice of a most likely slope, are much easier after a few iterations than they would have been initially.

In Section 10.5.1 we define a class of relaxation processes, and also discuss methods of determining the numerical parameters that are used in computing the increments at each iteration. Section 10.5.2 presents applications of this approach to thresholding, spectral classification, edge and curve detection. Sections 10.5.3–10.5.5 discuss some alternative types of relaxation methods, and give examples of their application to picture analysis.

10.5.1   Probabilistic Relaxation

Suppose that we have a set of objects $A_1, \ldots, A_n$ that we want to classify into m classes $C_1, \ldots, C_m$. (In most of our examples, the objects will be pixels, but they need not be.) Suppose further that the class assignments of the objects are interdependent; in other words, for each pair of class assignments $A_i \in C_j$ and $A_h \in C_k$, we have some quantitative measure of the *compatibility* of this pair, which we shall denote by $c(i, j; h, k)$. For concreteness, let us assume that positive values of $c(i, j; h, k)$ represent compatibility of $A_i \in C_j$ with $A_h \in C_k$; negative values represent incompatibility; and zero represents "don't care." We may further assume that the c's always lie in some fixed range, say $[-1, 1]$. Note that the c's need not be symmetric, i.e., we need not have $c(i, j; h, k) = c(h, k; i, j)$.

Let $p_{ij}^{(0)}$ be an initial estimate of the probability that $A_i \in C_j$, $1 \leqslant i \leqslant n$, $1 \leqslant j \leqslant m$. Thus for each i we have $0 \leqslant p_{ij}^{(0)} \leqslant 1$ and $\sum_{j=1}^{m} p_{ij}^{(0)} = 1$. We will now describe an interative method of computing successive "probability estimates" $p_{ij}^{(r)}$, $r = 1, 2, \ldots$, based on the initial probabilities and the compatibilities. We call these p's probability estimates because they satisfy $0 \leqslant p_{ij}^{(r)} \leqslant 1$ and $\sum_{j=1}^{m} p_i^{(r)} = 1$ for each i and r, but we make no claim that they are actually improved estimates in a decision-theoretic sense.

a.   The iteration scheme

Let us consider how p_{ij} should be adjusted at each iteration step, based on the current values of the other p's and on the values of the c's. Intuitively, if p_{hk} is high and $c(i, j; h, k)$ is positive, we want to increase p_{ij}, since it is compatible with the high-probability event $A_h \in C_k$. Similarly, if p_{hk} is high and $c(i, j; h, k)$ is negative, we want to decrease p_{ij}, since it is incompatible with $A_h \in C_k$, which has high probability. On the other hand, if p_{hk} is low, or if $c(i, j; h, k)$ is near zero, we do not want to change p_{ij} very much, since $A_h \in C_k$ either has low probability, or is essentially irrelevant to $A_i \in C_j$. A very simple way of defining an increment to p_{ij} that has these desired properties is to use the product $c(i, j; h, k) p_{hk}$.

How should these increments be combined for all the h's and k's? The simplest way of doing this is to linearly combine them. For each object A_h, let us add up the increments over the possible classes C_k, i.e.,

$$\sum_{k=1}^{m} c(i, j; h, k) p_{hk}.$$

This sum will be positive if the high-probability C_k's for A_h are compatible with $A_i \in C_j$, negative if they are incompatible, and near zero if they are irrelevant. Note that this sum is still in the range $[-1, 1]$ since $\sum_{k=1}^{m} p_{hk} = 1$.

We can then average these net increments for all objects $A_h \neq A_i$, i.e., we compute

$$q_{ij} \equiv \frac{1}{n-1} \sum_{\substack{h=1 \\ h \neq i}}^{n} \left(\sum_{k=1}^{m} c(i, j; h, k) p_{hk} \right)$$

This too is still in the range $[-1, 1]$.

It should be pointed out that there are many other possible ways of computing and combining the increments. Our simple averaging scheme does not take into account possible interrelationships among the compatibilities; it assumes that p_{ij} can be adjusted by considering each A_h separately. In other words, we are considering only pairwise constraints among the class assignments $A_i \in C_j$, and ignoring higher-order constraints. For example, we have no way of requiring that there exist a specified number of compatible A_h's, or that these A_h's be themselves related in specified ways. In fact, we cannot even force p_{ij} to be 0 in cases where $A_h \in C_k$ has probability 1 and is completely incompatible with $A_i \in C_j$, or to be 1 if $A_h \in C_k$ has probability 1 and implies $A_i \in C_j$. We shall consider here only the averaging scheme, but it must be emphasized that in many situations, other schemes might be preferable.

We must now apply the increment q_{ij} to the current estimate of p_{ij} in order to obtain a new estimate. We want the new estimates to be nonnegative and to satisfy $\sum_{j=1}^{m} p_{ij} = 1$. A simple way of insuring nonnegativeness (other ways are possible!) is to multiply p_{ij} by $1 + q_{ij}$; this quantity is nonnegative, since $q_{ij} \geq -1$. Note that if q_{ij} is negative, p_{ij} decreases, and if positive, it increases, as desired. Finally, we normalize the new estimates of p_{ij} by dividing each of them (for a given i) by their sum; this insures that they sum to 1. Our incrementation process is thus defined by

$$p_{ij}^{(r+1)} = \frac{p_{ij}^{(r)}(1 + q_{ij}^{(r)})}{\sum_{j=1}^{m} p_{ij}^{(r)}(1 + q_{ij}^{(r)})}$$

where

$$q_{ij}^{(r)} = \frac{1}{n-1} \sum_{\substack{h=1 \\ h \neq i}}^{n} \left(\sum_{k=1}^{m} c(i, j; h, k) p_{hk}^{(r)} \right)$$

Note that if any $p_{ij}^{(r)}$ is 0, it can never become nonzero, and if it is 1, it cannot change.

The iteration scheme just defined is appropriate when we use c's that can be either positive or negative. Suppose instead that we use nonnegative c's, where 0 represents high incompatibility and high values represent high compatibility. Then we could simply define a new estimate $p_{ij}^{(r+1)}$, based on the p_{hk}'s for a given A_h, as

$$p_{ij}^{(r+1)} = \frac{p_{ij}^{(r)} \sum_{k=1}^{m} c(i, j; h, k) p_{hk}^{(r)}}{\sum_{j=1}^{m} \sum_{k=1}^{m} c(i, j; h, k) p_{ij}^{(r)} p_{hk}^{(r)}}$$

Here the sum in the numerator is high if high p_{hk}'s are highly compatible with p_{ij}, and low otherwise; thus p_{ij}'s for which this is true will increase relative to other p_{ij}'s. These estimates can then be averaged over the A_h's, as before. This iteration scheme yields results comparable to those obtained using the one previously defined.

b.   The coefficients

In applying this approach to probabilistic pixel classification, we will generally take into account only the interactions between a point and its neighbors; thus in computing q_{ij}, we use only a few h's. This is equivalent to assuming that $c(i, j; h, k) = 0$ whenever point h is not a neighbor of point i. We will also ordinarily assume that the c's are space-invariant, i.e., they depend only on the position of point h relative to point i, and not on absolute position. This implies that the relaxation computation involves only $8m^2$ c's, if we use the usual eight neighbors of a point.

The coefficients $c(i, j; h, k)$ can be defined in a variety of ways; the exact definition does not seem to matter greatly in practice. The following is an example of how we might define them in the smooth curve detection application. Let p_{ij} be the probability of a curve of slope θ_j at point A_i, for $1 \leqslant j < m$, and let θ_{ih} be the direction from point A_i to point A_h; then we can define $c(i, j; h, k)$ as, e.g., $|\cos(\theta_j - \theta_{ih})||\cos(\theta_k - \theta_{ih})|$. This product of cosines is 1 if both θ_j and θ_k are collinear with θ_{ih}, and drops to 0 as one or both of them become perpendicular to θ_{ih}; note that it can never be negative. (If we were detecting edges rather than curves, we could use negative compatibilities between edges that have similar directions but opposite senses; the edge detection example will be discussed further in Section 10.5.2.) If we are considering pairs of points that are not necessarily immediate neighbors, we should also decrease $c(i, j; h, k)$ as the distance between point A_i and point A_h increases, corresponding to the concept that the slope probabilities are less relevant to one another for points that are farther apart. Next, let p_{im} be the probability that there is no curve at point A_i; then we can define $c(i, m; h, m)$ as a decreasing function of the distance between A_i and A_h, independent of direction. We can also define $c(i, m; h, k)$ as $-\cos 2(\theta_k - \theta_{ih})$; this corresponds to the fact that a curve of slope θ_k in direction θ_k from A_h is incompatible with no curve at A_i, while a curve of slope θ_k in the perpendicular direction positively supports the no-curve decision at A_i. Analogously we can define $c(i, j; h, m)$ as $-|\cos(\theta_j - \theta_{ih})|$; this corresponds to the fact that no curve in direction θ_j from A_i is incompatible with a curve of slope θ_j at A_i, but no curve in the perpendicular direction tells us nothing about the probability of such a curve at A_i. These c's too can drop off with the distance between A_i and A_h. Many variations on these definitions are possible; rather than using cosines, we can use any set of coefficients that behave similarly as functions of slope and direction.

The c's defined in the preceding paragraph are problem-dependent, i.e., they are specific to the smooth curve detection problem. We now describe a general method of defining c's in terms of the *a priori* and conditional probabilities of the events $A_i \in C_j$ and $A_h \in C_k$. Let $R(i, j; h, k)$ be the ratio $p(A_i \in C_j | A_h \in C_k)/p(A_i \in C_j)$. Intuitively, when $A_h \in C_k$ is compatible with $A_i \in C_j$, R should be greater than 1—i.e., knowing that $A_h \in C_k$ should increase the likelihood that $A_i \in C_j$. Conversely, when they are incompatible, R should be less than 1; and when they are irrelevant (i.e., independent), R should be 1. Thus $\log R$ has the desirable properties of a compatibility coefficient; it is positive when the two events are compatible, negative when they are incompatible, and zero when they are irrelevant. (Unfortunately, $\log R$ is not restricted to the range $[-1, 1]$, but this can be taken care of by truncation or scaling.) Since we always have

$$p(A_i \in C_j | A_h \in C_k) = \frac{p(A_i \in C_j, A_h \in C_k)}{p(A_h \in C_k)}$$

where the numerator is the joint probability of the two events, it follows that $R(i, j; h, k) = p(A_i \in C_j, A_h \in C_k)/p(A_i \in C_j)p(A_h \in C_k)$; thus the R's defined in this way are always symmetric. The quantity $\log R$ is called the *mutual information* of the two events. The prior and joint probabilities can be estimated by measuring the individual and joint frequencies of the events $A_i \in C_j$ in an ensemble of pictures; in practice, estimates obtained from a single picture are good enough. Other methods of defining c's in terms of the prior and joint probabilities can be formulated, but we will not describe these here.[§] In some cases, very crude c's can be used, e.g., a positive constant when the two classes are the same, and a negative constant when they are different, for all i and h. Some further examples of c's defined in various ways will be given in Section 10.5.2.

c.   Performance evaluation

Ideally, we would like a probability adjustment scheme of the type defined above to exhibit the following type of behavior: During the first few iterations,

---

[§] If we use the scheme mentioned at the end of (a) above, based on nonnegative c's, we can simply take $c(i, j; h, k)$ to be $R(i, j; h, k)$ itself, since this is high when $A_i \in C_j$ and $A_h \in C_k$ are highly compatible, and low when they are incompatible.

appreciable changes in some of the estimates should occur, as "noisy" initial estimates are brought into line with the consensus of evidence from their neighbors. Once this has happened, there should be little further change; the estimates should be relatively stable. We would also expect these "final" estimates to be less ambiguous than the initial ones.

Quantitatively, we expect the sum of absolute probability differences $\sum_{i,j}|p_{ij}^{(r)} - p_{ij}^{(r+1)}|$ to become small after a few iterations. At the same time, the final probabilities should not be too far away, on the average, from the initial ones; we would not be satisfied with the process if it converged to an arbitrary set of final probabilities unrelated to the initial ones. Thus $\sum_{i,j}|p_{ij}^{(r)} - p_{ij}^{(0)}|$ should not become very large. In addition, we expect the *entropy* of our probabilistic classification to decrease; in other words, we expect $-\sum_{i,j} p_{ij}^{(r)} \log p_{ij}^{(r)}$ to be smaller than $-\sum_{i,j} p_{ij}^{(0)} \log p_{ij}^{(0)}$. As we shall see in Section 10.5.2, these expectations seem to be fulfilled in practice.

Some theoretical studies of the convergence properties of relaxation schemes have been conducted, but we will not summarize these here. In any case, we are not normally interested in reaching a limit point, but only in carrying the process through a few iterations (typically, less than ten) so as to correct initial errors and reduce initial ambiguities.

After carrying out the desired number of iterations, we finally make our classification decisions based on the resulting probability estimates $p_{ij}^{(r)}$, e.g., we assign each A_i to the class C_j for which $p_{ij}^{(r)}$ is greatest. These classifications should be significantly more reliable than those that we would have obtained if we had classified on the basis of the initial probability estimates $p_{ij}^{(0)}$.

10.5.2 Examples

a. Thresholding

Let us first consider how relaxation methods might be applied to classifying pixels into dark and light classes based on their gray levels, i.e., to thresholding. This example is especially simple for two reasons:

(1) There are only two classes, corresponding to high and low gray levels.

(2) The compatibilities between neighboring pairs of points should all be alike, independent of direction, unless the picture is directionally biased.

Thus for each point A_i we have only two probabilities, p_{i1} and p_{i2}, where $p_{i2} = 1 - p_{i1}$. Moreover, for all pairs of neighboring points A_i, A_h we have only four compatibilities, $c(i, 1, h, 1)$, $c(i, 1, h, 2)$, $c(i, 2, h, 1,)$ and $c(i, 2, h, 2)$.

Since these are independent of i and h, we shall denote them for brevity by c_{11}, c_{12}, c_{21}, and c_{22}.

We can make the initial probability estimates $p_i^{(0)}$ in various ways. One very simple possibility is to let $p_{i1}^{(0)} = z_i/z_{max}$ where z_i is the gray level of A_i and z_{max} is the greatest possible gray level in the picture; this is certainly a reasonable measure of the "probability" that z_i is high. A more elaborate scheme would be to approximate the picture's histogram by a linear combination of two Gaussian probability densities, and then define $p_{i1}^{(0)}$ and $p_{i2}^{(0)}$ as the probabilities of gray level $f(A_i)$ given by these two densities. We shall use the z_i/z_{max} estimates in the following discussion.

The c's can also be chosen in a number of ways; the following is one simple possibility: Let p and $1 - p$ be the *a priori* probabilities of the high and low classes, and p' the conditional probability $p(h|h)$ that a point is in the high class given that a neighboring point is also high. Now $p(l|h) = 1 - p(h|h) = 1 - p'$. If we assume that $p(h|l) = p(l|h)$, this implies that $p(l|l) = 1 - p(h|l) = p'$. The R's of Section 10.5.1b are thus $p'/p, p'/(1 - p), (1 - p')/p$, and $(1 - p')/(1 - p)$, which give the compatibilities of high with high, low with low, high with low, and low with high, respectively. For example, suppose $p = \frac{1}{2}$, and that $p' > \frac{1}{2}$ (pairs that are both high or both low are more common than mixed pairs); then $p'/p = p'/(1 - p) > 1$ and $(1 - p')/p = (1 - p')/(1 - p) < 1$, so that $c_{11} = c_{22} > 0$ and $c_{12} = c_{21} < 0$.

At the first iteration step, the increments q_{i1} and q_{i2} are proportional to $\sum_h [c_{11}z_h + c_{12}(z_{max} - z_h)]$ and $\sum_h [c_{21}z_h + c_{22}(z_{max} - z_h)]$, respectively, where the sums are taken over the neighbors of A_i. Thus the increments are essentially weighted averages of the neighboring gray levels. If z_h is large compared to $z_{max} - z_h$, neighbor A_h will receive positive weight in q_{i1} (since c_{11} is positive and c_{12} negative) and negative weight in q_{i2}, and vice versa; this is consistent with the idea that a high gray level at A_h should give positive support to a high gray level at A_i, and vice versa. This discussion indicates that the relaxation process is analogous to, though somewhat more complicated than, the iterated weighted local averaging schemes used for gray level smoothing in Section 6.4. Unlike these schemes, however, the relaxation process is designed to drive the high and low gray levels to opposite ends of the gray scale, so that thresholding becomes trivial; it does not simply smooth the gray levels.

The effect of applying this process to a picture should be as follows: If a point has gray level $z_i > z_{max}/2$, its high gray level probability will initially be $> \frac{1}{2}$. If its neighbors also tend to have such gray levels, its high gray level probability should increase when the process is iterated; but if they tend to have low gray levels, the opposite may happen. The reverse is true for points having low gray level. Thus if we thresholded the initial probabilities, we would simply be thresholding the picture at the midpoint of the gray scale;

but if we iterate the process before thresholding, the results should be less noisy.

We can display the results of this process by treating the probability $p_{i1}^{(r)}$ at each iteration as a gray level (i.e., multiplying it by z_{max}). Note that if most of the gray levels in the picture lie on one side of the grayscale midpoint, the picture will become solid black or white; this process should only be applied to pictures that have substantial gray level populations on both sides of the midpoint. (If this is not the case, the gray scale can be adjusted, or the alternative process based on Gaussian fitting can be used.)

Figures 41a–41c show the results of applying the process to the pictures used in Fig. 1. The compatibilities used were obtained from mutual information estimates, as in Section 10.5.1b; they are tabulated in Figs. 41d–41f. Figures 42a–42c show the average rates of change of the probabilities at each iteration for these examples, and Figs. 42d–42f show the corresponding entropies. We see that both the rate of change and the entropy decrease significantly in the course of the iterations.

Some comments may be made about the compatibilities shown in Figs. 41d–41f.

(1)   When the numbers of light and dark points are very different, so are the (light, light) and (dark, dark) compatibilities; the rarer class of points yields higher compatibilities (the dark class, for the handwriting and the chromosomes). For this reason, there is some tendency for the dark points to expand under the relaxation process.

(2)   Directional biases in the input can be detected in the coefficients. For example, in the case of the handwriting, the (light, dark) and (dark, light) compatibilities are higher for horizontal neighbors than for those in the other directions; this corresponds to the fact that the strokes of the writing are usually narrowest in the horizontal direction.

(3)   The compatibilities between pairs of labels on the same point ("neighbor 8") are similar to those for neighboring points in the chromosome picture, where the gray level varies slowly. For the other two pictures, the (light, dark) and (dark, light) compatibilities for the point and its neighbors are less negative than those for the point and itself, reflecting the fact that {light, dark} neighboring pairs of points do occur.

b.   Spectral classification

More generally, we can use relaxation in classifying pixels based on their spectral (or local property) characteristics. As an example, consider the house picture whose red, green, and blue components are shown in Figs. 43a–43c (which are the same as Figs. 10a–10c). Figure 43d shows a segmentation of this picture into five classes, corresponding to sky, brick, shadowed brick,

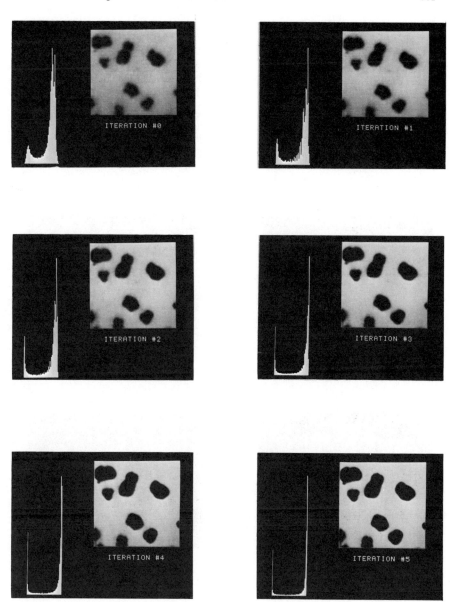

(a)

Fig. 41   (a)–(c) Results of applying the light/dark relaxation process to the pictures of Figs. 1a–1c (five iterations). At each iteration, the picture (probabilities displayed as gray levels) and its histogram are shown.

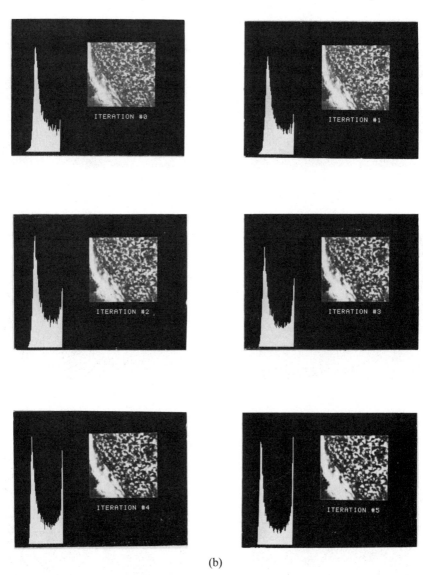

(b)

Fig. 41 (*Continued*)

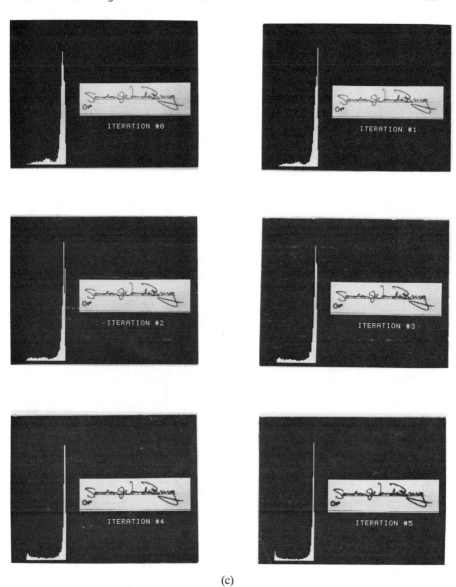

(c)

Fig. 41   (*Continued*)

	0	1	2	3	4	5	6	7	8
(d) Chromosomes									
Light, light	0.021	0.021	0.021	0.020	0.021	0.021	0.022	0.022	0.021
Light, dark	-0.061	-0.059	-0.059	-0.057	-0.059	-0.060	-0.062	-0.062	-0.061
Dark, light	-0.060	-0.060	-0.061	-0.061	-0.061	-0.059	-0.060	-0.058	-0.061
Dark, dark	0.095	0.095	0.096	0.096	0.096	0.094	0.094	0.093	0.096
(e) Clouds									
Light, light	0.033	0.028	0.033	0.027	0.033	0.029	0.033	0.026	0.043
Light, dark	-0.035	-0.030	-0.035	-0.028	-0.036	-0.030	-0.035	-0.027	-0.049
Dark, light	-0.035	-0.029	-0.035	-0.028	-0.035	-0.029	-0.035	-0.027	-0.049
Dark, dark	0.027	0.023	0.027	0.022	0.027	0.023	0.027	0.022	0.036
(f) Handwriting									
Light, light	0.006	0.004	0.004	0.003	0.006	0.005	0.005	0.004	0.008
Light, dark	-0.041	-0.024	-0.024	-0.018	-0.041	-0.033	-0.034	-0.027	-0.054
Dark, light	-0.041	-0.024	-0.025	-0.018	-0.041	-0.023	-0.024	-0.017	-0.054
Dark, dark	0.146	0.101	0.105	0.083	0.146	0.098	0.101	0.078	0.174

Fig. 41 (d)–(f) Compatibility coefficients used in a–c. The neighbors are

```
3  2  1
4  8  0
5  6  7
```

(i.e., the eighth neighbor is the point itself).

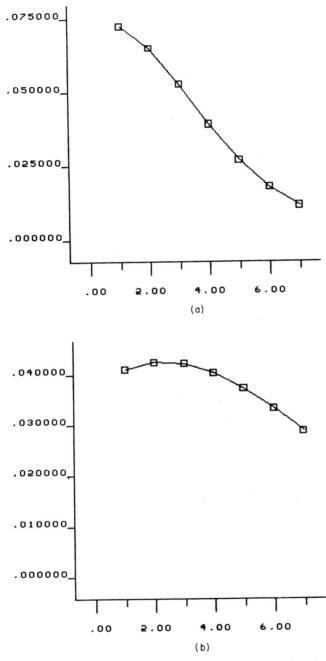

Fig. 42 (a)–(c) Average rate of change of the probability vectors, as a function of iteration number, for the relaxation processes in Figs. 41a–41c. (d)–(f) Average entropies for these processes.

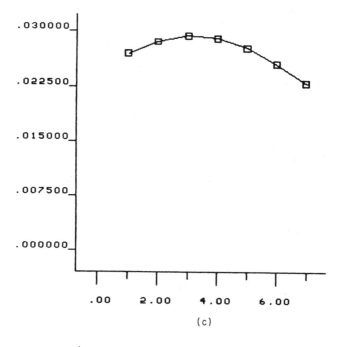

(c)

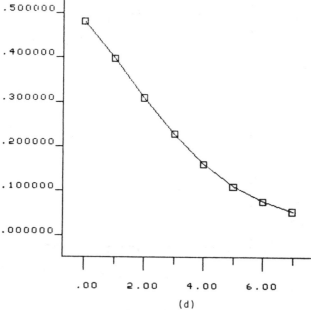

(d)

Fig. 42   (*Continued*)

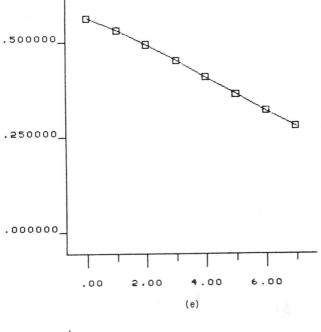

(e)

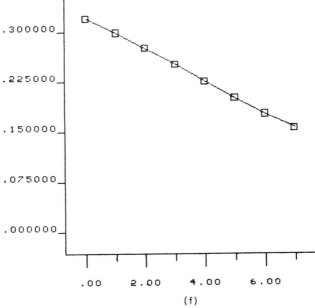

(f)

Fig. 42 (*Continued*)

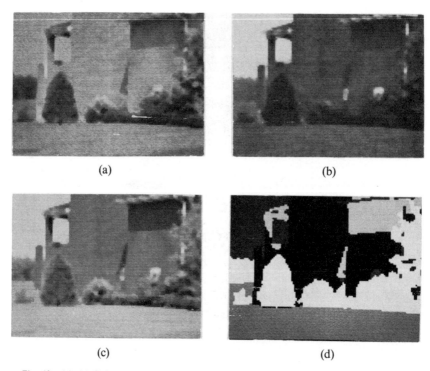

(a)   (b)

(c)   (d)

Fig. 43   (a)–(c) Color components of the house picture (same as Figs. 10a–10c). (d) Hand segmentation into five classes.

bushes, and grass; this was constructed by hand. These classes define clusters in (red, green, blue) space. Initial estimates of the probabilities of membership in these clusters were made by assuming the clusters to be normally distributed, and using the hand segmentation to define the prior probabilities of the classes.

Compatibility coefficients were then derived in the form of mutual information estimates, as in Section 10.5.1b; they are shown in Fig. 44. The relaxation process was then applied. The maximum-probability classifications at various steps ($r = 0, 5, 10, 15, 20$) are shown in Figs. 45a–45e. The initial error rate is 4.60%, while that after 20 iterations is 3.66%. Thus, over 20% of the errors in the initial classification have been corrected.

For comparison, two iterative noise cleaning schemes were applied to the same picture, one prior to classification, the other subsequent to classification, in order to determine whether such preprocessing or postprocessing would be equally effective in eliminating the errors. In the preprocessing approach, each point was averaged with its six most similar neighbors

Neighbor	Class compatibilities				
0	0.23	−0.56	−1.00	−0.31	−0.41
	−0.36	0.41	−1.00	−0.26	−0.49
	−1.00	−0.99	0.30	−1.00	−0.42
	−0.39	−0.30	−0.95	0.44	−0.11
	−0.43	−0.45	−0.46	−0.14	0.23
1	0.22	−0.49	−1.00	−0.23	−0.42
	−0.34	0.41	−1.00	−0.22	−0.49
	−0.97	−0.87	0.29	−0.87	−0.26
	−0.39	−0.19	−0.99	0.43	−0.09
	−0.34	−0.29	−0.46	−0.13	0.22
2	0.23	−0.65	−1.00	−0.33	−0.53
	−0.65	0.42	−1.00	−0.33	−0.52
	−0.98	−1.00	0.29	−0.92	−0.27
	−0.53	−0.23	−1.00	0.45	−0.11
	−0.39	−0.33	−0.45	−0.12	0.23
3	0.22	−0.33	−1.00	−0.27	−0.45
	−0.50	0.41	−0.99	−0.21	−0.50
	−0.96	−1.00	0.29	−0.82	−0.27
	−0.31	−0.16	−1.00	0.43	−0.11
	−0.34	−0.30	−0.43	−0.11	0.22
4	0.23	−0.36	−1.00	−0.40	−0.44
	−0.56	0.41	−0.99	−0.30	−0.45
	−1.00	−1.00	0.30	−0.96	−0.46
	−0.32	−0.26	−1.00	0.44	−0.14
	−0.42	−0.46	−0.43	−0.11	0.23
5	0.22	−0.35	−0.97	−0.41	−0.35
	−0.49	0.41	−0.90	−0.20	−0.32
	−1.00	−1.00	0.30	−1.00	−0.42
	−0.23	−0.24	−0.91	0.43	−0.14
	−0.43	−0.50	−0.26	−0.10	0.22
6	0.23	−0.65	−0.98	−0.55	−0.40
	−0.65	0.41	−1.00	−0.24	−0.35
	−1.00	−1.00	0.30	−1.00	−0.42
	−0.33	−0.38	−0.92	0.44	−0.13
	−0.53	−0.52	−0.27	−0.12	0.23
7	0.22	−0.50	−0.94	−0.32	−0.33
	−0.33	0.40	−1.00	−0.17	−0.33
	−1.00	−0.99	0.29	−1.00	−0.40
	−0.27	−0.23	−0.82	0.43	−0.11
	−0.44	−0.51	−0.27	−0.11	0.22
8	0.24	−1.00	−1.00	−0.76	−0.77
	−1.00	0.43	−1.00	−0.64	−0.66
	−1.00	−1.00	0.30	−1.00	−0.50
	−0.76	−0.64	−1.00	0.47	−0.17
	−0.77	−0.66	−0.50	−0.17	0.24

Fig. 44 Compatibility coefficients for the classes in Fig. 43. In each 5 × 5 table, the row index is the class of the center point, and the column index is the class of the neighbor. The neighbor numbering is the same as in Figs. 41d–41f.

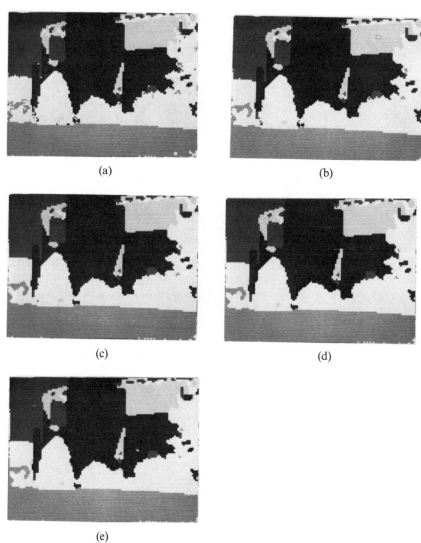

(a)

(b)

(c)

(d)

(e)

Fig. 45 Results of applying relaxation to the house picture, using the compatibilities of Fig. 44. (a)–(e) Maximum-probability classifications at iterations 0, 5, 10, 15, 20.

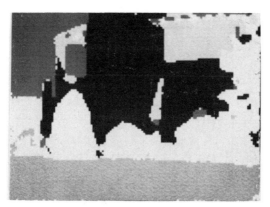

Fig. 46 Maximum-probability classification for the same picture after the second iteration of preprocessing; on subsequent iterations, the error rate increases.

(Section 6.4.3); at each iteration, class probabilities were estimated, and each point was assigned to the most likely class. The best result is shown in Fig. 46; the error rates are 4.60, 4.39, and 4.34, and begin to increase after the third iteration. In the postprocessing approach, the points were first classified on the basis of the initial probability estimates; each point was then reclassified if six of its neighbors belonged to a different class, and this process was iterated. The best result is shown in Fig. 47; the error rates are 4.60 and 4.46, and begin to increase after the second iteration. Thus neither pre- nor postprocessing was as effective as relaxation in reducing the classification errors.

Figure 48a shows the average rate of change of the probabilities at each iteration for this example, and Fig. 48b shows the corresponding entropies.

Fig. 47 Result of one iteration of postprocessing; on subsequent iterations, the error rate increases.

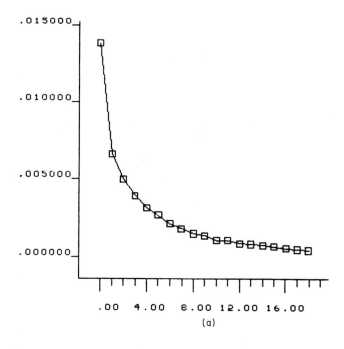

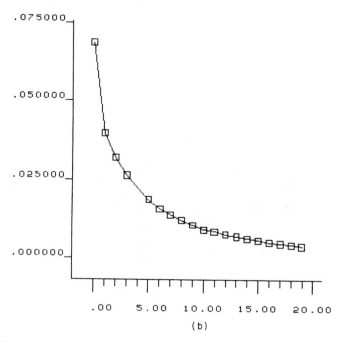

Fig. 48 (a) Average rate of change of the probability vectors as a function of iteration number for Fig. 45. (b) Average entropy for the same example.

 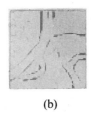  

(a)   (b)   (c)   (d)

Fig. 49 Results of applying curve enhancement relaxation to the edge strength values in Fig. 31a. (a)–(d) Iterations 0–3. At each iteration, if the highest probability at a point is the no-curve probability, that point is left blank; if it is one of the eight curve probabilities, that probability is displayed as a gray level (black = 1). Note how the weak values mutually reinforce.

We see that both the rate of change and the entropy decrease by an order of magnitude during the first few iterations, and quickly level off at low values.

c. Edge and curve detection

As already indicated, relaxation can be used as an aid in detecting smooth edges and curves. Figure 49 shows a curve detection example; the original picture is the same as in Fig. 31. The initial probabilities were estimated from the outputs of the nonlinear local operations of Section 10.3, in eight directions, at each point, and the compatibilities (Fig. 50) were based on mutual information. For each iteration ($r = 0, 1, 2 \ldots$) Fig. 49 displays those points at which a curve probability (in some orientation) is highest, with gray level proportional to that probability. It is seen that the curve probabilities for points on curves become stronger, while noise responses are eliminated. The average rate of change and the entropy for this example are shown in Fig. 51.

The relaxation process for edges is analogous to that for curves,[§] except that we must take into account the sense of the edge (i.e., which is its darker side) as well as its direction. We can do this by measuring edge direction (modulo 360°) so that the dark side is on the left when facing in the direction of the edge. Using this definition, we can modify the cosine coefficients described in Section 10.5.1b as follows: For edge/edge compatibility we use $\cos(\theta_j - \theta_{ih}) \cos(\theta_k - \theta_{ih})$; this is +1 when the edges are collinear in the same sense, 0 when they are perpendicular, and −1 when they are collinear in opposite senses. The edge/no-edge, no-edge/edge, and no-edge/no-edge coefficients can be the same as in Section 10.5.1b. Alternatively, we can use mutual information coefficients; if we do so, we should suppress nonmaxima in the direction across the edges, since otherwise parallel edges will have high

---

[§] Strictly speaking, edge probabilities should be associated with pairs of adjacent points rather than with single points, but we will ignore this distinction here.

Neighbor

Class compatibilities

0

```
0     0.00  -0.03  -0.01  -0.02  -0.04  -0.14  -0.06   0.00  -0.04
     -0.04   0.43  -1.00  -1.00  -1.00  -0.73  -1.00  -1.00   0.42
      0.01  -1.00   0.38  -1.00  -1.00  -0.46  -1.00  -1.00  -1.00
     -0.02  -1.00  -1.00   0.59  -1.00  -1.00   0.25  -1.00  -1.00
     -0.06  -0.49   0.14   0.02   0.73   0.38  -1.00  -1.00  -0.17
     -0.15  -0.90  -0.62  -1.00   0.26   0.66   0.41   0.01  -1.00
     -0.02  -0.09  -1.00  -1.00  -0.39  -1.00   0.79  -1.00  -1.00
     -0.06  -1.00  -1.00  -1.00  -1.00  -1.00  -1.00   0.58   0.75
      0.00  -1.00   0.42   0.18  -0.27  -1.00  -1.00  -1.00  -0.19


1     0.00  -0.03  -0.07  -0.10  -0.06  -0.03  -0.01   0.01  -0.01
     -0.05   0.44   0.33  -1.00  -1.00  -0.49  -1.00  -1.00   0.32
     -0.02  -1.00   0.60   0.36  -0.07  -1.00  -1.00  -1.00  -1.00
     -0.10  -1.00   0.44   0.78   0.26  -1.00  -1.00  -1.00  -1.00
     -0.03  -0.00  -0.23   0.21   0.69  -1.00  -1.00  -1.00  -1.00
     -0.04  -0.53  -0.48  -1.00   0.39   0.46   0.32  -1.00  -1.00
     -0.00  -0.06   0.01   0.23  -0.10   0.06  -1.00  -1.00  -1.00
      0.01  -1.00   0.39  -1.00  -1.00  -1.00  -1.00  -1.00  -1.00
      0.01  -1.00   0.29   0.10  -0.14  -1.00  -1.00  -1.00   0.06


2     0.00  -0.16  -0.04  -0.03  -0.00  -0.02  -0.02  -0.01  -0.10
     -0.18   0.64   0.18  -1.00  -0.44  -0.74  -0.42  -1.00   0.39
     -0.07   0.36   0.69   0.37  -0.09  -1.00  -1.00   0.22   0.02
     -0.03  -1.00  -0.15   0.60   0.32  -1.00  -1.00  -1.00   0.15
      0.01  -0.71  -1.00  -1.00   0.29  -1.00  -1.00  -1.00  -0.42
     -0.03  -0.52  -0.23  -1.00  -1.00   0.44   0.34  -1.00  -1.00
     -0.02  -0.23   0.09   0.19   0.44  -0.09   0.59  -1.00  -1.00
     -0.05  -1.00  -1.00  -1.00  -1.00   0.04  -1.00   0.82   0.62
     -0.03  -1.00   0.06  -1.00  -1.00  -1.00  -1.00  -1.00   0.68


3     0.00  -0.05   0.01   0.01   0.01  -0.02  -0.03  -0.09  -0.01
     -0.03   0.43  -1.00  -1.00  -0.32  -0.65  -0.06  -1.00  -1.00
      0.01  -1.00   0.08  -1.00   0.01  -1.00  -1.00  -1.00  -1.00
      0.01  -1.00  -1.00  -1.00   0.17  -1.00  -1.00  -1.00  -1.00
      0.01  -0.35  -1.00  -1.00  -1.00  -1.00  -1.00  -1.00  -1.00
     -0.02  -0.61  -0.21  -0.29  -1.00   0.40   0.36  -1.00  -1.00
     -0.01   0.10  -1.00  -1.00   0.30  -1.00   0.67  -1.00  -1.00
     -0.05  -1.00  -1.00  -1.00  -1.00   0.07  -1.00   0.98  -1.00
     -0.08   0.47  -1.00  -1.00  -1.00  -1.00  -1.00   0.65   0.53


4     0.00  -0.04   0.01  -0.02  -0.06  -0.14  -0.02  -0.06   0.00
     -0.03   0.43  -1.00  -1.00  -0.49  -0.90  -0.09  -1.00  -1.00
     -0.01  -1.00   0.38  -1.00   0.14  -0.62  -1.00  -1.00   0.42
     -0.02  -1.00  -1.00   0.59   0.02  -1.00  -1.00  -1.00   0.18
     -0.04  -1.00  -1.00  -1.00   0.73   0.26  -0.39  -1.00  -0.27
     -0.13  -0.73  -0.46  -1.00   0.38   0.66  -1.00  -1.00  -1.00
     -0.06  -1.00  -1.00   0.25  -1.00   0.41   0.79  -1.00  -1.00
      0.00  -1.00  -1.00  -1.00  -1.00   0.01  -1.00   0.58  -1.00
     -0.04   0.42  -1.00  -1.00  -0.17  -1.00  -1.00   0.75  -0.19
```

Neighbor   **Class compatibilities**

5

0.00	-0.05	-0.02	-0.10	-0.03	-0.04	-0.00	0.01	0.01
-0.03	0.43	-1.00	-1.00	-0.00	-0.53	-0.06	-1.00	-1.00
-0.07	0.33	0.60	0.44	-0.23	-0.48	0.01	0.39	0.29
-0.10	-1.00	0.36	0.78	0.21	-1.00	0.23	-1.00	0.10
-0.06	-1.00	-0.07	0.26	0.69	0.39	-0.10	-1.00	-0.14
-0.03	-0.49	-1.00	-1.00	-1.00	0.45	0.06	-1.00	-1.00
-0.01	-1.00	-1.00	-1.00	-1.00	0.32	-1.00	-1.00	-1.00
0.01	-1.00	-1.00	-1.00	-1.00	-1.00	-1.00	-1.00	-1.00
-0.01	0.32	-1.00	-1.00	-1.00	-1.00	-1.00	-1.00	0.06

6

0.00	-0.18	-0.07	-0.03	0.01	-0.03	-0.02	-0.05	-0.03
-0.16	0.64	0.36	-1.00	-0.71	-0.52	-0.23	-1.00	-1.00
-0.04	0.18	0.69	-0.15	-1.00	-0.23	0.09	-1.00	0.06
-0.03	-1.00	0.37	0.60	-1.00	-1.00	0.19	-1.00	-1.00
-0.00	-0.44	-0.09	0.32	0.29	-1.00	0.44	-1.00	-1.00
-0.02	-0.74	-1.00	-1.00	-1.00	0.44	-0.09	0.04	-1.00
-0.02	-0.42	-1.00	-1.00	-1.00	0.34	0.59	-1.00	-1.00
-0.01	-1.00	0.22	-1.00	-1.00	-1.00	-1.00	0.82	-1.00
-0.10	0.39	0.02	0.15	-0.42	-1.00	-1.00	0.62	0.68

7

0.00	-0.03	0.01	0.01	0.01	-0.02	-0.01	-0.05	-0.08
-0.05	0.42	-1.00	-1.00	-0.35	-0.61	0.10	-1.00	0.47
0.01	-1.00	0.08	-1.00	-1.00	-0.21	-1.00	-1.00	-1.00
0.01	-1.00	-1.00	-1.00	-1.00	-0.29	-1.00	-1.00	-1.00
0.01	-0.32	0.01	0.17	-1.00	-1.00	0.30	-1.00	-1.00
-0.02	-0.65	-1.00	-1.00	-1.00	0.41	-1.00	0.07	-1.00
-0.03	-0.06	-1.00	-1.00	-1.00	0.36	0.67	-1.00	-1.00
-0.09	-1.00	-1.00	-1.00	-1.00	-1.00	-1.00	0.98	0.65
-0.01	-1.00	-1.00	-1.00	-1.00	-1.00	-1.00	-1.00	0.53

8

0.01	-0.21	-0.15	-0.13	-0.16	-0.18	-0.08	-0.14	-0.16
-0.21	0.68	0.16	-1.00	-0.51	-0.99	-0.43	-1.00	-1.00
-0.15	0.16	0.91	0.26	-0.08	-0.71	-1.00	-1.00	-1.00
-0.13	-1.00	0.26	0.83	0.01	-1.00	-1.00	-1.00	-1.00
-0.16	-0.51	-0.08	0.01	0.93	0.22	-0.40	-1.00	-0.23
-0.18	-0.99	-0.71	-1.00	0.22	0.70	-1.00	-1.00	-1.00
-0.08	-0.43	-1.00	-1.00	-0.40	-1.00	0.99	-1.00	-1.00
-0.14	-1.00	-1.00	-1.00	-1.00	-1.00	-1.00	1.00	-1.00
-0.16	-1.00	-1.00	-1.00	-0.23	-1.00	-1.00	-1.00	0.88

Fig. 50 Compatibility coefficients for Fig. 49. In each 9 × 9 table, the row and column indices represent the classes no line, $0°$, $22\frac{1}{2}°$, $45°$, $67\frac{1}{2}°$, $90°$, $112\frac{1}{2}°$, $135°$, and $157\frac{1}{2}°$. The neighbor numbering is the same as in Figs. 41 and 44.

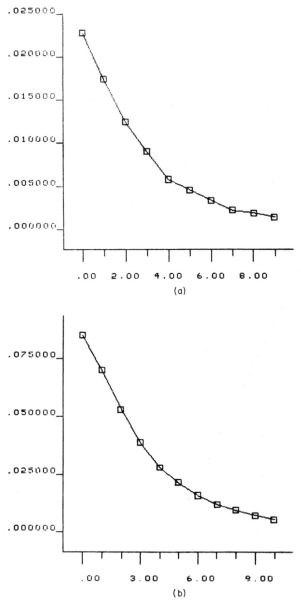

Fig. 51 (a) Average rate of change of the probability vectors as a function of iteration number for Fig. 49. (b) Average entropy for the same example.

compatibility, and the process will thicken the edges. The initial probabilities can be estimated from the digital gradient magnitude and direction at each point.

10.5.3 Fuzzy Relaxation

Up to now we have regarded the p_{ij}'s as probability estimates, and so have required them to sum to 1 for each object, i.e., $\sum_{j=1}^{m} p_{ij} = 1$ for all i. Another point of view is that the p's represent degrees of "fuzzy class membership," but need not sum to 1. This viewpoint is appropriate when the classes are not mutually exclusive or not exhaustive.

As an example in which fuzzy relaxation might be used, let us consider a simple point pattern matching problem. Let $A_1, \ldots, A_n$ and $B_1, \ldots, B_m$ be two sets of points; we want to find subsets of the A's (not necessarily disjoint) that match the set of B's. We regard "A_i is B_j" as an assignment of A_i to the class B_j; thus a correspondence between the patterns is equivalent to a classification of the A's as B's. Note, however, that the same A may correspond to several B's, or it may not correspond to any B. Thus it would not be reasonable to use probabilistic relaxation in this problem domain.

The compatibility $c(i, j; h, k)$ of two assignments, $A_i = B_j$ and $A_h = B_k$, is a function of how much the (actual) position of A_h relative to A_i differs from the (desired) position of B_k relative to B_j. If this difference has magnitude δ, we can define the compatibility as, e.g., $1/(1 + \delta^2)$, which is 1 for $\delta = 0$ and goes to zero as δ becomes large. We can take δ to be a relative, rather than absolute, difference, i.e., we can divide it by the distance from A_i to A_h. Note that these compatibilities are always nonnegative. Other formulas could be devised in which the compatibility becomes negative when the position discrepancy becomes sufficiently large, but this seems artificial.

We can still use $c(i, j; h, k)p_{hk}$ as a contribution to a new estimate of p_{ij}, but it no longer makes sense to treat this contribution as an increment, since the contributions are all positive. Rather, we can define the new estimate as an average of the previous estimate and the other points' contributions. In defining the net contribution of A_h to p_{ij}, it is reasonable to use the max, rather than the average, of the $c(i, j; h, k)p_{hk}$'s, since if any one of these terms is large there is strong support for p_{ij} from A_h, even if all the other terms are small. Thus a plausible relaxation formula in this situation is

$$p_{ij}^{(r+1)} = \frac{1}{n} \sum_{h=1}^{n} \left[\max_{k=1}^{m} c(i, j; h, k)p_{hk}^{(r)} \right]$$

We define $c(i, j; i, k) = 1$ if $j = k$, and 0 otherwise; thus the "self-support" term $(h = i)$ of the average is just $p_{ij}^{(k)}$. If desired, we could give this term greater weight relative to the other terms.

The initial estimates of the p's can be made in various ways. If the A's and B's are all indistinguishable, we can take all the $p_{ij}^{(0)}$'s to be 1; if not, we can define $p_{ij}^{(0)}$ in terms of some measure of similarity between A_i and B_j. If the patterns match exactly, the p's do not change under the iteration process, but otherwise they decrease at each iteration. However, if a good correspondence exists between (some of) the A's and (some of) the B's, those p_{ij}'s for which A_i corresponds to B_j should decrease slowly, since they have substantial support, while the other p_{ij}'s should decrease much more rapidly. An example of this process is given in Fig. 52; parts (a) and (b) show the two point patterns, and parts (c), (d), ... show the highest p_{ij}'s for each point ($\times 100$) at the first few iterations.

Other fuzzy relaxation formulas can be defined which may be more appropriate in some situations. For example, suppose we require, in point pattern matching, that each A *must* be one of the B's. We then might want to take the min, instead of the average, in combining the evidence from the neighbors, since p_{ij} cannot be stronger than the weakest of these pieces of evidence. This gives us the formula

$$p_{ij}^{(r+1)} = \min_{h=1}^{n} \left[\max_{k=1}^{m} c(i, j; h, k) p_{hk}^{(r)} \right]$$

In both this and the preceding formula, we can, if desired, take the average or min only over a subset of the A's (the "neighbors" of A_i), and regard the other A's as irrelevant to A_i. This is particularly sensible if the B pattern is smaller than the A pattern, so that only the A's close to A_i can contribute to p_{ij}.

Relaxation can be applied to many other types of picture matching problems. In hierarchical template matching, when we match parts of a template to a picture and then look for combinations of these matches in approximately the correct relative positions (see Section 9.4.3), we can use relaxation to iteratively reestimate the strengths of the part matches, based on the strengths of the other desired matches in the vicinity. Relaxation can also be used in relational structure matching; see Section 12.2.4a.

10.5.4 "Discrete Relaxation"

In some situations we cannot, or do not need to, assign probabilities to the class memberships $A_i \in C_j$; instead, we are concerned only with which memberships are possible and which are impossible. (We will give an example of such a situation shortly.) In such a case, the compatibilities $c(i, j; h, k)$ must also be qualitative rather than quantitative; i.e., $A_h \in C_k$ is either

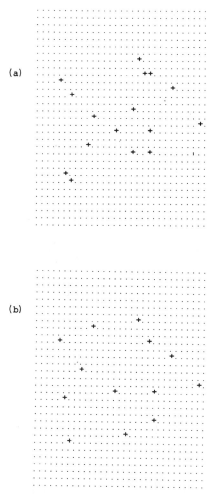

Fig. 52   Application of fuzzy relaxation to point pattern matching: (a) and (b) Point patterns. (c) Initial ratings of all the displacements that take a point of (a) into a point of (b), scaled to integer values. (d)–(g) Results of four iterations of fuzzy relaxation, showing substantial reduction in the ambiguity of the ratings.

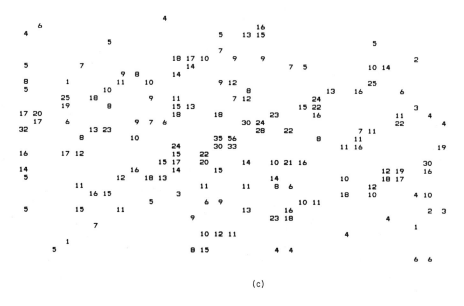

(c)

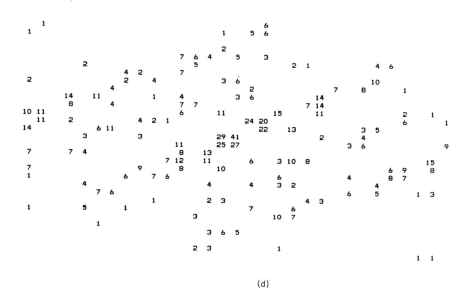

(d)

Fig. 52 (*Continued*)

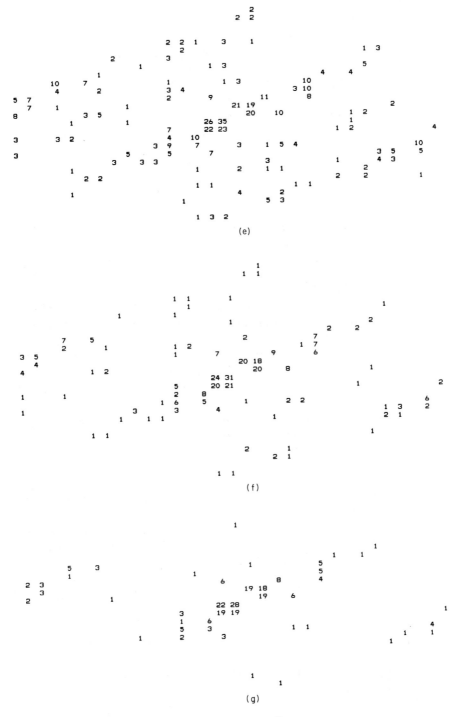

(e)

(f)

(g)

Fig. 52 (*Continued*)

compatible or incompatible with $A_i \in C_j$. Let us define $c(i, j; h, k)$ to be 1 in the former case and 0 in the latter. Similarly, we define p_{ij} to be 1 if $A_i \in C_j$ is possible, and 0 if it is impossible. Then for each A_h that is relevant to A_i, we must have at least one k for which $c(i, j; h, k) = 1$ and $p_{hk} = 1$; if there is no such k, $A_i \in C_j$ is impossible. Initially, we can take all the p_{ij}'s to be 1. At subsequent iterations, p_{ij} remains 1 if, for each relevant A_h, there exists k such that $c(i, j; h, k)p_{hk} = 1$; otherwise, p_{ij} becomes 0. It is easily seen that this is equivalent to

$$p_{ij}^{(r+1)} = \min_h \left[\max_{k=1}^{m} c(i, j; h, k)p_{hk}^{(r)} \right]$$

where the min is taken over the relevant h's; note that this is exactly the same as the second formula in Section 10.5.3.

The process defined by this "discrete relaxation" formula iteratively estimates possible class assignments for A_i by checking that each assignment is compatible with the surviving assignments of A_i's neighbors. This is a special type of constraint checking process in which only binary constraints are used. The concept can be generalized to higher-order constraints, but we will not pursue this here.

As an interesting example of binary constraint checking, we consider the problem of assigning three-dimensional interpretations to the lines in a line drawing of polyhedra. Any line in such a drawing can represent

(a, b)   A convex or concave dihedral angle, with both faces visible
(c, d)   An "occluding" dihedral angle with only one face visible; here too there are two cases, depending on which side of the line the visible face lies.

However, when two or more lines meet in a vertex, not all combinations of these interpretations are possible, and we can use these vertex constraints to eliminate interpretations.

As a simple illustration of these ideas, consider the drawing shown in Fig. 53. (The usual applications of vertex constraints deal with hidden-line drawings, but for simplicity we use a somewhat specialized set of constraints and give a wire-frame drawing example.) This drawing contains six "A" vertices (a, b, c, f, g, h) and two "Y" vertices (d, e) (see Fig. 39). Let us impose the following constraints on the lines that meet at such vertices:

(1)   At a "Y" vertex, either all three lines are convex or all three are concave.
(2)   At an "A" vertex, the outer lines are occluding and the inner line is either convex or concave. In the former case, the faces defined by the smaller

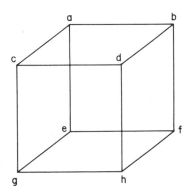

Fig. 53 Wire-frame drawing illustrating the use of "discrete relaxation" in applying vertex constraints (see text).

angles (e.g., *acd* and *gcd*) are in front of the faces defined by the large angle (*acg*); in the latter case, the reverse is true. We will call these two occluding cases O_1 and O_2.

(3) If a line is O_1 as seen from one of its ends, it is O_2 as seen from the other. For example, if *ca* is O_1, *acd* is in front of *acg*, which implies that *ac* is O_2, since *bac* (which is the same face as *acd*) must be in front of *cae* (which is the same face as *acg*). On the other hand, if a line is convex as seen from one end, it is also convex from the other, and similarly for concave.

Using these constraints, we can immediately identify lines *ab, bf, f h, hg, gc*, and *ca* as occluding (of one of the two types), and lines *db, dc, dh, ea, ef, eg* as either convex or concave. Thus each line has only two possible interpretations, rather than the four that it would have *a priori*. Moreover, as we shall now verify, only two combinations of these interpretations are possible, so that in this 12-line drawing we have a reduction from 4^{12} ($> 10^7$) to 2^{12} ($= 4096$) to 2 interpretations by applying constraints. To find the two legal interpretations, suppose that, say, *dc* is convex. By (1), *db* and *dh* must then also be convex; by (2), it follows that *ca, cg, ba, bf, hg*, and *hf* are O_1; by (3), this implies that *ac, ab, f b, f h, gc*, and *gh* are O_2; and by (2) again, this make *ea, ef*, and *eg* concave, which is consistent with (1). If we make the opposite assumption, that *dc* is concave, all of these interpretations are reversed.

It should be noted that if we impose the constraints only in parallel, we cannot eliminate the ambiguity; we can eliminate the O_1 and O_2 class assignments from *db, dc, dh, ea, ef, eg*, and convex and concave assignments from the other lines, but this still leaves two interpretations for each line. In order to eliminate all but the two consistent interpretations, we must make a choice for one of the lines; the effects of this choice then propagate, as just seen, to make all of the lines unambiguous.

10.5.5 Iterative Property Adjustment

We conclude this section by briefly discussing a class of iterative techniques that are closely related to, but somewhat simpler than, relaxation methods. In relaxation, we initially classify the objects (e.g., pixels) probabilistically on the basis of certain properties, and then iteratively adjust the class probabilities, based on the neighboring probabilities and compatibilities. An alternative idea is to iteratively adjust the property values, based on the neighboring values and "compatibilities," prior to classifying. We illustrate this idea with several examples.

In thresholding, we measure the gray level at each point and classify the point as (e.g.) light or dark on the basis of this measurement. In Section 10.5.2a we showed how to use relaxation to iteratively adjust the light and dark class probabilities, based on the neighboring probabilities. Alternatively, we can iteratively adjust the gray level values, based on the neighboring values and their compatibilities (e.g., similar gray levels are more compatible than dissimilar). This approach has already been described in Section 6.4; it is used in a variety of iterative noise cleaning techniques. Analogous ideas apply in the case of spectral (or spatial) classification, as we saw in Section 10.5.2b.

Similarly, edge or curve detection is based on measurements of the edge or curve strength (and direction) at each point. In Section 10.5.2c we applied relaxation to these tasks. Alternatively, we can iteratively adjust the strengths (and directions) at each point, based on the neighboring values and compatibilities. For example, let the strengths at points A_i and A_h be w_i and w_h, the directions θ_i and θ_h, and let the direction from A_i to A_h be θ_{ih}; then we can increment or decrement w_i by an amount proportional to w_h and to their compatibility, as defined by a cosine rule just as in Sections 10.5.1b and 10.5.2c. (If desired, we can also adjust our estimate of θ_i by comparing it with the neighboring θ's; we will not describe this process here in detail.) This will tend to increase the strengths at points that lie on smooth edges or curves, and decrease them elsewhere.

10.6 BIBLIOGRAPHICAL NOTES

Pavlidis [56] has defined segmentation formally as follows: Given a definition of "uniformity," a *segmentation* is a partition of the picture into connected subsets, each of which is uniform, but such that no union of adjacent subsets is uniform. The suggestion that the results of segmentation should be fuzzy subsets rather than ordinary subsets was first made by

Prewitt [59]. A discussion of the role of image models in segmentation can be found in Rosenfeld and Davis [66].

The use of histogram peaks and valleys in threshold selection is due to Prewitt and Mendelsohn [60]. For threshold selection based on busyness, on p-tiles, and on size see Weszka and Rosenfeld [82], Doyle [17], Bartz [8], and Tou [77]. Segmentation by averaging and thresholding is treated by Davis *et al.* [16]. On interpolation of local thresholds see Chow and Kaneko [12]; methods based on object "cores" are reviewed by Rutovitz [71] (see also the references cited below on region growing methods). A brief survey of threshold selection techniques is given by Weszka [81]; see also Lowitz [37] and Otsu [55].

On the use of spectral signatures for pixel classification, see, e.g., Reeves [61]. Recursive segmentation using one feature at a time is discussed by Ohlander *et al.* [54]. Segmentation based on local property values is treated by Coleman [13], Davis *et al.* [16], Schachter *et al.* [72], and Ahuja *et al.* [3]. The use of gray level difference values to improve thresholding is reviewed by Weszka and Rosenfeld [83]; on the use of neighboring gray levels or of local average gray level see Ahuja and Rosenfeld [4] and Kirby and Rosenfeld [34]. Conditional typicality was introduced by Haralick [26]; on clustering of gray levels based on conditional probabilities see Rosenfeld *et al.* [67] and Aggarwal [2].

The use of derivative operators, particularly the gradient, for edge detection dates back to Kovasznay and Joseph in the mid-1950's; see references [24] and [25] in Section 6.5. Many authors have defined digital approximations to these operators. A comparison of the orientation and noise sensitivities of many of these operators can be found in Abdou [1]. A review of edge detection methods through 1974 is given by Davis [15].

Modestino and Fries [44] define a recursive filter that yields a least mean-square estimate of the Laplacian. On the pseudo-Laplacian and mean-median difference see Schachter and Rosenfeld [73]. Differences of averages are used by Rosenfeld *et al.* [69, 70], who also discuss methods of automatically selecting a "best" amount of averaging to use at each point. On weighted averages see Argyle [5] and MacLeod [38]; for operators based on products see Rosenfeld [64], and compare Chen [11].

On the use of masks for edge detection see Prewitt [59], Kirsch [35], and Robinson [63]. Another approach to edge detection, due to Frei and Chen [21], is to observe that we can choose 3×3 edge masks that are mutually orthogonal; we can thus regard these masks as a basis for the space of edgelike 3×3 patterns. We then extend this basis, by adding additional masks, to obtain a basis for the entire nine-dimensional space of 3×3's. The "edge-likeness" of any 3×3 can then be defined in terms of its (angular) "closeness" to the edge subspace. Persoon [58] uses an edge detector based on differences of 2×5 averages in each of eight directions, but reduces the value of each difference if its 2×5's are not good fits to linear functions. Wechsler and Kidode [80] define an edge detector based on a max of forward, backward, and central difference operators of up to third order in four directions.

Step-fitting edge detection, using basis functions, is treated by Hueckel [28, 29], Nevatia [52], O'Gorman [53], Meró and Vássy [42], and Hummel [30]. Ramp and surface fitting for estimation of the gradient and Laplacian is discussed by Prewitt [59]; for more recent discussions see Brooks [10] and Beaudet [9] (the latter gives a table of derivative operators of orders up to 4 approximated on neighborhoods of size up to 8×8, and also discusses the uses of higher-order operators as feature detectors).

For an early use of decision theory to detect edges (or lines) see Griffith [23], as well as Yakimovsky [84]. A recent example, based on the sign test, is by Kersten and Kurz [33]. Nahi *et al.* [47, 48] have used Bayesian image estimation to detect the edges where one random-process image model replaces another. On "planning" see Kelly [32]; see also the discussion of hierarchical image representations in Section 11.1.5a. For examples of color edge detection techniques see Nevatia [51] and Robinson [62]. On detection of local property differences see Rosenfeld *et al.* [69, 70], as well as Thompson [76]; other statistical differencing techniques were used in early work on region growing, e.g., by Muerle and Allen [46]. Generalizations of edge detection to three (and more) dimensions are discussed by Herman and Liu [27, 36].

Nonlinear line, curve, streak, and spot detection operators are discussed in Rosenfeld *et al.* [69, 70]; on semilinear operators see VanderBrug [78]. For a review of Hough transforms, including a bibliography and an analysis of their performance, see Shapiro [74]. Edge linking into straight lines is discussed by Nevatia [50] and Dudani and Luk [19]. On edge linking along borders see Milgram [43] and Nakagawa and Rosenfeld [49]. On linking of curve segments into global curves see VanderBrug and Rosenfeld [79].

Raster tracking is an old technique that has been used in a number of applications, including bubble chamber picture processing. Omnidirectional tracking was used in early work on target detection and character recognition; a classical example of the latter is Greanias *et al.* [22]. An optimization approach to curve (or edge) tracking was introduced by Montanari [45], and a heuristic search approach by Martelli [40], who also presents some more recent results in [41]. Other algorithms for finding minimum-cost paths can also be applied to this problem; see, e.g., [6].

Region tracking was first introduced by Grimsdale *et al.* [24]. A review and discussion of region growing techniques, including a bibliography, is given by Zucker [85]. Partitioning techniques have been extensively studied by Pavlidis; his book [56] summarizes much of his work on this subject. The fuzzy and probabilistic relaxation models were introduced by Rosenfeld *et al.* [68]; see also Peleg [57]. A review of various relaxation applications is given by Rosenfeld [65].

REFERENCES

1. I. E. Abdou and W. K. Pratt, Quantitative design and evaluation of enhancement/thresholding edge detectors, *Proc. IEEE* **67**, 1979, 753–763.
2. R. K. Aggarwal, Adaptive image segmentation using prototype similarity, *Proc. IEEE Conf. Pattern Recognition Image Processing*, 1978, 354–359.
3. N. Ahuja, R. M. Haralick, and A. Rosenfeld, Neighbor gray levels as features in pixel classification, *Pattern Recognition* **12**, 1980, 251–260.
4. N. Ahuja and A. Rosenfeld, A note on the use of second-order gray level statistics for threshold selection, *IEEE Trans. Systems Man Cybernet.* **8**, 1978, 895–898.
5. E. Argyle, Techniques for edge detection, *Proc. IEEE* **59**, 1971, 285–286.
6. G. P. Ashkar and J. W. Modestino, The contour extraction problem with biomedical applications, *Comput. Graphics Image Processing* **7**, 1978, 331–355.
7. H. G. Barrow and J. M. Tenenbaum, MSYS: A system for reasoning about scenes, Technical Note 121, Artificial Intelligence Center, SRI International, Menlo Park, California, April 1976.
8. M. R. Bartz, Optimizing a video preprocessor for OCR, *Proc. Internat. Joint Conf. Artificial Intelligence, 1st* 1969, 79–90.
9. P. R. Beaudet. Rotationally invariant image operators, *Proc. Internat. Joint Conf. Pattern Recognition, 4th* 1978, 579–583.
10. M. J. Brooks, Rationalizing edge detectors, *Comput. Graphics Image Processing* **8**, 1978, 277–285.
11. C. H. Chen, Note on a modified gradient method for image analysis, *Pattern Recognition* **10**, 1978, 261–264.
12. C. K. Chow and T. Kaneko, Automatic boundary detection of the left ventricle from cineangiograms. *Comput. Biomed. Res.* **5**, 1972, 338–410.
13. G. B. Coleman and H. C. Andrews, Image segmentation by clustering, *Proc. IEEE* **67**, 1979, 773–785.
14. C. M. Cook and A. Rosenfeld, Size detectors, *Proc. IEEE* **58**, 1970, 1956–1957.
15. L. S. Davis, A survey of edge detection techniques, *Comput. Graphics Image Processing* **4**, 1975, 248–270.
16. L. S. Davis, A. Rosenfeld, and J. S. Weszka, Region extraction by averaging and thresholding, *IEEE Trans. Systems Man Cybernet.* **5**, 1975, 383–388.
17. W. Doyle, Operations useful for similarity-invariant pattern recognition, *J. ACM* **9**, 1962, 259–267.
18. R. O. Duda and P. E. Hart, "Pattern Classification and Scene Analysis." Wiley, New York, 1973.
19. S. A. Dudani and A. L. Luk, Locating straight-line segments on outdoor scenes, *Pattern Recognition* **10**, 1978, 145–157.
20. J. A. Feldman and Y. Yakimovsky, Decision theory and artificial intelligence: I. A semantics-based region analyzer, *Artificial Intelligence* **5**, 1974, 349–371.
21. W. Frei and C. C. Chen, Fast boundary detection: a generalization and a new algorithm, *IEEE Trans. Comput.* **26**, 1977, 988–998.

22. E. C. Greanias, P. F. Meagher, R. J. Norman, and P. Essinger, The recognition of hand-written numerals by contour analysis, *IBM J. Res. Develop.* 7, 1963, 14–21.

23. A. K. Griffith, Mathematical models for automatic line detection, *J. ACM* **20**, 1973, 62–80.

24. R. L. Grimsdale, F. H. Sumner, C. J. Tunis, and T. Kilburn, A system for the automatic recognition of patterns, *Proc. IEE* **106B**, 1959, 210–221.

25. A. Guzman, Decomposition of a visual scene into three-dimensional bodies, *Proc. Fall Joint Comput. Conf.*, 1968, 291–304.

26. R. M. Haralick, A resolution preserving textural transform for images, *Proc. IEEE Conf. Comput. Graphics Pattern Recognition and Data Structure* 1975, 51–61.

27. G. T. Herman and H. K. Liu, Dynamic boundary surface detection, *Comput. Graphics Image Processing* 7, 1978, 130–138.

28. M. F. Hueckel, An operator which locates edges in digitized pictures, *J. ACM* **18**, 1971, 113–125.

29. M. F. Hueckel, A local operator which recognizes edges and lines, *J. ACM* **20**, 1973, 634–647.

30. R. A. Hummel, Feature detection using basis functions, *Comput. Graphics Image Processing* 9, 1979, 40–55.

31. A. Kaufmann, "Introduction to the Theory of Fuzzy Subsets." Academic Press, New York, 1975.

32. M. D. Kelly, Edge detection in pictures by computer using planning, *in* "Machine Intelligence" (B. Meltzer and D. Michie, eds.), Vol. 6, pp. 377–396. Edinburgh Univ. Press, Edinburgh, Scotland, 1971.

33. P. Kersten and L. Kurz, Bivariate m-interval classifiers with application to edge detection, *Informat. Control* **34**, 1977, 152–168.

34. R. L. Kirby and A. Rosenfeld, A note on the use of (gray level, local average gray level) space as an aid in threshold selection, *IEEE Trans. Systems Man Cybernet.* 9, 1979, 860–864.

35. R. A. Kirsch, Computer determination of the constituent structure of biological images, *Comput. Biomed. Res.* **4**, 1971, 315–328.

36. H. K. Liu, Two- and three-dimensional boundary detection, *Comput. Graphics Image Processing* **6**, 1977, 123–134.

37. G. E. Lowitz, What a histogram can really tell the classifier, *Pattern Recognition* **10**, 1978, 351–357.

38. I. D. G. MacLeod, Comments on "Techniques for edge detection," *Proc. IEEE* **60**, 1972, 344.

39. D. Marr, Early processing of visual information, *Philos. Trans. Roy. Soc. London Ser. B* **275**, 1976, 483–524.

40. A. Martelli, Edge detection using heuristic search methods, *Comput. Graphics Image Processing* **1**, 1972, 169–182.

41. A. Martelli, An application of heuristic search methods to edge and contour detection, *Comm. ACM* **19**, 1976, 73–83.

42. L. Meró and Z. Vássy, A simplified and fast version of the Hueckel operator for finding optimal edges in pictures, *Proc. Internat. Joint Conf. Artificial Intelligence, 4th* 1975, 650–655.

43. D. L. Milgram, Region extraction using convergent evidence, *Comput. Graphics Image Processing* **11**, 1979, 1–12.

44. J. W. Modestino and R. W. Fries, Edge detection in noisy images using recursive digital filtering, *Comput. Graphics Image Processing* **6**, 1977, 409–433.

45. U. Montanari, On the optimum detection of curves in noisy pictures, *Comm. ACM* **14**, 1971, 335–345.

46. J. L. Muerle and D. C. Allen, Experimental evaluation of techniques for automatic segmentation of objects in a complex scene, *in* "Pictorial Pattern Recognition" (G. C. Cheng *et al.*, eds.), pp. 3–13. Thompson Books, Washington, D.C.
47. N. E. Nahi and M. H. Jahanshahi, Image boundary estimation, *IEEE Trans. Comput.* **26**, 1977, 772–781.
48. N. E. Nahi and S. Lopez-Mora, Estimation-detection of object boundaries in noisy images, *IEEE Trans. Automat. Control* **23**, 1978, 834–846.
49. Y. Nakagawa and A. Rosenfeld, Edge/border coincidence as an aid in edge extraction, *IEEE Trans. Systems Man Cybernet.* **8**, 1978, 899–901.
50. R. Nevatia, Locating object boundaries in textured environments, *IEEE Trans. Comput.* **25**, 1976, 1170–1175.
51. R. Nevatia, A color edge detector and its use in scene segmentation, *IEEE Trans. Systems Man Cybernet.* **7**, 1977, 820–826.
52. R. Nevatia, Evaluation of a simplified Hueckel edge-line detector, *Comput. Graphics Image Processing* **6**, 1977, 582–588.
53. F. O'Gorman, Edge detection using Walsh functions, *Artificial Intelligence* **10**, 1978, 215–223.
54. R. Ohlander, K. Price, and D. R. Reddy, Picture segmentation using a recursive region splitting method, *Comput. Graphics Image Processing* **8**, 1978, 313–333.
55. N. Otsu, A threshold selection method from gray level histograms, *IEEE Trans. Systems Man Cybernet.* **9**, 1979, 62–66.
56. T. Pavlidis, "Structural Pattern Recognition." Springer, New York, 1977.
57. S. Peleg, A new probabilistic relaxation scheme, *IEEE Trans. Pattern Analysis Machine Intelligence* **2**, 1980, 362–369.
58. E. Persoon, A new edge detection algorithm and its applications in picture processing, *Comput. Graphics Image Processing* **5**, 1976, 425–446.
59. J. M. S. Prewitt, Object enhancement and extraction, *in* "Picture Processing and Psychopictorics" (B. S. Lipkin and A. Rosenfeld, eds.), pp. 75–149. Academic Press, New York, 1970.
60. J. M. S. Prewitt and M. L. Mendelsohn, The analysis of cell images, *Ann. N.Y. Acad. Sci.* **128**, 1966, 1035–1053.
61. R. G. Reeves (ed.), "Manual of Remote Sensing," pp. 766–785. American Society of Photogrammetry, Falls Church, Virginia, 1975.
62. G. S. Robinson, Color edge detection, *Opt. Eng.* **16**, 1977, 479–484.
63. G. S. Robinson, Edge detection by compass gradient masks, *Comput. Graphics Image Processing* **6**, 1977, 492–501.
64. A. Rosenfeld, A nonlinear edge detection technique, *Proc. IEEE* **58**, 1970, 814–816.
65. A. Rosenfeld, Iterative methods in image analysis, *Pattern Recognition* **10**, 1978, 181–187.
66. A. Rosenfeld and L. S. Davis, Image segmentation and image models, *Proc. IEEE* **67**, 1979, 764–772.
67. A. Rosenfeld, H. K. Huang, and V. B. Schneider, An application of cluster detection to text and picture processing, *IEEE Trans. Informat. Theory* **15**, 1969, 672–681.
68. A. Rosenfeld, R. A. Hummel, and S. W. Zucker, Scene labeling by relaxation operations, *IEEE Trans. Systems Man Cybernet.* **6**, 1976, 420–433.
69. A. Rosenfeld and M. Thurston, Edge and curve detection for visual scene analysis. *IEEE Trans. Comput.* **20**, 1971, 562–569.
70. A. Rosenfeld, M. Thurston, and Y. H. Lee, Edge and curve detection: further experiments, *IEEE Trans. Comput.* **21**, 1972, 677–715.
71. D. Rutovitz, Expanding picture components to natural density boundaries by propagation methods—the notions of fall-set and fall-distance, *Proc. Internat. Joint Conf. Pattern Recognition*, *4th* 1978, 657–664.

72. B. J. Schachter, L. S. Davis and A. Rosenfeld, Some experiments in image segmentation by clustering of local feature values, *Pattern Recognition* **11**, 1978, 19–28.
73. B. J. Schachter and A Rosenfeld, Some new methods of detecting step edges in digital pictures, *Comm. ACM* **21**, 1978, 172–176.
74. S. D. Shapiro, Feature space transforms for curve detection, *Pattern Recognition* **10**, 1978, 129–143.
75. J. M. Tenenbaum and H. G. Barrow, IGS: A paradigm for integrating image segmentation and interpretation, *in* "Pattern Recognition and Artificial Intelligence" (C. H. Chen, ed.), pp. 472–507. Academic Press, New York, 1976.
76. W. B. Thompson, Textural boundary analysis, *IEEE Trans. Comput.* **26**, 1977, 272–276.
77. J. T. Tou, Zoom-thresholding technique for boundary determination, *J. Comput. Informat. Sci.* **8**, 1979, 3–8.
78. G. J. VanderBrug, Semilinear line detectors, *Comput. Graphics Image Processing* **4**, 1975, 287–293.
79. G. J. VanderBrug and A. Rosenfeld, Linear feature mapping, *IEEE Trans. Systems Man Cybernet.* **8**, 1978, 768–774.
80. H. Wechsler and M. Kidode, A new edge detection technique and its implementation, *IEEE Trans. Systems Man Cybernet.* **7**, 1977, 827–836.
81. J. S. Weszka, A survey of threshold selection techniques, *Comput. Graphics Image Processing* **7**, 1978, 259–265.
82. J. S. Weszka and A. Rosenfeld, Threshold evaluation techniques, *IEEE Trans. System Man. Cybernet.* **8**, 1978, 622–629.
83. J. S. Weszka and A. Rosenfeld, Histogram modification for threshold selection, *IEEE Trans. Systems Man Cybernet.* **9**, 1979, 38–52.
84. Y. Yakimovsky, Boundary and object detection in real world images, *J. ACM* **23**, 1976, 599–618.
85. S. W. Zucker, Region growing: childhood and adolescence, *Comput. Graphics Image Processing* **5**, 1976, 382–399.

Chapter 11

Representation

Segmentation decomposes a picture into subsets or regions. Geometrical properties of these subsets—connectedness, size, shape, etc.—are often important in picture description. As we shall see, there are many methods of measuring such properties; the preferred method usually depends on how the subsets are represented. This chapter discusses various representation schemes and their application to geometric property measurement.

In this chapter, Σ denotes a picture; subsets of Σ are denoted by $S, T, \ldots$, and points by $P, Q, \ldots$.

In general, a segmented picture Σ is partitioned into a collection of non-empty subsets $S_1, \ldots, S_m$ such that $\bigcup_{i=1}^{m} S_i = \Sigma$ and $S_i \cap S_j = \varnothing$ for all $i \neq j$. An important special case is $m = 2$, i.e., Σ consists of a set S and its complement $\bar{S}$. (In the general case, we can take any of the S_i as S, so that $\bigcup_{j=1, j \neq i}^{m} S_j = \bar{S}$.) Note that the S's are not necessarily connected; connectedness will be treated in Sections 11.1.7 and 11.3.1.

11.1 REPRESENTATION SCHEMES

Any subset S of a digital picture Σ can be represented by a binary picture (or "bit plane") χ_S of the same size as Σ, having 1's at the points of S and 0's elsewhere. If Σ is $n \times n$, the χ_S representation requires n^2 bits. χ_S can be regarded as the *characteristic function* of the subset S; this is the function that maps points of S into 1 and points of $\bar{S}$ into 0.

More generally, any partition of Σ into $S_1, \ldots, S_m$ can be represented by an m-valued picture having i's at the points of S_i, $1 \leqslant i \leqslant m$. In particular, if Σ is any picture, then in this sense, Σ represents its own partition into sets of constant gray level, i.e., S_i is the set of points of Σ having gray level i. For an $n \times n$ picture, this representation requires $n^2 \log_2 m$ bits.

The storage requirements of this trivial representation are the same for all partitions of Σ into a given number of sets. Our main interest in this chapter is to study representations which are more economical for "simple" partitions. In the following sections we will define a variety of such representations.

Exercise 1. If the subsets $S_1, \ldots, S_{m-1}$ consist of only a few points, they can be specified by listing the coordinates of these points. For example, if $m = 2$, and $S = S_1$ consists of k points, the list requires $2k \log_2 n$ bits ($2 \log_2 n$ bits for the coordinates of each point) for an $n \times n$ picture, as compared with n^2 bits for the bit plane representation, and is more economical if $k < n^2/2 \log_2 n$. How many bits are required for arbitrary m when there are k points? ∎

11.1.1   Rows

a.   Runs

Each row of a picture consists of a sequence of maximal runs of points such that the points in each run all have the same value. Thus the row is completely determined by specifying the lengths and values of these runs. If there are only a few runs, this representation is very economical; for this reason, *run length coding* is sometimes used for picture compression, as mentioned in Section 5.8. For example, suppose that the row has length n, and there are r runs. Since it takes $\log_2 n$ bits to specify the length of a run (it may have any length between 1 and n), the number of bits needed to specify all the run lengths is $r \log_2 n$. Thus if there are m possible values, this representation of the row requires $r(\log_2 n + \log_2 m)$ bits, as compared with the $n \log_2 m$ bits that are required when the row is treated as a string of length n.

When $m = 2$, we need only specify the value of the first run in the row, since the values must alternate. Thus the run length specification of a row of a binary picture requires $1 + r \log_2 n$ bits, as compared with the n bits required to represent the row as a bit string. Note that these savings are only one-dimensional; for an $n \times n$ picture, if the average number of runs in each row is r, the total number of bits required by the run length representation is $n(1 + r \log_2 n)$ as compared with n^2 in the binary case, or $nr(\log_2 m + \log_2 n)$ as compared with $n^2 \log_2 m$ in the general case.

b. Binary trees

Suppose, for simplicity, that the row length is a power of 2, say $n = 2^k$. We shall now describe a method of representing the row by a binary tree, each of whose leaves corresponds to a (not necessarily maximal) run of constant value whose length is a power of 2, say 2^i, and whose position coordinate is a multiple of 2^i.

The root node of the tree represents the entire row. If the row all has one value, we label the root node with that value and stop; in this case, the tree consists only of the root node. Otherwise, we add two descendants to the root node, representing the two halves of the row. The process is then repeated for each of these new nodes: if its half of the row has constant value, we label it with that value and do not give it any descendants; if not, we give it two descendants corresponding to the two halves of its half. In general, at level h in the tree (where the root is at level 0), the nodes (if any) represent pieces of the row of length 2^{k-h}, in positions which are multiples of 2^{k-h}. If a piece has constant value, its node is a leaf node (i.e., it has no descendants), labeled with that value; otherwise, that node has two descendants, corresponding to the two halves of the piece. At level k, the nodes (if any) correspond to single pixels, and are all leaf nodes, labeled with the values of their pixels.

A simple example of a string and the corresponding binary tree is shown in Fig. 1. Note that the blocks of constant value corresponding to the leaf nodes of the tree are not necessarily maximal runs; if the length of a run is not

1 0 0 1 0 1 1 0 1 1 1 1 1 1 1 1 0 1 0 0 1 0 1 1 1 1 1 1 1 0 1 1 0

(a)

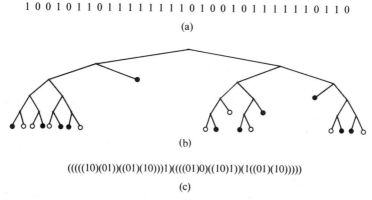

(b)

$$(((((10)(01))((01)(10)))1)((((01)0)((10)1))(1((01)(10)))))$$

(c)

Fig. 1   Binary tree representation of a string. (a) Binary string of length 32. (b) Tree; solid and open circles are leaf nodes corresponding to blocks of 1's and 0's, respectively. There are 39 nodes, 20 of which are leaf nodes. (c) Parenthesis string representation of the tree. The run length representation of this string is 1:1,2,1,1,2,1,8,1,1,2,1,1,6,1,2,1, where the initial 1 represents the fact that the string begins with a run of 1's; there are 16 runs.

a power of 2, or if its starting position is not a multiple of the proper power of 2, it is represented by more than one leaf node.

The space required to store the tree is proportional to the number of nodes. (For example, the tree can be represented by a parenthesis string in which each node maps into a pair of parentheses enclosing the substring representing the subtree rooted at that node, as illustrated in Fig. 1c. The leaf nodes can be represented by their associated values.) If the row consists of only a few runs, the tree will have relatively few nodes, but the exact number depends on the positions and lengths of the runs.

The binary tree representation of the rows of a picture does not seem to have been used in practice. It was presented here as a preliminary to the two-dimensional quadtree representation in Section 11.1.2b.

11.1.2   Blocks

a.   MATs

With each point P of the picture Σ, let us associate the set of upright squares of odd side lengths n centered at P. (Our discussion generalizes to any family of mutually similar shapes, but for simplicity we will assume that they are upright squares and that they all have P at their centers.) Let S_P be the largest such square that is contained in Σ and has constant value, and let r_P be the radius of S_P. There may exist other points Q such that S_P is contained in S_Q; if no such Q exists, we call S_P a *maximal block.*

It is easy to see that if we specify the set of centers P, radii r_P, and values v_P of the maximal blocks, Σ is completely determined, since any point of Σ lies in at least one maximal block. In fact, we need only do this for $m - 1$ of the values; the points not covered by any of these blocks must have the omitted value. Thus in the case where there are only two values, we need only specify the blocks for one value. The maximal blocks for a simple picture are shown in Fig. 2.

The set of centers and radii (and values) of the maximal blocks is called the *medial axis* (or *symmetric axis*) *transformation* of Σ, abbreviated MAT or SAT. It has this name because the centers are located at midpoints, or along local symmetry axes, of the regions of constant value in Σ. Intuitively, if a block S_P is maximal and is contained in the constant-value region S, it must touch the border of S in at least two places; otherwise we could find a neighbor Q of P that was farther away than P from the border of S, and then S_Q would contain S_P. A more rigorous treatment of these ideas will be given in Section 11.2.1.

Evidently, if Σ consists of only a few constant-value regions which have simple shapes (i.e., which are unions of only a few blocks), its MAT repre-

```
0  0  0  0  0  0  0  0

0  0  1  0  0  0  0  0

0  0  0  0  0  1  0  0

1  1  1  1  1  1  1  0
                                    x    y    r
1  1  1  1  1  1  1  0          ───────────────
                                    3    3    2
1  1  1  1  1  1  1  0          3    7    0
                                    4    3    2
1  1  1  1  1  1  0  0          6    4    1

1  1  1  1  1  1  0  0          6    6    0
```

(a)   (b)

Fig. 2 Simple example of a MAT representation. (a) 8 × 8 binary picture, with centers of MAT blocks (of 1's) underlined. (b) Center coordinates and radii of MAT blocks (lower left corner has coordinates (1,1)).

sentation will be quite compact. For an $n \times n$ picture, $2 \log_2 n$ bits are needed to specify the coordinates of each block center, and $(\log_2 n) - 1$ bits to specify the radius; thus if there are b blocks in the MAT, the total number of bits required to specify it is $b(3 \log_2 n + \log_2(m - 1) - 1)$.

Note that the MAT representation may still be redundant, i.e., some blocks may be contained in unions of others. There does not seem to be any simple way of reducing this redundancy without carrying out a lengthy search process.

b. Quadtrees

Maximal blocks can be of any size and in any position; they are analogous to runs in the one-dimensional case. (Maximal connected regions of constant value are also analogous to runs, but they are not good primitive elements for representation purposes, since they themselves cannot be specified compactly; a block, on the other hand, is defined by specifying its center and radius.) We next describe a two-dimensional representation based on trees of degree 4; it is analogous to the binary tree representation for rows. We assume for simplicity that the size of the picture Σ is $2^k \times 2^k$.

The root node of the tree represents the entire picture. If the picture has all one value, we label the root node with that value and stop. Otherwise, we add four descendants to the root node, representing the four quadrants of the picture. The process is then repeated for each of these new nodes; and so on. In general, the nodes at level h (if any) represent blocks of size $2^{k-h} \times 2^{k-h}$, in positions whose coordinates are multiples of 2^{k-h}. If a block has constant value, its node is a leaf node; otherwise, its node has four descendants at level

0	0	0	0	0	0	0	0
0	0	1	0	0	0	0	0
0	0	0	0	0	1	0	0
1	1	1	1	1	1	1	0
1	1	1	1	1	1	1	0
1	1	1	1	1	1	1	0
1	1	1	1	1	1	0	0
1	1	1	1	1	1	0	0

(a)

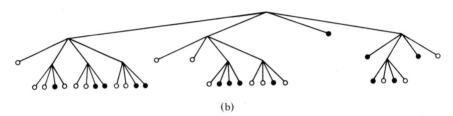

(b)

Fig. 3 Quadtree for the picture in Fig. 2. (a) Picture, with quadtree blocks marked. (b) Tree; notation is as in Fig. 1. There are 41 nodes, 31 of which are leaf nodes. The order of the sons on each row is NW, NE, SW, SE.

$h + 1$, corresponding to the four quadrants of the block. The nodes at level k, if any, are all leaf nodes corresponding to single pixels.

The tree constructed in this way is called a *quadtree*, since its nonleaf nodes all have degree 4. A simple example of a picture and its corresponding quadtree is shown in Fig. 3. The space required to store the tree is proportional to the number of nodes, as in the one-dimensional case. Note that, unlike the MAT, the quadtree representation is not redundant.

The chief disadvantage of the quadtree representation is that, unlike the nontree representations considered here, it is shift-variant. Two pictures that differ only by a translation may give rise to very different quadtrees. Thus it is hard to tell from their quadtree representations whether two pictures are congruent.

11.1.3 Borders

All of the representations considered up to now are based on maximal runs or blocks of constant value, possibly restricted as to size and position. These

A B

  C D

    E $A_t, B_t, B_r, D_t, D_r, E_r, F_r,$

    F A, B, D, E, F, E, C $F_b, F_l, E_l, C_b, C_l, A_b, A_l$

  (a)   (b)   (c)

Fig. 4 Methods of defining border sequences. (a) Set S; each point is labeled with a different letter. In this simple example there are no interior points. (b) Clockwise sequence of points around the border, beginning with A. Note that point E is visited twice. (c) Clockwise sequence of "cracks" around the border, beginning with the crack A_t at the top of A. The subscripts t, r, b, l denote top, right, bottom, and left, respectively. Note that if we delete the subscripts, and eliminate consecutive repetitions of the same point, we obtain the point sequence (b).

approaches represent each S_i as the union of the maximal runs or blocks that are contained in it.

Another class of approaches to representation makes use of the fact that the sets S_i are determined by specifying their borders. (Compare the discussion of contour coding in Section 5.8.) The border S' of a set S is the set of points of S that are adjacent to points of the complement $\bar{S}$. We can regard S' as consisting of a set of closed curves; each of these curves contains the points that belong to a particular connected component of S and are adjacent to a particular connected component of $\bar{S}$.[§] These concepts will be defined more precisely in Sections 11.1.7 and 11.2.2, and algorithms given for finding the border curves of a given S and for reconstructing S from its borders. In this section we will only discuss how to specify border curves.

A border curve is determined by specifying a starting point and a sequence of moves around the border. Figure 4 illustrates two ways in which this can be done; one approach moves along a sequence of border points of the set S, while the other moves along a sequence of "cracks" between the points of S and the adjacent points of $\bar{S}$.

In moving from point to point around a border, we always go from a point to one of its eight neighbors. Let us number the neighbors as follows:

$$3 \quad 2 \quad 1$$
$$4 \qquad 0$$
$$5 \quad 6 \quad 7$$

(mnemonic: neighbor i is in direction $45i°$, measured counterclockwise from the positive x-axis). Thus each move is defined by one of the digits $0, 1, \ldots, 7$,

---

[§] More generally, the points of any S_i that are adjacent to any given S_j consist of a set of arcs.

i.e., by an octal digit. For example, the sequence of moves used in Fig. 4b corresponds to 0766233. A sequence of moves represented by octal digits in this way is called a *chain code*. A border is thus defined by giving the co-ordinates of a starting point together with a chain code representing a sequence of moves.

If we follow the cracks around a border, at each move we are going either left, right, up, or down; if we denote direction $90i°$ by i, these moves can be represented by a sequence of 2-bit numbers (0, 1, 2, 3). For example, the sequence in Fig. 4c is represented by 00303332112121. We shall call this representation a *crack code*. A border is specified by giving the coordinates of a starting crack together with a crack code.

Each move in a crack code is represented by a 2-bit number, while chain code moves require 3-bit numbers; but the number of moves in crack following is somewhat greater than that in chain code, since several moves may be required to traverse the cracks around a single point. If we assume that at most half of the border points of S have two neighbors in $\bar{S}$ (i.e., are corners or "waists" of S), and that border points having three neighbors in $\bar{S}$ are rare, then the average number of cracks per border point is at most $1\frac{1}{2}$. Thus the number of bits required to represent a border by a crack code should be no greater than the number required to represent it by a chain code.

The storage requirements of these border code schemes depend on the total border length of the sets S_i. In an $n \times n$ picture, there are $2n(n + 1)$ cracks, including those around the edges of the picture. If fraction β of these cracks are border cracks, we need about $8\beta n(n + 1)$ bits to represent all the crack codes (two bits per move), since each border crack—ignoring the edges of the picture—belongs to two border curves.[§] As indicated in the preceding paragraph, chain codes would require a similar number of bits. In addition, for each border curve we need to specify starting point coordinates ($2 \log_2 n$ bits), as well as which set S_i lies on (say) the right as the border is traversed [$\log_2 m$ or $\log_2(m - 1)$ bits.] Thus if there are B borders, the total number of bits required for this type of representation is $8\beta n(n + 1) + B(\log_2 m + 2 \log_2 n)$. Note that B itself may be on the order of n^2, if there are many small connected components; in this case, border representations would not be economical.

For any given chain code, we can construct a *difference chain code* whose values represent the successive changes in direction, e.g., let 0 represent no turn, ± 1 represent 45° right or left turns, ± 2 and ± 3 similarly represent 90°

[§] It suffices to specify the borders for $m - 1$ of the subsets S_i, so that some of these border curves can be omitted. In particular, if $m = 2$, we need only specify the border curves of S, so that each border crack is on only one border curve, and the crack codes require only about $4\beta n(n + 1)$ bits.

and 135° turns, and 4 represent a 180° turn. Thus the difference chain code, like the chain code, has eight possible values and requires three bits per move. However, the values are no longer equally likely, e.g., 0 and 1 should be very common, while 4 is rare. (What does this imply about the compressibility of the difference code?) Evidently, a border is determined by specifying the coordinates of the starting point, the starting direction, and the difference chain code. The case of the *difference crack code* is analogous; here there are only three possible difference values, 0° and $\pm 90°$.

Exercise 2. If borders often contain long straight segments, their chain or crack codes can be further compressed by run length coding. Analyze the savings that can be obtained in this way. Also, discuss the possibilities for compressing difference codes using run length coding (note, in particular, that turns in a given direction cannot occur in long runs). ∎

Chain codes can also be used for the digital representation of plane curves. The chain code of a given curve C can be constructed as follows: Imagine a Cartesian grid superimposed on C. As we move along C, whenever C enters a grid square, we take the nearest corner of that square as a point on the digitization of C. (If C enters at the midpoint of a side of the square, use any standard rounding convention to pick the "nearest" corner.) When C enters and then leaves a grid square, the two successive grid points (= grid square corners) on the digitization are either the same or are neighbors in the grid, since they are corners of the same square. Thus the sequence of grid points constituting the digitization can be represented by a starting point and a chain code defining the sequence of moves from neighbor to neighbor. Chain codes of digital curves will be discussed further in Sections 11.1.5c and 11.2.3.

11.1.4 Representations of Derived Sets

We often need to define new subsets or partitions of a picture in terms of given ones, e.g., by performing set-theoretic operations on the given ones, or by resegmenting them. If we are using a particular type of representation, we would like to be able to derive the representations of the new sets directly from the representations of the original sets. Methods of doing this in particular situations will be described during the course of this chapter; e.g., see Sections 11.3.1a and 11.3.1b on connected component labeling using various representations. In this section we briefly discuss the problem of performing set-theoretic operations on representations of a given type. For simplicity, we deal only with partitions into a set S and its complement.

If S and T are represented as binary arrays χ_S and χ_T it is trivial to obtain $\bar{S}, S \cup T, S \cap T$, etc., by pointwise Boolean operations on these arrays. Such operations can be done in a single parallel step on a cellular array computer, or they can be done in a single scan of χ_S and χ_T on a conventional computer.

Given the run length representation of S (and T), we get that of $\bar{S}$ by simply reversing the designation of the first value on each row. In the remainder of this paragraph we describe an algorithm that creates the run list L for $S \cap T$ on a given row from the lists L_1 and L_2 of S and T. Analogous algorithms can be given for other Boolean functions. If L_1 and L_2 both begin with runs of 1's, so does L; otherwise, L begins with 0. We will now describe how to successively add runs to L by examining the initial parts of L_1 and L_2; each time we do this, we delete or truncate the initial run(s) in L_1 and L_2. The process is then repeated using the shortened L_1 and L_2. When they are empty, we have the desired L. A counter is also associated with L, and is initially set at 0; its role will be explained in step (b) of the algorithm.

(a)   Suppose L_1 and L_2 (currently) both begin with runs of 1's, ρ_1 and ρ_2, say of lengths $|\rho_1| \leqslant |\rho_2|$. In this case we add a run of 1's of length $|\rho_1|$ to b; delete the initial run from L_1; and shorten the initial run of L_2 to length $|\rho_2| - |\rho_1|$ (or delete it, if $|\rho_1| = |\rho_2|$). Note that at least one of L_1 and L_2 now begins with a run of 0's.

(b)   Suppose L_1 begins with a run ρ_1 of 0's, and L_2 begins with a run of 1's or with a shorter run of 0's. Add up the run lengths in L_2 until $|\rho_1|$ is reached, i.e.,

$$|\rho_{21}| + |\rho_{22}| + \cdots + |\rho_{2,k-1}| \leqslant |\rho_1| < |\rho_{21}| + |\rho_{22}| + \cdots + |\rho_{2k}|$$

(b1)   If ρ_{2k} is a run of 1's, truncate it to length

$$|\rho_{2k}| - (|\rho_1| - |\rho_{21}| - |\rho_{22}| - \cdots - |\rho_{2,k-1}|);$$

delete ρ_1 from L_1; add a run of 0's of length $|\rho_1| + C$ to L, where C is the value in the counter; and reset the counter to 0.

(b2)   If ρ_{2k} is a run of 0's, truncate it as above; delete ρ_1 from L_1; and add $|\rho_1|$ to the counter. In this case L_2 still begins with a run of 0's, so we are still in case (b), possibly with the roles of L_1 and L_2 reversed.

Given the MATs of S and T, we get a redundant MAT for $S \cup T$ by simply taking their union. A procedure for constructing a MAT for $S \cap T$ by finding maximal intersections of blocks in the MATs of S and T is described in [49]; the details will not be given here.

The quadtree of $\bar{S}$ is the same as that of S with "black" leaf nodes (= nodes corresponding to blocks of 1's) changed to "white" and vice versa. To get the quadtree of $S \cup T$ from those of S and T, we traverse the two trees simultaneously. Where they agree, the new tree is the same. If S has a gray (= nonleaf) node where T has a black node, the new tree gets a black node; if T has a white node there, we copy the subtree of S at that gray node into the new tree; if S has a white node and T a black node, the new tree gets a black node. The algorithm for $S \cap T$ is exactly analogous, with the roles of black and white reversed. The time required for these algorithms is proportional to the number of nodes in the smaller of the two trees [27, 72].

The crack codes of the borders of $\bar{S}$ are the same as those of S (but in reverse order, if we want to maintain a convention as to the order of border following). Constructing chain codes of the borders of $\bar{S}$ from those of S is somewhat more complex; the details will be omitted here. For either crack or chain codes, it is quite complicated to construct the codes of $S \cup T$ of $S \cap T$ from those of S and T; in particular, this involves finding which pairs of borders surround one another, and also finding all the intersections of each pair of borders. Border representations are not well suited for performing set-theoretic operations.

11.1.5   Approximate Representations

The representations considered so far in this section are all exact, i.e., they completely determine the given partition of Σ. In the following paragraphs we discuss methods of obtaining approximations to a partition based on these representations. Here again, for simplicity, we will usually assume that the partition consists of a set S and its complement.

It should be pointed out that the representations of a set S considered in this chapter are quite sensitive to noise. For example, if we make a tiny hole in S, its various types of maximal-block representations may change substantially (see Fig. 5), and it also acquires a new border. To minimize this problem, one can noise-clean the picture (Section 6.4) before segmenting it; one can noise-clean S by a shrinking and expanding process (see Section 11.3.2d); or one can use approximate representations of S, which should be less sensitive to the presence of noise.

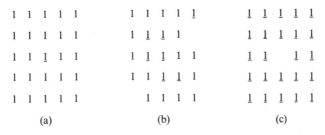

Fig. 5   Effects of noise on the MAT representation: A simple object (a) and two noisy versions (b) and (c), with centers of MAT blocks underlined.

a. Arrays

The simplest method of approximation is simple resolution reduction. We can reduce the resolution of a picture by resampling it coarsely; this is equivalent to digital demagnification (see Section 9.3 on geometric transformations of digital pictures). We can either demagnify Σ, obtaining Σ' (say), and then segment Σ' the same way that we segmented Σ; or we can demagnify the segmented Σ (i.e., χ_S). Note that in Σ', each gray level is (in general) a weighted average of a block of gray levels of Σ; thus if we demagnify χ_S, the result may no longer be binary, but we can make it binary again by thresholding. If desired, we can remagnify to make the simplified picture the same size as the original one for display purposes.

On the advantages of using reduced-resolution pictures to obtain preliminary information about which parts of Σ to analyze, see Sections 9.4.4 and 10.2.3a. A "pyramid" of successive reductions, e.g., each half the size of the preceding (n by n, $n/2$ by $n/2$, $n/4$ by $n/4$, ...) provides a wide range of resolutions, at least one of which should be approximately right for any desired purpose. The total storage space required for this pyramid is less than $1\frac{1}{3}$ times that required for Σ alone $(1 + \frac{1}{4} + (\frac{1}{4})^2 + \cdots = 1\frac{1}{3})$.

Other types of picture approximation, not involving reduced resolution, were discussed in Section 10.4.3 in connection with picture partitioning.

b. Blocks

When S is represented by maximal blocks (or runs, etc.) we can simplify it by eliminating some of the blocks. For example, if we eliminate small blocks from the MAT, we retain the gross features of S and lose only some of the details (see Fig. 6). A related method of simplifying S by shrinking or expanding it will be discussed in Section 11.3.2d. Eliminating low-level nodes from a quadtree representation may result in substantial simplification, since it may lead to the elimination of nodes at higher levels.

Another possibility is to use a representation based on blocks whose values are only approximately constant (e.g., have variances less than some threshold). To generalize the MAT [1], we consider the set of squares centered at each point P, and let S_P be such that all the smaller squares have below-threshold variances, while the next larger square does not. Maximal

```
1   1   1              1   1   1
1   1   1   1          1   1   1
1   1   1   1          1   1   1
        (a)                    (b)
```

Fig. 6 Simplification by elimination of small MAT blocks. (a) Object, with centers of MAT blocks underlined. (b) Simplified object resulting from ignoring blocks having radius 0.

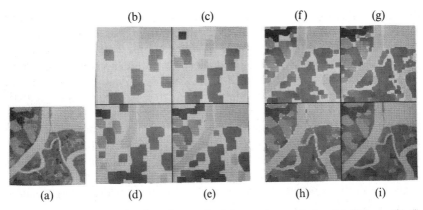

Fig. 7 Picture approximations by maximal low-variance blocks. (a) Picture; (b)–(i)
blocks of radii 7, 7 and 6, 7 and 6 and 5, . . . , 7 and 6 and ⋯ and 0, displayed with their mean
gray levels. When blocks having different radii overlap, the mean gray level of the smaller block
is used; when blocks of the same radius overlap, the maximum of their gray levels is used.

blocks defined in this way are illustrated in Fig. 7. The other maximal-block
representations (runs, trees) can be generalized analogously.

If S is not too noisy, the points of its MAT (in brief: its *medial axis*, MA)
will tend to lie along a set of arcs or curves. Thus we can represent S, at least
approximately, by specifying these curves (e.g., by chain codes) and defining
a "radius function" along each curve [7]. This defines a set of "generalized
ribbons" (i.e., arbitrarily curved ribbons of varying width) whose union is
S. Unfortunately, it is not obvious how to find a simple set of curves that
contain the MA; see Section 11.2.1a.

c.   Borders and curves

Borders can be approximated piecewise linearly, i.e., by polygons (or,
analogously, using pieces of higher-order curves) in various ways. Some
simple methods of constructing and refining polygonal approximations to a
border or curve are briefly described in Section 11.3.3c.

Given a set of successive approximations to a border, these approximations
define a tree structure in which each node represents a polygon side, and the
sons of a node are its immediate refinements; the sides of the coarsest polygon
are the sons of the root node. If desired, we can associate with each polygon
side a rectangular strip that just contains the arc; these strips define the zone
in which the border might lie [5].

A *line drawing* is a picture that can be segmented into a set of everywhere
elongated subsets and a background. (This is a rather imprecise definition;
elongatedness will be defined more precisely in Section 11.3.4b.) In this case

the subsets themselves can be approximated by unions of arcs and curves; this is a special case of the "generalized ribbon" representation in which the radius function is constant (or piecewise constant, if the lines have different thicknesses). "Thinning" processes that reduce elongated regions to unions of arcs and curves will be described in Section 11.2.3. These unions can then be further separated into individual arcs and curves by breaking them apart at junctions, and the individual curves can be represented by chain codes. (Here crack codes are not appropriate, since we are representing a one pixel thick curve, not a region border; but chain codes are applicable, since successive points on a curve are neighbors.) Alternatively, the curves can be approximated by polygons (etc.), as above; or one can use generalized chain codes based on moves from neighbor to neighbor in which a larger neighborhood is allowed.

11.1.6   Representation of Three-Dimensional Objects

In analyzing the three-dimensional structure of a scene from one or more pictures, it is often necessary to represent how the scene is divided into objects of various types and empty space. In principle, this could be done using a 3-d array in which the values represent the types of objects; for example, in a binary array 1's could represent points occupied by objects and 0's could represent unoccupied points. However, except for very small sizes, 3-d arrays require a prohibitively large amount of storage; even a $64 \times 64 \times 64$ binary array requires $2^{18} > 250,000$ bits. Thus compact representations and approximations become especially important in the 3-d case.

One approach to 3-d representation is to regard the 3-d array as a sequence of 2-d arrays ("slices" or "serial sections"), just as in the two-dimensional case we regarded a picture as a sequence of rows. Any of the representation schemes described in this chapter can then be used to represent the individual slices.

Another possibility is to use 3-d blocks. The 3-d analog of the MAT is based on maximal upright cubes (or blocks of some other standard shape) of constant value centered at each point. The analog of the quadtree is the "octree," obtained by recursive subdivision of the 3-d array (which we assume to be $2^k \times 2^k \times 2^k$) into octants until octants of constant value are reached.

A class of approximate representations known as *generalized cones* (or *cylinders*) has been widely used to represent three-dimensional objects [42]. A generalized cone is defined by an axis, a cross-section shape, and a size function; it is the volume swept out as the shape moves along the axis (perpendicular to it, or at some other fixed angle, and with the axis passing through it at a specified reference point), changing size from position to position as specified by the size function. (Compare the remarks on "generalized ribbons" in Section 11.1.5b.)

The border of a 3-d region consists of a set of surfaces. The slope of a surface at a point is defined by a pair of angles that specify how the normal to the surface at that point is oriented. Thus the analog of chain or crack code, in the case of a surface, would be an array of quantized values representing the pair of slopes for each unit patch of the surface. On 3-d chain codes for space curves see [19].

Exercise 3. Generalize as many as possible of the techniques in this chapter to the 3-d case. ∎

Various types of three-dimensional information about a scene can be represented in the form of two-dimensional arrays. For example, given a picture of a scene, the range from the observer to each visible point, and the surface slope at each visible point, can be represented as arrays in register with the picture. Note that the slope array is an array of vectors; Marr has called this array the "$2\frac{1}{2}$-d sketch." It should be pointed out that the three-dimensional resolution of such an array varies from point to point of the scene; as the slope of the surface becomes very oblique, the size of the surface patch that maps into a unit area on the picture becomes very large, so that the range and slope information at that point of the array become unreliable.

On the extraction of three-dimensional information about a scene from pictures of the scene, e.g., slope from shading or perspective, and relative range from occlusion cues, see the Appendix to Chapter 12.

11.1.7 Digital Geometry

This section introduces some of the basic geometric properties of subsets of a digital picture. Earlier we mentioned the concept of connectedness for such subsets; we now define this concept more precisely. We also define the borders of a subset, and indicate why they can be regarded as closed curves. Some other concepts of digital geometry, including some useful digital distance functions such as "city block" and "chessboard" distance, will also be introduced.

The results presented in this section are not all self-evident; many of them require rather lengthy proofs. We will not give such proofs here; for a mathematical introduction to the subject see [55].

A point $P = (x, y)$ of a digital picture Σ has four horizontal and vertical neighbors, namely the points

$$(x - 1, y), \qquad (x, y - 1), \qquad (x, y + 1), \qquad (x + 1, y)$$

We will call these points the 4-*neighbors* of P, and say that they are 4-*adjacent* to P. In addition, P has four diagonal neighbors, namely

$$(x - 1, y - 1), \qquad (x - 1, y + 1), \qquad (x + 1, y - 1), \qquad (x + 1, y + 1)$$

These, together with the 4-neighbors, are called 8-*neighbors* of P (8-*adjacent* to P). Note that if P is on the border of Σ, some of its neighbors may not exist. If S and T are subsets of Σ, we say that S is 4- (8-) adjacent to T if some point of S is 4- (8-) adjacent to some point of T.

Exercise 4.   Define the 6-*neighbors* of (x, y) as the 4-neighbors together with

$$(x - 1, y - 1) \quad \text{and} \quad (x - 1, y + 1), \qquad \text{if } y \text{ is odd}$$
$$(x + 1, y - 1) \quad \text{and} \quad (x + 1, y + 1), \qquad \text{if } y \text{ is even}$$

These can be regarded as the neighbors of (x, y) in a "hexagonal" array constructed from a square array by shifting the even-numbered rows half a unit to the right. Develop "hexagonal" versions of all the concepts in this section. ∎

a.   Connectedness

A *path* π of length n from P to Q in Σ is a sequence of points $P = P_0$, $P_1, \ldots, P_n = Q$ such that P_i is a neighbor of $P_{i-1}, 1 \leqslant i \leqslant n$. *Note that there are two versions of this and the following definitions, depending on whether "neighbor" means "4-neighbor" or "8-neighbor."* Thus we can speak of π being a 4-*path* or an 8-*path*.

Let S be a subset of Σ, and let P, Q be points of S. We say that P is (4- or 8-) *connected* to Q in S if there exists a (4- or 8-) path from P to Q consisting entirely of points of S. For any P in S, the set of points that are connected to P in S is called a connected *component* of S. If S has only one component, it is called a connected set. For example, if we denote the points of S by 1's, the set

1

1

is 8-connected but not 4-connected.

It is easily seen that "is connected to" is reflexive, symmetric, and transitive, and so is an equivalence relation, i.e., for all points P, Q, R of S:

(a)   P is connected to P (hint: a path π can have length $n = 0$);
(b)   if P is connected to Q, then Q is connected to P;
(c)   if P is connected to P and Q to R, then P is connected to R.

Thus two points are connected to each other in S iff they belong to the same component of S.

b. Holes and surroundness

Let $\bar{S}$ be the complement of S. We will assume, for simplicity, that the border Σ' of Σ (i.e., its top and bottom rows and its left and right columns) is in $\bar{S}$. The component of $\bar{S}$ that contains Σ' is called the *background* of S. All other components of $\bar{S}$, if any, are called *holes* in S. If S is connected and has no holes, it is called *simply connected*; if it is connected but has holes, it is called *multiply connected*.

In dealing with connectedness in both S and $\bar{S}$, it turns out to be desirable to use opposite types of connectedness for S and $\bar{S}$, i.e., if we use 4- for S, then we should use 8- for $\bar{S}$, and vice versa. This will allow us to treat borders as closed curves; see Subsection (c). We will adopt this convention from now on.

Let S and T be any subsets of Σ. We say that T *surrounds* S if any path from any point of S to the border of Σ must meet T, i.e., if for any path P_0, $P_1, \ldots, P_n$ such that P_0 is in S and P_n is on the picture border, some P_i must be in T. Evidently the background of S surrounds S. On the other hand, S surrounds any hole in S; if it did not, there would be a path from the hole to the picture border that did not meet S, contradicting the fact that the hole and the border are in different components of $\bar{S}$. The type of path here (4- or 8-) must be the same as the type of connectedness used for $\bar{S}$.

Exercise 5.

(a) Prove that any S surrounds itself, and that if W surrounds V and V surrounds U, then W surrounds U.

(b) Can S and T surround each other without being the same? (Hint: Can a proper subset of S surround S?)

(c) Prove that if S and T are disjoint, then only one of them can surround the other. (Thus for disjoint sets, "surrounds" is a strict partial order relation, i.e., is irreflexive, antisymmetric, and transitive.) ∎

c. Borders

As indicated in Section 11.1.3, the *border* S' of a set S is the set of points of S that are adjacent to $\bar{S}$. We will assume, for simplicity, that "adjacent" here means "4-adjacent." The set of nonborder points of S is called the *interior* of S.

Let C be a component of S, and let D be a component of $\bar{S}$ that is adjacent to C. Note that if C and D are 8-adjacent, they are also 4-adjacent; indeed, consider the pattern

$$P \quad X$$
$$Y \quad Q$$

where P is in C and Q in D. If X is in S, it is in C, and if it is in $\bar{S}$, it is in D, and similarly for Y; thus in any case C and D are 4-adjacent.

The set C_D of points of C that are adjacent to D is called the D-*border* of C. Similarly, the set D_C of points of D that are adjacent to C is called the C-*border* of D. It is these component borders that can be regarded as closed curves, and for which we can define border following algorithms, as we shall see in Section 11.2.2.

It can be shown that if C is adjacent to several components of $\bar{S}$, then exactly one of those components, say D_0, surrounds C, and the others are surrounded by C. The D_0-border of C is called its *outer border*, and the other D-borders of C, if any, are called *hole borders*. This surroundness property is what allows us to treat the borders as closed curves. It is true ónly if we use opposite types of connectedness for S and $\bar{S}$. For example, suppose that we use 4-connectedness for both S and $\bar{S}$, and that the points of S are

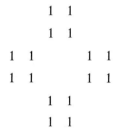

Then each block C of 1's is a 4-component of S, and the block D of 0's that they surround is a 4-component of $\bar{S}$, but C_D and D_C are not closed curves, and D does not surround the C's nor does any one of them surround it. Similarly, if we use 8-connectedness for both S and $\bar{S}$, then S and $\bar{S}$ each consists of a single cómponent but $S_{\bar{S}}$ and $\bar{S}_S$ are not closed curves. On the other hand, if we use 4-connectedness for S and 8- for $\bar{S}$, then each block C of S is a component, all of $\bar{S}$ is a single component, and each $C_{\bar{S}}$ and $\bar{S}_C$ is a closed curve. Similarly, if we use 8- for S and 4- for $\bar{S}$, then S consists of a single component, $\bar{S}$ has the surrounded block D of 0's as a component, and S_D and D_S are closed curves.

Exercise 6. Show that when we use 8-connectedness for S, the outer border and a hole border of S can be the same. Can two borders of a 4-component be the same? Can one be a subset of another? On how many different borders of C can a given point of C lie? ∎

Alternatively, we can define the (C, D)-*border* (of C or D) as the set of pairs (P, Q) such that P is in C, Q is in D, and P, Q are 4-adjacent. This is the same as the set of "cracks" between the points of C and the adjacent points of D. It is evident that any two such borders must be disjoint.

The surroundness and closed curve properties hold only when C and D are components of a set S and its complement, respectively. If we take C and D to be arbitrary connected subsets of Σ, we can say nothing about the D-border of C; C and D can touch in many distinct places, neither of them need surround the other, and the border need not be anything like a closed curve.

d. Distance

The *Euclidean distance* between two points $P = (x, y)$ and $Q = (u, v)$ is

$$d_e(P, Q) = \sqrt{(x - u)^2 + (y - v)^2}$$

It is sometimes convenient to work with simpler "distance" measures on digital pictures. In particular, the *city block distance* between P and Q is defined as

$$d_4(P, Q) = |x - u| + |y - v|$$

and the *chessboard distance* between them is

$$d_8(P, Q) = \max(|x - u|, |y - v|)$$

It can be verified that all three of these measures, d_e, d_4, and d_8, are *metrics*, i.e., for all P, Q, R we have

$$d(P, Q) \geqslant 0, \qquad \text{and} = 0 \qquad \text{iff} \quad P = Q$$
$$d(P, Q) = d(Q, P)$$
$$d(P, R) \leqslant d(P, Q) + d(Q, R)$$

In other words, each of these d's is *positive definite, symmetric,* and satisfies the *triangle inequality*.

It is not hard to see that the points at city block distance $\leqslant t$ from P form a diamond (i.e., a diagonally oriented square) centered at P. For example, if we represent points by their distances from P (so that P is represented by 0), the points at distances $\leqslant 2$ are

$$
\begin{array}{ccccc}
 & & 2 & & \\
 & 2 & 1 & 2 & \\
2 & 1 & 0 & 1 & 2 \\
 & 2 & 1 & 2 & \\
 & & 2 & & \\
\end{array}
$$

In particular, the points at distance 1 are just the 4-neighbors of P. It can be shown that $d_4(P, Q)$ is equal to the length of a shortest 4-path from P to Q. (Note that there may be many such paths!)

Analogously, the points at chessboard distance $\leqslant t$ from p form an upright square centered at P; e.g., the points at distances $\leqslant 2$ are

$$
\begin{array}{ccccc}
2 & 2 & 2 & 2 & 2 \\
2 & 1 & 1 & 1 & 2 \\
2 & 1 & 0 & 1 & 2 \\
2 & 1 & 1 & 1 & 2 \\
2 & 2 & 2 & 2 & 2
\end{array}
$$

so that the points at distance 1 are the 8-neighbors of P. In general, $d_8(p, q)$ is the length of a shortest 8-path from P to Q, and there are many such paths.

Exercise 7. d_4 and d_8 are integer-valued, but d_e is not. Is $\lfloor d_e \rfloor$ (the greatest integer $\leqslant d_e$) a metric? What about the closest integer to d_e? ▮

The distance between a point P and a set S is defined to be the shortest distance between P and any point of S. The *diameter* of a set is the greatest distance between any two of its points.

11.2 CONVERSION BETWEEN REPRESENTATIONS

Some of the conversions between one representation and another are quite straightforward. For example, it is obvious how to convert from arrays to runs: scan the array row by row; increment a counter as long as the value remains constant; when the value changes, append the (old) value and count to the run list and reset the counter. Conversely, to convert from runs to arrays, we create the array row by row, putting in the current value and decrementing the current run length until it reaches zero, at which point we switch to the next value and length on the run list.

In this section we give conversion algorithms involving blocks (MAT, quadtree) and border representations. We also describe algorithms for thinning line drawings into sets of arcs and curves. As before, we will usually deal only with the case where the given picture Σ is partitioned into a set S and its complement.

Conventionally, the algorithms described here would be implemented on an ordinary (single-processor) digital computer, which would process the given picture row by row. On the other hand, some of the algorithms are very well suited for implementation on a cellular array computer, in which a processor is assigned to each pixel, so that local operations can be performed at all the pixels simultaneously. Cellular computers of reasonable size (100 by

100 processors or more) currently exist or are being built, and it is expected that they will play a major role in picture processing in future years.

11.2.1 Blocks

a. MATs

In Section 11.1.2a we defined the MAT in terms of the set of maximal constant-value blocks (= upright squares of odd side length) in Σ. When Σ is partitioned into S and $\bar{S}$, we can speak of the MAT of S (defined by the maximal blocks that are contained in S) and the MAT of $\bar{S}$.

Let us first observe that the MA of S (= the set of centers of the MAT blocks) consists of those points of S whose chessboard (= d_8) distances from $\bar{S}$ are local maxima. Recall that the points within a given chessboard distance of P form an upright square centered at P. If $d_8(P, \bar{S}) = d$, then the square $S_P^{(d-1)}$ of radius $d - 1$ centered at P cannot intersect $\bar{S}$ (since there would then be a point of $\bar{S}$ at distance $\leqslant d - 1$ from P), but the square of radius d does intersect $\bar{S}$. Suppose $d_8(P, \bar{S})$ is not a local maximum; then P has a neighbor Q such that $d_8(Q, \bar{S}) \geqslant d + 1$. Thus a square S_Q of radius $\geqslant d$ centered at Q does not intersect $\bar{S}$; but S_Q contains $S_P^{(d-1)}$, so that $S_P^{(d-1)}$ is not a maximal block. Conversely, if $S_P^{(d-1)}$ is not maximal, it is not hard to see that it must be contained in a square S_Q of constant value centered at some neighbor of P, and S_Q has radius at least d; thus $d_8(Q, \bar{S}) \geqslant d$, so that $d_8(P, \bar{S})$ is not maximal.

Analogously, if we define the MAT using diagonally oriented rather than upright squares, we can show that the MA of S consists of those points of S whose city block (d_4) distances from $\bar{S}$ are local maxima. In this case "local maximum" means that no 4-neighbor of the point has greater distance from $\bar{S}$, whereas in the previous case the MA points were 8-neighbor distance maxima. Figure 8 shows the sets of city block and chessboard distances to $\bar{S}$ when S is a rectangle or a diamond; the local maxima in each case are underlined.

These remarks imply that we can construct the MAT of S by computing the distances to $\bar{S}$ from all points of S and discarding nonmaxima. In the next two subsections we will present some algorithms for computing these distances from a given binary array or row-by-row representation.

Since the MA is a set of local distance maxima, it is generally quite disconnected; indeed, two maxima cannot be adjacent unless their values are equal. We can attempt to make the MA connected by keeping some of the nonmaxima. For example, in the d_8 case, we might keep P if at most one of its 8-neighbors has a larger distance to $\bar{S}$, and if it has a 4-neighbor that is a distance maximum. [The first condition implies that the layer of points at

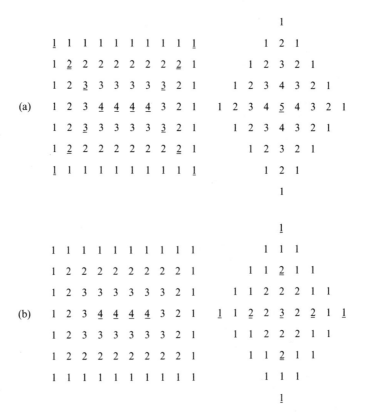

Fig. 8 Distances to $\bar{S}$ for the points of a rectangle and a diamond; the MA points are underlined. (a) d_4; (b) d_8. Note that the d_4 and d_8 values for the rectangle are the same. The MAT radii
are 1 less than the distances.

distance $d_8(P)$ from $\bar{S}$ turns sharply at P.] Of course, when we discard fewer
points, we have a less economical representation of S. However, if we can
find a set of arcs and curves containing the MA, we can represent these curves
by chain codes, rather than having to list all the MA points individually, so
that the economy of the representation may be improved.

b. Distance computation

We first give a cellular array computer algorithm for distance computation.
This algorithm works for any metric d that is integer-valued and has the
following property:

For all P, Q such that $d(P, Q) \geqslant 2$, there exists a point R, different from P
and Q, such that $d(P, Q) = d(P, R) + d(R, Q)$.

Such metrics will be called *regular*. It can be shown that d is regular if and only if, for all distinct P, Q, there exists a point R such that $d(P, R) = 1$ and $d(P, Q) = d(R, Q) + 1$.[§]

Given χ_S, which is 1 at the points of S and 0 elsewhere, we define $\chi_S^{(m)}$ inductively for $m = 1, 2, \ldots$ as follows:

$$\chi_S^{(m)}(P) = \chi_S^{(0)}(P) + \min_{d(Q, P) \leqslant 1} \chi_S^{(m-1)}(Q) \tag{1}$$

where $\chi_S^{(0)} = \chi_S$. Thus $\chi_S^{(m)}$ can be computed by performing a local operation on the pair of arrays $\chi_S^{(0)}$ and $\chi_S^{(m-1)}$ at every point.

To see how (1) works, note first that 0's (i.e., points of $\bar{S}$) remain 0's, since if $\chi_S^{(0)}(P) = 0$ we have

$$\min_{d(Q, P) \leqslant 1} \chi_S^{(m-1)}(Q) = \chi_S^{(m-1)}(P) = 0 \qquad \text{for} \quad m = 1, 2, \ldots$$

Similarly, 1's at distance 1 from $\bar{S}$ (i.e., border 1's) remain 1's, since for such points too, the min is 0. On the other hand, on the first iteration of (1), all 1's at distance > 1 from $\bar{S}$ (i.e., interior 1's) become 2's, since for such points the min is 1. On the second iteration, all interior 2's ($=$ all 2's at distance > 2 from $\bar{S}$) become 3's, since the min for such points is 2. On the third iteration, all interior 3's become 4's, and so on. Thus if $d(P, \bar{S}) = k > 0$, the values of $\chi_S^{(m)}(P)$ for $m = 1, 2, \ldots, k$ are $1, 2, \ldots, k$, respectively, and the value remains k at all subsequent iterations. It follows that if m is the greatest distance between $\bar{S}$ and any point in the picture, we have $\chi_S^{(m)}(P) = d(P, \bar{S})$ for all P. It certainly suffices to iterate (1) a number of times equal to the picture's diameter; for an $n \times n$ picture, this is $2(n - 1)$ for d_4 and $n - 1$ for d_8.

If we want to find a shortest path from each P to $\bar{S}$, then at the iteration when $\chi_S^{(m)}(P)$ stops increasing, we create a pointer from P to one of its neighbors which has the minimum value in (1). This neighbor must be on a shortest path from P to $\bar{S}$, so that if we follow the pointers starting at P, we must move along a shortest path to $\bar{S}$ [61]. Note, incidentally, that P is an MA point iff it does not lie on a shortest path from any other point to S to $\bar{S}$. Indeed, if P were on a shortest path from Q to $\bar{S}$, its predecessor on this path would be a neighbor of P and would have higher distance to $\bar{S}$ than P.

Exercise 8. For a multivalued picture, define the *gray-weighted distance* [61] between P and Q as the smallest possible sum of values along any path from P to Q. (Here a "path" means a sequence of points such that the distance between consecutive points is 1.) Prove that if we apply (1) to a multivalued picture that contains 0's, the value at any point P eventually becomes equal

---

[§] Regularity is a necessary and sufficient condition for a metric to be the distance on a graph; see F. Harary, *Graph Theory*, Addison-Wesley, Reading, Massachusetts, 1969, p. 24, Exercise 2.8.

to the gray-weighted distance between P and the set of 0's. Note that if there are only two values, 0 and 1, the gray-weighted distance to the set of 0's is the same as the ordinary distance. We can also define a gray-weighted MA [33] as the set of points whose gray-weighted distance to the 0's is a local maximum, or equivalently, which do not lie on a minimum-sum path from any other point to the 0's. Is a multivalued picture reconstructible from its gray-weighted MAT? ▮

Metrics d_4 and d_8 are quite non-Euclidean; the set of points at distance $\leqslant k$ from a given point is a square, rather than a circle, for these metrics. We can compute a distance in which this set is an octagon by using d_4 and d_8 at alternate iterations in (1). The points at "octagonal distance" $\leqslant 2$ are

$$
\begin{array}{ccccc}
  &   & 2 & 2 & 2 &   &   \\
  & 2 & 2 & 1 & 2 & 2 &   \\
  & 2 & 1 & 0 & 1 & 2 &   \\
  & 2 & 2 & 1 & 2 & 2 &   \\
  &   & 2 & 2 & 2 &   &   \\
\end{array}
$$

To get the MAT for this distance, we can use 4-neighbor maxima when the distance is odd, and 8-neighbor maxima when it is even, to insure that the argument given in Subsection (a) remains valid.

Algorithm (1) is quite efficient on a cellular array computer, since each iteration can be performed at all points in parallel, and the number of iterations is at most the diameter of the picture. On a conventional computer, however, each iteration requires a number of steps proportional to the area of the picture. We now present an algorithm that computes all the distances (d_4 or d_8) to $\bar{S}$ in only two scans of the picture, so that a large number of iterations is not required. We assume that the border of the picture consists entirely of 0's. Let $N_1(P)$ be the set of (4- or 8-) neighbors that precede P in a row-by-row (left to right, top to bottom) scan of the picture, and let $N_2(P)$ be the remaining (4- or 8-) neighbors of P; i.e., $N_1(x, y)$ consists of the 4-neighbors $(x - 1, y)$ and $(x, y + 1)$, as well as the 8-neighbors $(x - 1, y + 1)$ and $(x + 1, y + 1)$. Then the algorithm is as follows:

$$
\chi_S'(P) = \begin{cases} 0 & \text{if } P \in \bar{S} \\ \min_{Q \in N_1} \chi_S'(Q) + 1 & \text{if } P \notin \bar{S} \end{cases}
$$

$$
\chi_S''(P) = \min_{Q \in N_2} [\chi_S'(P), \chi_S''(Q) + 1] \tag{2}
$$

Thus we can compute χ_S' in a single left to right, top to bottom scan of the picture, since for each P, χ_S' has already been computed for the Q's in N_1.

Similarly, we can compute χ_S'' in a single *reverse* scan (right to left, bottom to top). Then for all P we have $\chi_S''(P) = d_4(P, \bar{S})$ or $d_8(P, \bar{S})$, depending on whether we use 4- or 8-neighbors in the algorithm. As a very simple example, let χ_S be

$$
\begin{array}{cccc}
0 & 1 & 1 & 1 \\
1 & 1 & 1 & 0 \\
1 & 1 & 1 & 1
\end{array}
$$

surrounded by 0's. Then χ_S' in the 4-neighbor case is

$$
\begin{array}{cccc}
0 & 1 & 1 & 1 \\
1 & 2 & 2 & 0 \\
1 & 2 & 3 & 1
\end{array}
$$

and χ_S'' is

$$
\begin{array}{cccc}
0 & 1 & 1 & 1 \\
1 & 2 & 1 & 0 \\
1 & 1 & 1 & 1
\end{array}
$$

c. Shrinking and expanding

Any S can be "shrunk" by repeatedly deleting its border, and "expanded" by repeatedly adding to it the border of its complement $\bar{S}$. In general, by the border of S we mean the set of points that are at distance 1 from $\bar{S}$. Let S' be the border of S in this general sense; then the successive stages in shrinking S are $S^{(0)} \equiv S$, $S^{(-1)} \equiv S - S'$, $S^{(-2)} \equiv S^{(-1)} - S^{(-1)'}$, and so on. Similarly, the successive stages in expanding S are $S^{(1)} \equiv S \cup (\bar{S})'$; $S^{(2)} \equiv S^{(1)} \cup (\overline{S^{(1)}})'$; and so on. It is easily seen that for any k we have $S^{(k)} = \overline{(\bar{S})^{(-k)}}$, or equivalently $S^{(-k)} = \overline{(\bar{S})^{(k)}}$. Given χ_S, we get $\chi_{S^{(1)}}$ by changing 0's to 1's if they have any 1's as neighbors (i.e., at distance 1), and we get $\chi_{S^{(-1)}}$ by changing 1's to 0's if they have any 0's as neighbors.

It should be pointed out that shrinking and expanding do not commute with one another; $(S^{(m)})^{(-n)}$ is not necessarily the same as $(S^{(-n)})^{(m)}$, and neither of them is the same as $S^{(m-n)}$. For example, if S consists of a single point P, then $S^{(1)}$ consists of P and its neighbors, and $(S^{(1)})^{(-1)} = S = \{P\}$; but $S^{(-1)}$ is empty, and so is $(S^{-1})^{(1)}$. However, it can be shown that $(S^{(-n)})^{(m)} \subseteq S^{(m-n)} \subseteq (S^{(m)})^{(-n)}$. In particular, we have $(S^{(-k)})^{(k)} \subseteq S \subseteq (S^{(k)})^{(-k)}$ for all k.

In Sections 11.3.2d and 11.3.4b we will show how combinations of shrinking and expanding can be used to define and detect elongated parts, isolated parts, and clusters of parts of a set S and remove "noise" from S. In this section we are primarily concerned with how shrinking and expanding are related to distance and to the MAT. Since S' is the set of points of S that are at distance 1 from $\bar{S}$, readily $S^{(t)}$ is the set of points at distance $\leqslant t$ from S, and $S^{(-t)}$ is the set of points at distance $> t$ from $\bar{S}$. Thus $\chi_S^{(m)} = \chi_S + \chi_{S^{(-1)}} + \cdots + \chi_{S^{(-m)}}$, which gives us an algorithm for computing the distances to $\bar{S}$ by repeated shrinking and adding.

To obtain the MA of S by shrinking and expanding, we proceed as follows: Let $S_k = S^{(-k)} - (S^{(-k-1)})^{(1)}$ (note that the first of these sets contains the second). Any point P of S_k is at distance exactly $k + 1$ from $\bar{S}$, since if it were farther away it would be in $S^{(-k-1)}$, hence in $(S^{(-k-1)})^{(1)}$. Furthermore, P has no neighbors whose distances from $\bar{S}$ are greater than $k + 1$, since any such neighbor would be in $S^{(-k-1)}$, so that P would be in $(S^{(-k-1)})^{(1)}$. Thus P's distance from $\bar{S}$ is a local maximum, making it an MA point; in fact, S_k is just the set of MA points that are at distance $k + 1$ from $\bar{S}$.

Shrinking and expanding can make use of arbitrary neighborhoods [70]. Let $N(P)$ denote the "neighborhood" of P; this could consist of P together with any set of points having given positions relative to P. Then we can define the N-expansion of S as $\bigcup_{P \in S} N(P)$, and the N-contraction of S as $\{P \in S \mid N(P) \cap \bar{S} = \varnothing\}$. Readily, if neighborhoods are symmetric [i.e., $Q \in N(P)$ iff $P \in N(Q)$], these operations are complementary, i.e. the N-contraction of S is the complement of the N-expansion of $\bar{S}$, and vice versa. These operations can be iterated or combined in various ways, as before. Note that if $N(P)$ consists of the points at distance $\leqslant 1$ from P, these definitions reduce to the ones given earlier.

The analogs of shrinking and expanding for multivalued pictures [41] are local min and local max operations defined over a given neighborhood. Note that if there are only two values, 0 and 1, local min shrinks the 1's, while local max expands them. Just as shrinking and expanding can be used to delete noise from a segmented (two-valued) picture, so local min and max operations can be used to smooth multivalued pictures; compare Section 6.4.4 and Section 11.3.2d.

A single step of shrinking or expanding can be done in a single parallel operation on a cellular array computer, but requires a complete scan of the picture on a conventional computer. On each row, points are marked for changing (from 1 to 0 or from 0 to 1), but they are not actually changed until the next row has been marked. If we want to do a sequence of k steps without having to scan the entire picture k times, we can operate on the rows in a sequence such as $1; 2, 1; 3, 2, 1; 4, 3, 2, 1; \ldots$ (Once we have operated on the second row, we can do a second operation on the first row; once we have operated on the third row, we can do a second operation on the second row, and then a third operation on the first row; and so on.) Once we have reached the kth row, and done the sequence $k, (k - 1), \ldots, 2, 1$, we are finished with the first row, since it has been processed k times. The next sequence is $(k + 1)$, $k, \ldots, 3, 2$, after which we are finished with the second row; the first row can now be dropped. The next sequence is $(k + 2), (k + 1), \ldots, 4, 3$, after which the second row can be dropped; and so on. Thus only $k + 2$ rows (those currently being processed, plus one before and one after them) need to be available at a time.

d.   Reconstruction from MATs

Given the MAT of S, we can construct χ_S by creating solid squares of 1's having the specified centers and radii. Alternatively, we can represent the MAT by a picture Σ_S whose values at the distance maxima are equal to their distances, and consisting of 0's elsewhere. We can then reconstruct the entire set of distances to $\bar{S}$ by applying an iterative algorithm to Σ_S [compare (1)]:

$$\Sigma_S^{(m)}(P) = \max\left[0, \ \max_{d(P,Q)=1} \Sigma_S^{(m-1)}(Q) - 1\right] \quad \text{provided} \quad \Sigma_S^{(m-1)}(P) = 0$$
$$= \Sigma_S^{(m-1)}(P) \qquad\qquad\qquad\qquad \text{otherwise} \qquad\qquad (3)$$

The number of iterations required is one less than the value of the largest distance maximum. As an example, if Σ_S is

```
    1               1

      2           2

          3   3   3

      2           2

    1               1
```

(see Fig. 8a), where 0's are represented by blanks, then two iterations of (3) using d_4 give

```
1   1           1   1           1   1   1   1   1   1   1

1   2   2   2   2   2   1        1   2   2   2   2   2   1

    2   3   3   3   2      and   1   2   3   3   3   2   1

1   2   2   2   2   2   1        1   2   2   2   2   2   1

1   1           1   1           1   1   1   1   1   1   1
```

as in Fig. 8a. Alternatively, we can reconstruct the distances using the following two-pass algorithm [compare (2)]:

$$\Sigma_S'(P) = \max_{Q \in N_1}(\Sigma_S(P), \Sigma_S'(Q) - 1)$$
$$\Sigma_S''(P) = \max_{Q \in N_2}(\Sigma_S'(P), \Sigma_S''(Q) - 1) \qquad\qquad (4)$$

Here we first compute Σ' in a left to right, top to bottom scan, and then compute Σ'' in a right to left, bottom to top scan. In the example given above, using d_4, Σ_S' and Σ_S'' are

1	0	0	0	0	0	1		1	1	1	1	1	1	1
0	2	1	0	0	2	1		1	2	2	2	2	2	1
0	1	3	3	3	2	1	and	1	2	3	3	3	2	1
0	2	2	2	2	2	1		1	2	2	2	2	2	1
1	1	1	1	1	1	1		1	1	1	1	1	1	1

respectively. It can be shown that the MAT is the smallest set of distance values from which all of the distance values can be reconstructed using (3) or (4). S itself is then just the set of points that have nonzero values after (3) or (4) has been applied.

e.   Quadtrees

In this subsection we briefly describe methods of converting between the quadtree and binary picture representation of a given S. The details of the algorithms can be found in three papers by Samet [63,65,67].

In Section 11.1.2b we described a method of constructing the quadtree corresponding to a given binary picture by recursively subdividing the picture into blocks that are quadrants, subquadrants, ... If a block consists entirely of 1's or 0's, it corresponds to a "black" or "white" leaf node in the tree; otherwise, it corresponds to a "gray" nonleaf node, which has four sons corresponding to its four quadrants. If we do this in a "top–down" fashion, i.e., first examine the entire picture, then its quadrants, then their quadrants, etc., as needed, it may require excessive computational effort, since parts of the picture that contain finely divided mixtures of 0's and 1's will be examined repeatedly.

As an alternative, we can build the quadtree "bottom up" by scanning the picture in a suitable order, e.g., in the sequence

1	2	5	6	17	18	21	22
3	4	7	8	19	20	23	24
9	10	13	14	25	26	29	30
11	12	15	16	27	28	31	32
33	⋯						

where the numbers indicate the order in which the points are examined. As we discover maximal blocks of 0's or of 1's, we add leaf nodes to the tree, together with their needed ancestor gray nodes. This can be done in such a way that leaf nodes are never created until they are known to be maximal, so that it is never necessary to merge four leaves of the same color and change their common parent node from gray to black or white. For the details of this algorithm see [65].

Bottom-up quadtree construction becomes somewhat more complicated if we want to scan the picture row by row. Here we add leaf nodes to the tree as we discover maximal 1×1 or 2×2 blocks of 0's or 1's; if four leaves with a common father all have the same color, they are merged. The details can be found in [67].

Given a quadtree, we can construct the corresponding binary picture by traversing the tree and, for each leaf, creating a block of 0's or 1's of the appropriate size in the appropriate position. A more complicated process can be used if we want to create the picture row by row. Here we must visit each quadtree node once for each row that intersects it (i.e., a node corresponding to a $2^k \times 2^k$ block is visited 2^k times), and, for each leaf, output a run of 0's or 1's of the appropriate length (2^k) in the appropriate position. For the details, see [63].

11.2.2 Borders

We recall that the border S' of a set S is the set of points of S that are 4-adjacent to $\bar{S}$. To find all the border points of S, we can scan χ_S and check the four neighbors of each 1 to see if any of them is 0 (or vice versa). On a cellular array computer, we can find the border by performing a Boolean operation at each point P of χ_S. Let P_n, P_s, P_e, P_w be the four neighbors of P; then we compute

$$P \wedge (\bar{P}_n \vee \bar{P}_s \vee \bar{P}_e \vee \bar{P}_w)$$

where the overbars denote logical negation ($\bar{1} = 0, \bar{0} = 1$). Evidently, this yields 1's at border points of S and 0's elsewhere.

Exercise 9. A point of S is called (4- or 8-) *isolated* if it has no neighbors in S. Define a Boolean operation on χ_S that detects 4- or 8- isolated points. ∎

In this section we describe algorithms for finding, following, and coding the individual borders between components of S and $\bar{S}$, and for reconstructing S from these border codes. These algorithms provide the means for converting between the bit plane and border code representations. Conversion between quadtrees and border codes will also be briefly discussed.

We will continue to assume here that S does not touch the edges of the picture. This will make it unnecessary for our algorithms to handle special cases involving neighbors that are outside the picture.

a. Crack following

Let C, D be components of S, $\bar{S}$, respectively, and let P, Q be 4-adjacent points of C and D, so that (P, Q) defines one of the cracks on the (C, D)-border. Let U, V be the pair of points that we are facing when we stand on the crack between P and Q with P on our left, with U, V 4-adjacent to P, Q respectively, i.e.,

$$\begin{matrix} P & U \\ Q & V \end{matrix}, \quad \begin{matrix} U & V \\ P & Q \end{matrix}, \quad \begin{matrix} V & Q \\ U & P \end{matrix}, \quad \text{or} \quad \begin{matrix} Q & P \\ V & U \end{matrix}$$

(Defining U, V in terms of P on the left implies that outer borders will be followed counterclockwise and hole borders clockwise; the reverse would be true if we had P on the right.) Then the following rules define the next crack (P', Q') along the (C, D)-border:

(1)   If we use 8-connectedness for C

U	V	P'	Q'	Turn
$-$	1	V	Q	right
1	0	U	V	none
0	0	P	U	left

(2)   If we use 4-connectedness for C

U	V	P'	Q'	Turn
0	$-$	P	U	left
1	0	U	V	none
1	1	V	Q	right

The algorithm stops when we come to the initial pair again.

To generate the crack code corresponding to this traversal of the border, note that the four initial configurations of P, Q, U, and V given above correspond to crack codes of 0, 1, 2, and 3, respectively. To determine the code for (P', Q') from that for (P, Q), we simply add 1 (modulo 4) if we turned left, subtract 1 (modulo 4) if we turned right, and use the same code if we made no turn. (Alternatively, the turning rules give us the difference crack code directly.)

b. Border following

The algorithms for following the D-border of C are somewhat more complicated. Let P, Q be as before, and let the eight neighbors of P, in counterclockwise order starting from Q, be $Q = R_1, R_2, \ldots, R_8$. (Using counterclockwise order implies that outer borders will be followed counterclockwise and hole borders clockwise.) We assume that C does not consist solely of the isolated point $\{P\}$. The following rules define a new pair P', Q':

(1)   If we use 8-connectedness for C:

Let R_i be the first of the R's that is 1 (such an R exists since P is not isolated). Then $P' = R_i, Q' = R_{i-1}$.

(2)   If we use 4-connectedness for C:

Let R_i be the first 4-*neighbor* that is 1 (i.e., the first of R_3, R_5, R_7). If $R_{i-1} = 0$, take $P' = R_i, Q' = R_{i-1}$; if $R_{i-1} = 1$, take $P' = R_{i-1}, Q' = R_{i-2}$.

The algorithm stops when we come to the initial P again, provided that we find the initial Q (as one of the R's) before finding the next P'. This last condition is necessary because a border can pass through a point twice, as we will see in the example immediately below.

As an example, let C, P, and Q be as in Fig. 9a. The R's and the new P, Q at the first two steps of the algorithm are shown in Figs. 9b and 9c. If we use 8-connectedness for C, the next three steps are shown in Figs. 9d–9f; if we

```
 Q  P  1            R  1  1              1  1
    1              R  Q  P           R  1
    1                 1              Q  P
       1                 1                 1
      (a)               (b)               (c)


    1  1              1  1              1  1
       1                 1              P  Q
 R  1              P  Q  R              1  R
 R  Q  P           R  1  R                 1
                   R  R  R
      (d)               (e)               (f)


                                     Q  R  R
    1  1           1  P  Q           P  1  R
       P  Q           1  R              1
 R  1  R              1                 1
 R  R  1 = R          1                 1
      (f')               (g)               (h)
```

Fig. 9 Border following, showing the initial P and Q in (a), and the R's and new P and Q at successive steps (b)–(h). For detailed explanation see text.

use 4-connectedness, the second step (Fig. 9c) is followed directly by Fig. 9f'. Note that P is the same in Figs. 9e and 9c, but the algorithm would not stop even if this had been the initial P, since the Q of Fig. 9c is not one of the R's encountered at the next step before finding a 1. Analogous remarks apply to Figs. 9b and 9f'. In both the 4- and 8-cases, the last two steps are as shown in Figs. 9g–9h. After step 9h the algorithm stops, since P is the same as the initial P, and the initial Q is one of the R's encountered at the next step before finding a 1. An equivalent stopping criterion would be as follows: The algorithm stops when it finds two successive P's that it had previously found in the same consecutive order.

Any two successive P's found by the algorithm are always 8-neighbors, and in the case where C is 4-connected, they are also 4-connected through 1's; thus all the P's belong to C. Any two successive Q's are always 8-connected through 0's (when C is 4-connected, they may not be 4-connected, since one of the intermediate R's may be a 1 diagonally adjacent to P), and when C is 8-connected, they are also 4-connected through 0's; thus all the Q's belong to D, so that the P's are all on the D-border of C. It is much harder to prove that they constitute the entire D-border; see [55] for a detailed treatment of this. Since successive P's are 8-neighbors, the sequence of P's defines the chain code of the border.

Exercise 10. Define borders using 8-adjacency rather than 4-adjacency, and devise a border following algorithm (for the case where C is 4-connected) in which successive P's are always 4-neighbors. ∎

Exercise 11. Let P_1, P_2 be 4-adjacent points of C, D, respectively, and define P_i for $i > 2$ as follows [13, 36]: If $P_{i-1} = 0$, turn left (relative to the direction from P_{i-2} to P_{i-1}); if $P_{i-1} = 1$, turn right. Work out an example of the operation of this algorithm. How does it handle 8-adjacency? (Hint: Try it on 8-adjacencies in two perpendicular diagonal directions.) ∎

An alternative method of obtaining the successive points of the D-border of C is as follows: Use the crack following algorithm to generate the successive cracks (P_i, Q_i) of the (C, D)-border. If a given point P occurs two or more times in succession as the first term of a crack, eliminate the repetitions. [This situation arises when a border point of C has several consecutive 4-neighbors in D, corresponding to several successive cracks on the (C, D)-border.] The sequence of first terms that remain after these eliminations is just the sequence of points of C_D. In fact, we can generate the chain code of C_D directly from the crack following algorithm. For example, if the current point pairs are

$$P \quad U$$
$$Q \quad V$$

and we next turn right, we add 7 to the chain; if we make no turn, we add 0; while if we turn left, we add nothing to the chain, since the new first term is still P. Since the algorithm for border following is more complicated than that for crack following, it may be more economical to use crack following to generate the sequence of border points and the chain code, as just described, rather than using border following.

c.   Border finding

Suppose that we want to follow all of the borders of S, one at a time, in order to obtain all their chain codes. (The discussion for crack codes is analogous.) To accomplish this, we might scan the picture systematically, say row by row; when we hit a border (e.g., when we find a 1 immediately following a 0), follow it around; when it has been completely followed, resume the scan. This process will certainly find all the borders, since on any border there must be a 1 with a 0 immediately to its left (e.g., a leftmost point of an outer border or a rightmost point of a hole border). However, it will find the same border repeatedly (on every row that contains points of the component or of the hole), so that each border will be followed many times, which is certainly undesirable. We can eliminate this problem by marking each border while following it, and initiating border following only when we hit an unmarked border; this insures that no border will be followed twice. However, some borders may now be missed; in fact, a hole border may be a subset of an outer border (the simplest examples are

```
    1               1  1  1

1       1    and    1       1

    1               1  1  1
```

—see Exercise 6), or at least all its points with 0's on their left may be on an outer border, so that these points get marked when the outer border is followed, and the hole border does not itself get followed. To avoid this problem, we should mark only those border points (e.g., on the D-border of C) that have points of D as left-hand neighbors; this insures that we will never hit the same border by a transition from 0 to an unmarked 1, but it does not interfere with detecting other borders that share points with the given one.

If we want to distinguish between outer borders and hole borders, we can label the connected components of $\bar{S}$ (see Section 11.3.1) before searching for borders; this gives the background a distinctive label, so that outer borders are identifiable by their adjacency to points having that label. Alternatively, we use the following modified process of scanning, border following, and marking: We use two marks, l and r, for points of the D-border that have D on their left and right, respectively; and we initiate border following at any transition from a 0 to a 1 not marked l, or from a 1 not marked r to a 0. Now an outer border is always first encountered as a transition from 0 to 1 (at the leftmost of the uppermost points of C), and similarly a hole border is always first encountered as a transition from 1 to 0; hence this scheme allows us to identify them as soon as they are encountered.

d. Border tracking

Crack and border following have the disadvantage that they require access to the points of the picture in an arbitrary order, since a border may be of any shape and size. In the following paragraphs we describe a method of constructing crack or chain codes of all the borders of S in a single row-by-row scan of the picture. This allows us to convert a row-wise representation of S (e.g., runs) directly into a border representation. Conversion from borders to runs will also be discussed.

We will now show how to convert runs into crack codes; conversion to chain codes is quite similar and is left as an exercise for the reader.

(1)   For each run ρ on the first row, or each run ρ not adjacent to a run on the preceding row, we initialize a code string of the form $1\,2 \cdots 2\,3$, where the number of 2's is the length of ρ; this represents the top and sides of ρ, which we know must be on a border. The end of this string is associated with the left end of ρ, and its beginning with the right end.

(2)   In general, a run ρ on a given row has one end of a code string associated with each of its ends (we will see shortly how the strings at the two ends can be different). Let these strings be α_ρ and β_ρ, where the end of α_ρ is associated with the left end of ρ, and the beginning of β_ρ is associated with its right end. Note that α_ρ always ends with 3 and β_ρ always begins with 1. If no run on the following row is adjacent to ρ, we link the end of α_ρ to the beginning of β_ρ by a string of the form $0 \cdots 0$, where the number of 0's is the length of ρ. This string represents the bottom of ρ, which must be on a border.

(3)   If just one run ρ' on the following row is adjacent to ρ, we add to the end of α_ρ a string of the form $2 \cdots 2\,3$, if ρ' extends to the left of ρ; a single 3, if ρ' and ρ are lined up at their left ends; and a string of the form $0 \cdots 0\,3$, if ρ extends to the left of ρ'. The end of the extended α_ρ is now associated with the left end of ρ'. Similarly, we add to the beginning of β_ρ a string of the form $1\,2 \cdots 2$, if ρ' extends to the right of ρ; a single 1, if ρ' and ρ are lined up at their right ends; and a string of the form $1\,0 \cdots 0$, if ρ extends to the right of ρ'. The beginning of the extended β_ρ is associated with the right end of ρ'.

(4)   If several runs, say $\rho_1', \ldots, \rho_k'$, on the following row are adjacent to ρ, we add strings to the end of α_ρ and beginning of β_ρ just as in (3), depending

on the positions of ρ_1' relative to the left end of ρ and of ρ_k' relative to its right end. The end and beginning of these strings are associated with the left end of ρ_1' and the right end of ρ_k', respectively. In addition, for each consecutive pair of runs ρ_{i-1}', ρ_i' ($1 < i \leqslant k$), we create a new string of the form $1\,0\cdots0\,3$, representing the right side of ρ_{i-1}', the bottom portion of ρ between ρ_{i-1}' and ρ_i', and the left side of ρ_i'. The beginning of this new string is associated with the right end of ρ_{i-1}', and its end with the left end of ρ_i'.

(5)   Suppose finally that run ρ' is adjacent to several runs, say $\rho_1, \ldots, \rho_k$, on the preceding row. In this case, we add strings to the end of α_{ρ_1} and the beginning of β_{ρ_k} just as in (3); the forms of these strings depend on the positions of ρ' relative to the left end of ρ_1 and the right end of ρ_k. The end and beginning of these extended strings are associated with the left and right ends of ρ', respectively. In addition, for each consecutive pair of runs ρ_{i-1}, ρ_i ($1 < i \leqslant k$), we link the end of α_i to the beginning of β_{i-1} by a string of the form $2\cdots2$, representing the top portion of ρ' between ρ_{i-1} and ρ_i.

When all rows have been processed, we are left with a set of closed crack codes representing all the borders of S, with outer borders represented counterclockwise and hole borders clockwise. A simple example of the operation of the algorithm is shown in Fig. 10.

If we want to convert crack or chain codes into runs, we must store them in such a way that it is easy to access all the code strings relating to a given row. For example, we can break each code into a set of substrings each of which represents a monotonically nondecreasing or nonincreasing sequence of y-coordinates, and record the coordinates of the endpoints of these segments. We can then generate a row-by-row run representation by scanning the appropriate substrings (in forward order, if y is nonincreasing; in reverse order, if it is nondecreasing) to determine the run ends. Further details will not be given here; see [8].

Fig. 10   Border tracking. (a) Input, with rows numbered. (b) Crack codes at successive rows; new segments on each row are underlined.

e. Reconstruction from borders

Suppose that we are given all the borders of S, e.g., we are given $\chi_{S'}$, which has 1's at border points of S and 0's elsewhere. Assume that we are using 8-connectedness for S and 4-connectedness for $\bar{S}$.[§] We might then try to reconstruct S along the following lines: (1) Label all the 4-components of 0's (see Section 11.3.1 on connected component labeling); let b be the label of the background component. (2) Mark 1's adjacent to b's, say with primes. (3) If $l \neq b$ is any label that occurs adjacent to a primed 1, change all points labeled l to primed 1's. (4) Mark 1's adjacent to primed 1's with stars. (5) If l is any label that occurs adjacent to a starred 1, change all points labeled l to b's. ("Adjacent" in all of these steps means "4-adjacent".) Steps (2)–(5) are repeated until all labels of 0's have been changed to either b's or primed 1's. Unfortunately, this process breaks down if any hole border and outer border of S have a point in common; that point will be adjacent to b's at some stage, and so will become primed, which will result in the 0's inside the hole turning into primed 1's. To avoid this problem, we must be given more than just $\chi_{S'}$. (In fact, $\chi_{S'}$ alone does not determine S; for example, if S' is

$$
\begin{array}{ccc}
 & 1 & \\
1 & & 1 \\
 & 1 & \\
\end{array}
$$

is S

$$
\begin{array}{ccc}
 & 1 & \\
1 & 1 & 1 \\
 & 1 & \\
\end{array}
$$

or is S the same as S'?) In particular, suppose that 1's that lie on more than one border are specially marked. When such 1's are adjacent to b's, we given them stars rather than primes; labels occurring adjacent to them are then turned into b's, not into primed 1's. The successive steps in this algorithm are illustrated schematically in Fig. 11.

If S has only a few borders, a border-following method of reconstructing it from its borders may be more appropriate than the approach just described. Suppose we are given the crack code of the (C, D)-border of S, and the coordinates of the initial pair (P, Q); based on the convention that P is always on the left, this allows us to find all the successive pairs. As we find them, we mark the P's as 1's and the Q's as 0's (on an initially blank array). When this

---

[§] If we are using 4-connectedness for S, the borders used here must be "thick," i.e., must include all points of S that are 8-adjacent to $\bar{S}$. Otherwise. e.g., the interior points of S that are 8-adjacent to the background component (see below) will be labeled as part of it.

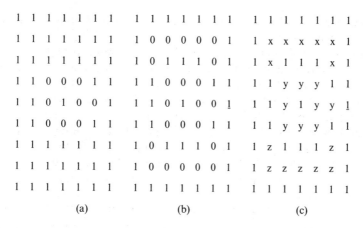

Fig. 11 Reconstruction from borders. (a) S (surrounded by 0's). (b) S' (surrounded by 0's); the 1 that lies on two borders is underlined. (c) All 4-components of $\bar{S}'$ (surrounded by b's) are labeled. (d) 1's adjacent to b's are primed; the previously underlined 1 is starred. (e) The labels x and z occur adjacent to primed 1's; all x's and z's become primed 1's. (f) 1's adjacent to primed 1's are starred. (g) The label y occurs adjacent to starred 1's; all y's become b's. Step (d) is now repeated, and the center 1 is primed; if it were on the outer border of a component of 1's having holes, repetition of steps (e)–(g) would turn that component's interior into primed 1's, its hole borders into starred 1's, and its holes into b's.

process is complete, the D-border of C has been completely marked with 1's, and the C-border of D with 0's. After all the borders of S have been marked in this way, it is easy to "fill in" the rest of S (and $\bar{S}$) by expanding 1's and 0's into 4-neighboring blanks (but not into each other); on a cellular array processor, the number of expansion steps required is less than the diameter of the picture. On a conventional computer, we can "fill in" S and $\bar{S}$ by doing a single row-by-row scan of the picture, turning consecutive blanks into 0's starting at any 0 or at the picture border, and turning consecutive blanks into 1's starting at any 1.

The reconstruction process is somewhat more complicated if we are given chain codes rather than crack codes. Suppose we know the chain code of the D-border of C and the coordinates of the starting point P; then the chain code gives us the next point P'. Let us first consider the case where we use 8-connectedness for C. We know that the 8-neighbor of P preceding P' (in counterclockwise order) is Q', and that (at least) the 4-neighbor of P preceding P' is in D (it is either Q, or is one of the R's visited before finding a 1; note that if P' is an 8-neighbor of P, this 4-neighbor is the same as Q'). We mark P and P' as 1, and this 4-neighbor—call it Q—and Q' as 0 (on the initially blank array). The chain code now gives us the next point P'', and we know that the 8-neighbor of P' preceding P'' is Q'', and that all the 8-neighbors of P' between Q' and Q'' are also in D. We mark P'' as 1 and all these 8-neighbors (including Q'') as 0's. This process is repeated until we reach the end of the chain; this should get us to P again, say from P^*. At this step, the 8-neighbors of P between Q^* and Q must all be 0's. Note that we may visit a point more than once, but different neighbors will be 0's at each visit. The D-border of C and C-border of D have now all been marked with 1's and 0's, respectively. The procedure is analogous when C is 4-connected, except that now only the 4-neighbors of P_i between Q_i and Q_{i+1} become 0's, since the intermediate 8-neighbors may actually be 1's. Furthermore, if P_{i+1} is an 8-neighbor of P_i, we know that the 4-neighbor following it must also be in C, and we make both of these neighbors 1's. A simple example of reconstruction from a border chain code is given in Fig. 12.

Expanding border 1's and 0's into blanks can be used to "fill in" regions even when the borders do not form perfectly closed curves [75]. For example, suppose that we have detected a set of edges that lie on a region border; we mark the points on the dark sides of the edges as 1's, and the points on the light sides as 0's, as illustrated in Fig. 13a. We then expand both 0's and 1's into blanks; when a blank has both 0's and 1's as neighbors, we mark it 0. Figure 13b shows the results of the first expansion step; it is evident that in a few more steps, the region will be filled with 1's and the background with 0's. This technique can only be used if the edges "surround" the border adequately, and there are no noise edges in the interior or background. If the border is not well surrounded, 1's will "escape" into the background; if noise edges are present, the expansion will create regions of 0's in the interior or 1's in the background.

f. Quadtrees and borders

In this subsection we briefly describe methods of converting between quadtree and border representations. The details can be found in two papers

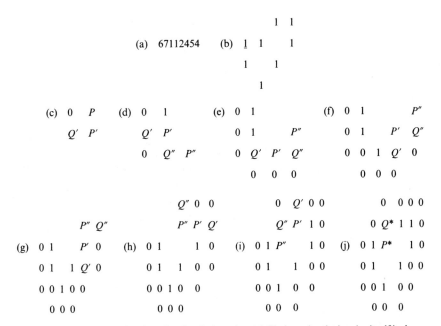

Fig. 12 Reconstruction from border chain codes. (a) Chain code; the border itself is shown in (b), with the starting point underlined. (c) Initial step; (d)–(j) successive steps. At the last step, with $P^* = 1$ and $Q^* = 0$, it is straightforward to fill in the interior by expanding 1's into blanks. The process shown here treats the 1's as 8-connected.

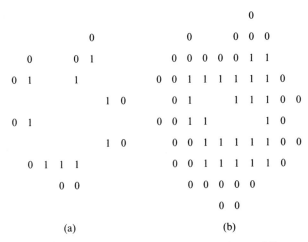

Fig. 13 Reconstruction from incomplete borders. (a) Initial 0's and 1's representing points on the light and dark sides of a border. (b) Result of 4-neighbor expansion of 0's and 1's into blanks (one step); blanks adjacent to both 0 and 1 become 0 (cf. near the lower left). Evidently, on successive steps the hole will fill with 1's and the background with 0's.

by Samet *et al.* [16, 64]. Conversion between MAT and border representations will not be treated here; on construction of a continuous MAT from a polygonal border see Montanari [38].

In order to determine, for a given leaf node M of a quadtree, whether the corresponding block is on a border, we must visit the leaf nodes that correspond to 4-adjacent blocks and check whether they are black or white. To find the nodes corresponding to, e.g., right-hand neighbor blocks, we move upward from M in the tree until we reach some ancestor node from its northwest or southwest son. (If we reach the root node before this happens, M is on the east edge of the picture and has no right-hand neighbor blocks.) As soon as this occurs, we go back down the tree making the mirror images of the moves made on the way up, i.e., the first move down is to the northeast or southeast son, and the following moves are to northwest or southwest sons. If a leaf node is reached by the time we come to the end of this move sequence, its block is at least as large as M's block, and so is M's sole right-hand neighbor. Otherwise, the nonleaf node reached at the end of the sequence is the root of a subtree whose leftmost leaf nodes correspond to M's right-hand neighbors, and we can find these nodes by traversing that subtree.

Let M, N be black and white leaf nodes whose blocks are 4-adjacent. Thus the pair M, N defines a common border segment of length 2^k (the smaller of the side lengths of M and N) which ends at a corner of M or of N (or both). To determine the next segment along this border, we must find the other leaf P whose block touches the end of this segment:

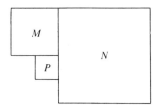

If the segment ends at a corner of both M and N, we must find the other two leaves P, Q whose blocks meet at that corner:

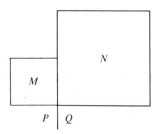

This can be done by an ascending and descending procedure similar to that described in the preceding paragraph; see [16] for the details. The next border segment is then the common border defined by M and P if P is white, or by N and P if P is black. [In the common corner case, the pair of blocks defining the next border segment is determined exactly as in the crack following algorithm of Subsection (a) above, with M, N, P, Q playing the roles of P, Q, U, V, respectively.] This process is repeated until we come to M, N again, at which stage the entire border has been traversed. The successive border segments constitute a crack code, broken up into pieces whose lengths are powers of 2. The time required for this process is on the order of the number of border nodes times the tree height.

Using the methods described in the last two paragraphs, we can traverse the quadtree, find all borders, and generate their crack codes. As in Subsection (c) above, we should mark each border as we follow it, so that we will not follow it again from another starting point; note that the marking process is complicated by the fact that a node's block may be on many different borders.

To generate a quadtree from a set of crack codes, we first traverse each code and create pairs of leaf nodes having the given border segments, together with the necessary nonleaf nodes. We then generate leaf (and nonleaf) nodes corresponding to the interior blocks. At any stage, if four leaves with a common father all have the same color, they are merged. The details of this algorithm will not be given here; see [64]. The time required is on the order of the perimeter (= total crack code length) times tree height.

11.2.3 Curves

This section discusses methods of "thinning" a given S into a set of arcs and curves. We will assume that S consists entirely of elongated parts, so that the resulting arcs and curves constitute a reasonable approximation to S (Section 11.1.5c). For other types of S's the results of thinning would not be particularly meaningful. (For a definition of elongatedness see Section 11.3.4b.) The result of thinning S will be called the *skeleton* of S.

The MA of an everywhere elongated S can be regarded as a skeleton, but it has two defects. At places where S has even width, its MA is two points thick, since the MA is the set of maxima of the distance to $\bar{S}$, as shown in Section 11.2.1a. Also, as pointed out there, the MA tends to be disconnected, and we would like connected pieces of S to be thinned into connected arcs or curves. In this section we will describe a thinning scheme that yields thin, connected skeletons.

a. Thinning

Our thinning algorithm is a specialized shrinking process which deletes from S, at each iteration, border points whose removal does not locally disconnect their neighborhoods; it can be shown [54] that this guarantees that the connectedness properties of S do not change, even if all such points are deleted simultaneously. To prevent an already thin arc from shrinking at its ends, we further stipulate that points having only one neighbor in S are not deleted.

Unfortunately, if we delete all such border points from S, and S is only two points thick, e.g.,

$$1 \quad 1 \quad 1 \quad 1 \quad 1$$

$$1 \quad 1 \quad 1 \quad 1 \quad 1$$

then S will vanish completely. We could avoid this by using an algorithm that examines more than just the immediate neighbors of a point, but such an algorithm would have to be quite complicated. Instead, we delete only the border points that lie on a given side of S, i.e. that have a specific neighbor (north, east, south, or west) in $\bar{S}$, at a given iteration. To insure that the skeleton is as close to the "middle" of S as possible, we use opposite sides alternately, e.g., north, south, east, west . . . . (It is possible to devise algorithms that remove border points from two adjacent sides at once, e.g., north and east, then west and south; but this approach is somewhat more complicated and will not be described here in detail. Another possibility is to check the neighbors of a point on two sides to determine whether they too will be deleted, and if so, not to delete the given point.)

In order to state the algorithm more precisely, we must give the exact conditions under which a border point can be removed. The border point P of S is called *simple* if the set of 8-neighbors of P that lie in S has exactly one component that is adjacent to P. This last clause means that if we are using 4-connectedness for S, we care only about components that are 4-adjacent to P. If we are using 8-connectedness, the last clause can be omitted. For example, P is 4-simple if its neighborhood is

$$0 \quad 1 \quad 1$$

$$0 \quad P \quad 0$$

$$1 \quad 0 \quad 0$$

since in this case only one 4-component of 1's is 4-adjacent to P; but P is not 4-simple if its neighborhood is

$$0 \quad 1 \quad 1 \qquad\qquad 0 \quad 1 \quad 0$$

$$0 \quad P \quad 0 \quad\; \text{or} \quad\; 0 \quad P \quad 1$$

$$0 \quad 1 \quad 0 \qquad\qquad 0 \quad 0 \quad 0$$

On the other hand, P is 8-simple in the third case, but not in the first two cases.

It is easily seen that deleting a simple point from S does not change the connectedness properties of either S or $\bar{S}$; $S - \{P\}$ has the same components as S, except that one of them now lacks the point P, and $\bar{S} \cup \{P\}$ has the same components as $\bar{S}$, except that P is now in one of them. Note that an isolated point (having no neighbor in S) is not simple, and that an end point (having exactly one neighbor in S) is automatically simple.

Our thinning algorithm can now be stated as follows: Delete all border points from a given side of S, provided they are simple and not end points. Do this successively from the north, south, east, west, north, ... sides of S until no further change takes place. A simple example of the operation of this algorithm is shown in Fig. 14.

The deletion of border points from a given side of S should be done "in parallel," i.e. the conditions for deletion of a point should be checked before any other points are deleted. (Suppose we did not do this, but simply performed the deletion row by row. When we deleted north border points, we would strip away layer after layer from the top of S, and the resulting skeleton would not be symmetric; e.g., if S were an upright rectangle, nothing would be left but its bottom row after the first operation.) Thus each iteration of the algorithm can be done as a simple parallel operation on a cellular array computer.

The algorithm can be implemented on a conventional computer, using one scan of the picture for each iteration. On each row, points are marked for deletion, but are not deleted until the points on the following row have been marked. We can avoid repeated scanning of the entire picture by operating on the rows in sequence $1; 2, 1; 3, 2, 1; \ldots$ as in Section 11.2.1c. After k steps, where $2k + 1$ is the maximum width of S, no more thinning is needed on the first row, so it can be dropped; thus only about k rows need to be available at a time.

Exercise 12 [3]. Prove that a north border point (=having north neighbor 0) is 8-simple iff

$$w\bar{s}e + \bar{w}(nw)\bar{n} + \bar{n}(ne)\bar{e} + \bar{e}(se)\bar{s} + \bar{s}(sw)\bar{w} = 0$$

where n, e, s, w, (nw), (ne), (se), (sw) denote the north, south, east, west, northwest, northeast, southeast, and southwest neighbors of the point, respectively, and the overbars denote negation ($\bar{0} = 1$, $\bar{1} = 0$). Can you formulate an analogous Boolean condition for 4-simplicity? ∎

b.   Alternative thinning schemes

Simplified approaches to thinning can sometimes be used. For example, if S has everywhere essentially constant thickness (see Section 11.3.2d), say

$2k + 1$, it can be thinned (at least roughly) by shrinking it k times. Note, however, that the resulting skeleton may occasionally be thick or broken. The thinning processes described in (a) will handle S's of variable width. Another method [32] that can be used if S is an arc of constant thickness is to initiate two border-following processes at one end of S that traverse the opposite sides of S. (Analogously, if S is a closed curve of constant thickness, we initiate two border-following processes at points just across the width of S from each other, which traverse the two borders of S.) If the distance between the border followers gets significantly larger than the width, we stop one of them until the other one catches up; thus they always remain approximately

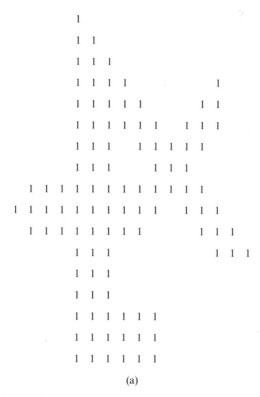

(a)

Fig. 14   Thinning. (a) Original S; (b)–(e) results of thinning successively from north, south, east, and west, treating S as 8-connected. At the next north step, the uppermost point of (e) will be deleted; at the next south step, the leftmost of the lowermost points; and at the next west step, the leftmost of the uppermost points. There will be no other changes. Note how the angle at the upper left shrinks down to the same height as the diagonal line at the upper right. (b′)–(e′) Analogous results treating S as 4-connected, and defining end points as having only one 4-neighbor in S. One more point will be removed at the next north step; there will be no other changes.

Fig. 14 (Continued)

```
        1
        1
        1
        1   1                               1
        1   1   1                           1
        1   1   1   1   1               1   1
        1   1           1   1       1   1
        1   1                   1   1   1
        1   1                   1       1
1   1   1   1   1   1   1   1   1   1       1   1
        1   1                           1   1
        1   1                               1   1   1
        1   1
        1   1
        1   1
        1   1   1   1   1   1
```

(d')

```
        1
        1
        1
        1   1                               1
            1   1                           1
            1   1   1   1               1   1
            1           1   1       1   1
            1                   1   1   1
            1                   1       1
1   1   1   1   1   1   1   1   1   1       1   1
            1                           1   1
            1                               1   1   1
            1
            1
            1
            1   1   1   1   1
```

(e')

Fig. 14 (*Continued*)

```
         1
         1
         1
         1  1                        1
         1  1  1                     1
         1  1  1  1  1          1  1
         1  1           1  1      1  1
         1  1              1  1  1
         1  1              1     1
   1  1  1  1  1  1  1  1  1  1     1  1
         1  1                     1  1
         1  1                        1  1  1
         1  1
         1  1
         1  1
         1  1  1  1  1  1
```

(d′)

```
         1
         1
         1
         1  1                        1
            1  1                     1
            1  1  1  1          1  1
            1           1  1      1  1
            1              1  1  1
            1              1     1
   1  1  1  1  1  1  1  1  1  1     1  1
            1                     1  1
            1                        1  1  1
            1
            1
          ˇ 1
            1  1  1  1  1
```

(e′)

Fig. 14 (*Continued*)

alongside one another. The midpoint of the line segment joining the border followers thus traces out a skeleton of S.

Thinning algorithms can be defined for multivalued pictures in various ways. One approach [15] is to generalize the definition of a simple point as follows: Define the strength of a path $P_1, \ldots, P_n$ as the minimum value of any point on the path, and the degree of connectedness of P and Q as the maximum strength of any path from P to Q. We call P "simple" if replacing it by the minimum of its neighbors does not decrease the degree of connectedness of any pair of points within its 8-neighborhood. It can be verified that this generalizes the two-valued definition given in (a). Thinning is then defined as a specialized local minimum operation: we repeatedly replace points by the minima of their neighbors, provided they are simple and have more than one higher-valued neighbor (this generalizes the condition that isolated and end points are not deleted). At each iteration we do this from only one side, i.e., we do it only to points that have a lower-valued neighbor on a specific side (north, south, ...). The results of applying this process to a set of pictures are shown in Fig. 15.

The output of edge detection operators is often thick (see Section 10.2). It can be thinned by suppressing nonmaxima in the gradient direction at each point; this discards all but the steepest edge value on each cross section of the edge, but does not allow points along the edge to compete with one another. Another possibility is to increase or decrease the value at each point in proportion to how much greater or smaller it is than its neighbors in the gradient direction. This causes the maximum value on each edge cross section to grow at the expense of the lower values, so that eventually it absorbs all of the responses on its cross section [17].

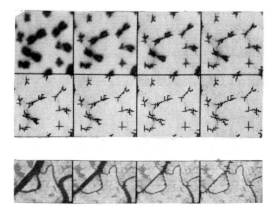

Fig. 15 Examples of gray-weighted thinning.

c. The results of thinning

Ideally, we call a subset A of the set T a simple digital *arc* if it is a connected component of T each point of which has exactly two neighbors in T, except for the two end points, which have one neighbor each. Note that this is two definitions in one, depending on whether we mean 4- or 8-neighbors. Note also that an arc cannot branch, cross itself or even touch itself, since otherwise it would contain points having more than two neighbors in S. A simple digital closed *curve* is a connected component C of T each point of which has exactly two neighbors in T; here too we have two definitions. Note that a border is not always a simple closed curve, since it may pass through some points twice.

Exercise 13. Show that A cannot be both a 4-arc and an 8-arc unless it is a horizontal line segment, and that C can never be both a 4-curve and an 8-curve. ▮

The goal of thinning S is to produce a T which is a union of such arcs and curves, perhaps after a few crossing or branching points—i.e., points having more than two neighbors—have been deleted. Note that such points will often occur in clusters; e.g., consider

```
    1                 1       1

    J                 J   J

1   J   J   J   1   and   J   J

    J                 1       1

    1
```

where we have indicated the junction points by J's. If many curve branchings and intersections occur close together, it becomes difficult to identify and classify individual junctions. A thinned S may even have interior points; an 8-connected example is

```
        1       1       1

            1   1   1

        1   1   1   1   1

            1   1   1

        1       1       1
```

in which every border point is either an end point or nonsimple, so that thinning has no effect on it. The results of thinning depend on orientation; for example, when we 8-thin

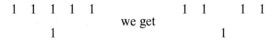

but when we 8-thin

$$
\begin{array}{ccccc}
1 & & & & \\
1 & 1 & 1 & 1 & 1
\end{array}
\quad \text{we get} \quad
\begin{array}{ccccc}
1 & 1 & 1 & 1 & 1
\end{array}
$$

(Under 4-thinning, neither of these patterns changes.)

Once points having more than two neighbors have been eliminated, it is straightforward to construct the chain code of an arc by moving from neighbor to neighbor, starting at one of the end points and ending at the other; or of a closed curve by beginning at an arbitrary starting point and moving in one direction until the starting point is reached again.

Thinning sometimes produces "spiky" skeletons, especially when S has "hairy" borders that contain many end points, or when end points arise prematurely in the course of the thinning process; this is especially likely in the 4-connected version of the algorithm. One way [56] to detect an inappropriate skeleton branch is to consider the distances of the points of the branch from $\bar{S}$, or, equivalently, the iterations at which they become border points. On a true skeleton branch these distances should be approximately constant, but on a spike branch they should increase steadily as we move in from the tip of the spike.

11.3 GEOMETRIC PROPERTY MEASUREMENT

In this section we describe how to measure geometrical properties of and relationships among subsets of a digital picture, using the various representations introduced earlier. We also discuss methods of deriving new segmentations from a given one, based on geometrical properties or relations.

11.3.1 Topology

We first consider properties and relationships involving the concepts of adjacency and connectedness. These properties are "topological" in the sense that they do not depend on size or shape; in the continuous case, they remain unchanged under arbitrary "rubber-sheet" distortions. For the basic definitions of these concepts see Section 11.1.7.

a.   Component labeling of arrays

We often want to treat the individual connected components of a set S as separate objects. The array representation of these objects is a multivalued picture in which the points of each component have a unique nonzero label, and the points of $\bar{S}$ have value zero. In the following paragraphs we discuss how to construct this representation from a given representation of S.

If S is represented by a binary picture, we can label its components by performing two row-by-row scans. Let us first assume that we want to label 4-connected components. During the first scan, for each point P having value 1, we examine the upper and left-hand neighbors of P; note that if they exist, they have already been visited by the scan, so that if they are 1's, they have already been labeled. If both of them are 0's, we give P a new label; if only one is 0, we give P the other one's label; and if neither is 0, we give P (say) the left one's label, and if their labels are different, we record the fact that they are equivalent, i.e., belong to the same component. When this scan is complete, every 1 has a label, and no label has been assigned to points that belong to different 4-components; but many different labels may have been assigned to points in the same component (see Figs. 16a and 16b). We now sort the equivalent pairs into equivalence classes, and pick a label to represent each class. Finally, we do a second scan of the picture and replace each label by the representative of its class (Fig. 16c); each component has now been uniquely labeled.

To label 8-connected components, we also examine the two upper diagonal neighbors of each 1; these have also already been visited by the scan. If all four neighbors are 0, the current point P gets a new label; if one of them is 1, P gets the same label; if two or more of them are 1, P gets one of their labels, and the equivalences are noted. (We need only note those equivalences that have not already been detected in previous positions of P; the details are left to the reader.) The equivalence processing and relabeling are done just as in the 4-connected case. Alternatively, we can examine the left, upper, and upper left neighbors of every point P, whether P is 0 or 1; if P is 1, we proceed as above, and if P is 0, but its upper and left neighbors are 1, we note the equivalence of their labels. Figures 16d and 16e show results of the first scans using these two schemes.

The component labeling process seems to be basically sequential. Algorithms can be devised for labeling components using a cellular array computer, but they are not very efficient. For example, suppose that we first give each 1 a unique label, e.g., its coordinates in the picture; this can be done

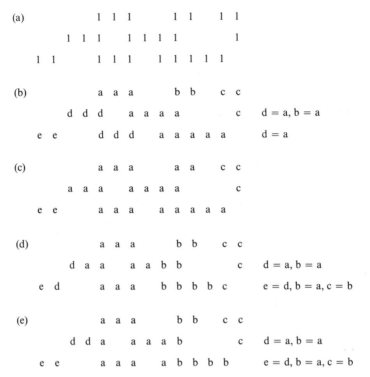

Fig. 16   Component labeling. (a) Input S. (b) Result of first scan, using 4-connectedness for S; equivalences discovered on each row are indicated. (c) Result of second scan, replacing all equivalent labels by a representative one. (d)–(e) Results of first scan using two versions of the 8-connectedness algorithm (see text).

in a number of steps equal to the picture diameter. We can then repeatedly perform a local maximum operation in parallel, where the maximum is defined by the lexicographic ordering of the coordinate pairs; points labeled 0 remain 0. We use the 4-neighbor maximum if we want to label 4-components, and the 8-neighbor for 8-components. When this is iterated until no further change takes place, every point of a given component is labeled with the coordinates of the uppermost of its rightmost points. However, the number of iterations required may be on the order of the picture area (consider a snakelike component).

Component labeling can be carried out simultaneously for the components of S and $\bar{S}$, or in fact for the components of the sets in any partition $S_1, \ldots, S_m$ (e.g., the partition of an arbitrary picture into sets of constant gray level), using 4- or 8-connectedness for the S_i's in any combination.

b.   Component labeling in other representations

A row-by-row approach to labeling the components of S (or of an arbitrary partition) can easily be defined in the case where S is given by its run length representation. On the first row, each run of 1's gets a new label. On subsequent rows, we compare the position of each run ρ with those of the runs on the preceding row. If ρ is adjacent (4- or 8-) to no run on the preceding row, it gets a new label. If it is adjacent to just one such run, it gets that run's label. If it is adjacent to two or more such runs, it gets (say) the first of their labels, and we note that all of their labels are equivalent. When the rows have all been processed, we sort out the equivalences and then do a second scan to give the runs their final labels.

If S is represented by a quadtree, we can label its components by traversing the tree in a standard order, say NW, NE, SW, SE. Whenever we come to a black leaf node, we visit the leaf nodes whose blocks adjoin M's block on its south and east sides (or at its southeast corner, if we are labeling 8-components); see Section 11.2.2f on how to find these nodes. If we find unlabeled black leaf nodes, we give them the same label as M; if we find black leaf nodes that already have labels, we note that their labels are equivalent. When the traversal is complete, we sort out the equivalences, retraverse the tree, and give the black leaf nodes their final labels. The time required is on the order of the number of nodes in the tree times the tree height. For the details of this algorithm see [66].

To label components based on the MAT representation, we must check all pairs of blocks that overlap or are adjacent (i.e., whose centers are no farther apart than the sum of their radii) in order to determine label equivalences. The details of this process are left to the reader. If we are given a border representation, and we know which are the outer borders, we can mark the borders in an array as in Section 11.2.2e, and give the points of each outer border a unique label; these labels can then be expanded into the interiors.

c.   Component counting

Once we have labeled the components of S, we know how many components it has, since this is just the number of final labels used. In this subsection we describe a method of counting components without labeling them, based on a parallel shrinking operation [34, 37] suitable for implementation on a cellular array computer.

Suppose that we use 4-connectedness for S and 8- for $\bar{S}$. If the right, lower, and lower right neighbors of P are

$$P \quad X$$
$$Y \quad Z$$

we define the operation Ψ as taking P into 1 iff $P + X = 2$, $P + Y = 2$, or $X + Y + Z = 3$. Readily, this is equivalent to saying that

$$\text{if } P = 1, \quad \text{we have } \Psi(P) = 0 \quad \text{iff } X = Y = 0$$

$$\text{if } P = 0, \quad \text{we have } \Psi(P) = 1 \quad \text{iff } X = Y = Z = 1$$

For any subset T of the picture, let T_1 be the set of 1's that are either in T or immediately above and to the left of T after Ψ is applied, and let T_0 be the set of such 0's. It can be shown that if C is a 4-component of 1's, so is C_1, and if D is an 8-component of 0's, so is D_0 (unless C or D consists of only one point, in which case C_1 or D_0 is empty). Moreover, if C is 4-adjacent to D, then C_1 is 4-adjacent to D_0. Thus, except for components consisting of single points, Ψ preserves the connectedness properties of both S and $\bar{S}$.

When Ψ is applied repeatedly, it can be shown that any component C of S shrinks to the single point (x_c, y_c), where x_c is the coordinate of the leftmost point of C and y_c is the coordinate of its uppermost point. This single point then vanishes. The number of steps required for this to happen is just the largest city block distance between (x_c, y_c) and any point of C. The same is true for any component D of $\bar{S}$ other than the background component. A component can shrink to a point even if it originally has holes, because the holes shrink to points (which then vanish) before the component does. An example of the operation of Ψ is shown in Fig. 17.

If we use 8-connectedness for S and 4- for $\bar{S}$, Ψ is defined to take P into 1 iff $P + Z = 2$ or $P + X + Y \geqslant 2$. Readily, this is equivalent to

$$\text{if } P = 1, \quad \text{we have } \Psi(P) = 0 \quad \text{iff } X = Y = Z = 0$$

$$\text{if } P = 0, \quad \text{we have } \Psi(P) = 1 \quad \text{iff } X = Y = 1$$

The results of applying this Ψ repeatedly are exactly analogous to those above. Of course, we could have defined Ψ using other 2×2 neighborhoods of P; such Ψ's would shrink components toward the northeast, southeast, or southwest rather than toward the northwest. A symmetric shrinking algorithm, based on a 3×3 neighborhood, is described in [51].

To count components using a Ψ operation, we modify it so that, instead of vanishing, isolated points turn into special marks, which do not interfere with the effect of Ψ on 1's and 0's. These marks shift leftward and upward to a counter at the northwest corner of the picture. The number of iterations required to shrink all components to points and count them in this way is less than the diameter of the picture.

d. The genus

If all the components of S are simply connected (i.e., have no holes), special methods can be used to count the components.[§] More generally, for any S, special methods can be used to compute the number of components of S minus the number of holes; this number is called the *genus* or *Euler number* of S.

If we repeatedly delete simple points from S in parallel, it can be shown that any simply connected component either shrinks to a single point or vanishes. (For example, any component of the form

$$1 \quad 1 \quad 1 \quad \cdots \quad 1$$
$$1 \quad 1 \quad 1 \quad \cdots \quad 1$$

vanishes immediately, since all its points are simple.) By making the deletion direction-dependent, we can guarantee that all simply connected components shrink to single points; the details will not be given here. Thus we can count the simply connected components of S by applying this process until there is no further change and then counting the resulting isolated points by shifting them to the upper left corner.

There are several simple methods of computing the genus $g(S)$ by counting local patterns of various types in the picture Σ. Let us use the following notation (blanks can be either 0's or 1's):

Pattern	No. of those patterns in Σ	Pattern	No. of those patterns in Σ
1	v	$\begin{matrix}1 & 0\\0 & 0\end{matrix}$, $\begin{matrix}0 & 1\\0 & 0\end{matrix}$, $\begin{matrix}0 & 0\\1 & 0\end{matrix}$, or $\begin{matrix}0 & 0\\0 & 1\end{matrix}$	v'
$\begin{matrix}1\\1\end{matrix}$ or $1 \ 1$	e	$\begin{matrix}1 & 0\\0 & 1\end{matrix}$ or $\begin{matrix}0 & 1\\1 & 0\end{matrix}$	d'
$\begin{matrix}1 & \\ & 1\end{matrix}$ or $\begin{matrix} & 1\\1 & \end{matrix}$	d	$\begin{matrix}0 & 1\\1 & 1\end{matrix}$, $\begin{matrix}1 & 0\\1 & 1\end{matrix}$, $\begin{matrix}1 & 1\\0 & 1\end{matrix}$, or $\begin{matrix}1 & 1\\1 & 0\end{matrix}$	t'
$\begin{matrix}1 & 1\\1 & 1\end{matrix}$, $\begin{matrix}1 & 1\\1 & 1\end{matrix}$, $\begin{matrix}1 & 1\\1 & \end{matrix}$, or $\begin{matrix}1 & 1\\ & 1\end{matrix}$	t		
$\begin{matrix}1 & 1\\1 & 1\end{matrix}$	q		

[§] One should be careful about assuming that all components are simply connected; a single pinhole adjacent to the border is just as much of a hole as a completely hollowed out interior.

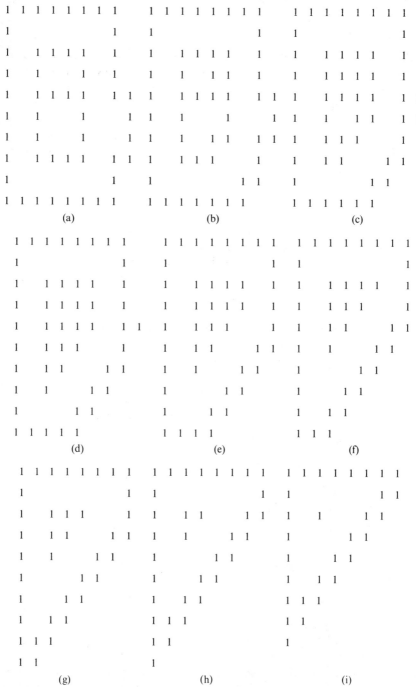

Fig. 17 Shrinking operation Ψ. (a) Input S. (b)–(q) Results of successive applications.

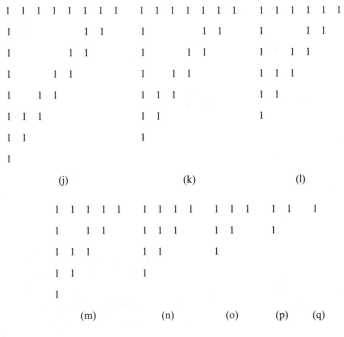

Fig. 17   (*Continued*)

If we use 4-connectedness for S and 8- for $\bar{S}$, then it can be shown [23, 37] that

$$g(S) = v - e + q = \tfrac{1}{4}(v' - t' + 2d')$$

In the reverse case we have

$$g(S) = v - e - d + t - q = \tfrac{1}{4}(v' - t' - 2d')$$

These patterns can be counted in a row-by-row scan by comparing each row with the preceding one, or they can be converted to special marks by local parallel logical operations, and the marks can then be shifted and counted.

The primed formulas just given provide a very simple method of computing the genus from the crack codes of the borders of S. In fact, both formulas imply that $4g(S)$ is the number of convex corners of S minus the number of concave corners. To see this, note that each v' pattern has a convex corner at its center point; each t' pattern has a concave corner; and each d' pattern has two convex corners if diagonally adjacent 1's are not connected, but two concave corners if they are connected. Of course, if we know which borders of S are outer borders and which are hole borders, we can obtain

$g(S)$ by simply subtracting the number of hole borders from the number of outer borders.

The unprimed formulas can be generalized to the case where S is represented by a quadtree [14]. In particular, let v be the number of black leaf nodes; e the number of pairs of such nodes whose blocks are horizontally or vertically adjacent; and q the number of triples or quadruples of such nodes whose blocks meet at and surround a common point, e.g.,

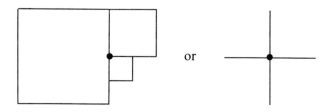

Then $g(S) = v - e + q$. These adjacencies can be found (see Section 11.2.2f) by traversing the tree; the time required is on the order of the number of nodes in the tree times the tree height.

Finally, we give a very simple formula for computing the genus from the run length representation of S. For each run ρ, let $k(\rho)$ be the number of runs on the preceding row to which ρ is adjacent; then $g(S) = \sum_{\rho} (1 - k(\rho))$.

e.   Adjacency, surroundedness, and nesting

Given any partition of a picture Σ, say into $S_1, \ldots, S_m$, we define the (4- or 8-) *adjacency graph* of the partition as the graph whose nodes are the S_i's, and in which nodes S_i and S_j are joined by an arc iff S_i is adjacent to S_j. It is straightforward to construct this graph by scanning Σ and checking the neighbors of each point to find all the adjacencies. This information is also easy to derive from the run length representation: check all consecutive pairs of runs on each row, and check all pairs of dissimilar-valued runs on successive rows for possible adjacency. Given the MAT representation, we check all pairs of dissimilar-valued MAT points for adjacency (distance between centers equal to sum of radii). In the quadtree representation, for each leaf node, we check the leaf nodes whose blocks are adjacent to it (Section 11.2.2f). In a border representation the information would be immediately available, since for each border arc we must specify which pair of regions defines it.

In the preceding paragraph the sets S_i may or may not be connected. (Given any partition $S_1, \ldots, S_m$ we can define a refined partition $C_1, \ldots, C_M$ consisting of the connected components of the S_i's.) An important special case is that in which the S_i's are the connected components of a set S and its complement $\bar{S}$, where we use 4-connectedness for S and 8- for $\bar{S}$ or vice versa. Construction of the graph is still straightforward, once we have labeled all the components. It makes no difference whether we use 4- or 8-adjacency between components, since as already pointed out, if a component of S and a component of $\bar{S}$ are 8-adjacent, they are also 4-adjacent. In this case, as we have seen, the borders are all closed curves. Moreover, it can be shown [55] that in this case the adjacency graph is a *tree*. (*Corollary*: If C_1, C_2 are any two components of S, and D_1, D_2 any two components of $\bar{S}$, we cannot have C_1, C_2 both adjacent to each of D_1, D_2, since otherwise the graph would contain the cycle C_1, D_1, C_2, D_2.)

We recall (Section 11.1.7b) that T *surrounds* S if any path from S to the border of Σ meets T. We also saw (Section 11.1.7c) that if C, D are adjacent components of S, $\bar{S}$, respectively, then one of them surrounds the other. Hence the adjacency tree of the components of S and $\bar{S}$ becomes a directed tree if we use the additional relationship of surroundness. Evidently, this tree is rooted at the background component of $\bar{S}$, which contains the border of Σ and so surrounds all the other components of S and $\bar{S}$. Just below the root we have "continent" components of S that are adjacent to the background; just below them, "lake" components of $\bar{S}$; just below them, "island" components of S; and so on.

We often need to know whether one of two sets surrounds the other, and in particular, whether a given set S surrounds a given point P of $\bar{S}$. (In cartography, this is known as the *point-in-polygon* problem.) If we are given the array representation of S, this problem can be solved for all P by simply labeling the background component of $\bar{S}$, say with b's; now, if $P = 0$ it is surrounded by S, and if $P = b$ it is not. Maximal-block representations of S (runs, MAT, quadtree) do not seem to be very useful in connection with this problem.[§] For border representations, on the other hand, there are various algorithms [44] for deciding whether P is inside or outside a given closed curve; using such algorithms, we can test whether P is inside the outer border of each component of S. We mention here only one standard approach: Move to the right (say) from P, and count the number of times that the curve is crossed (but do not count times that the curve is only touched without being crossed), until the border of Σ is reached. If the number of crossings is odd, P is inside the curve; if it is even, P is outside.

If $S_1, \ldots, S_m$ and $T_1, \ldots, T_n$ are partitions of Σ, we say that $T_1, \ldots, T_n$ is a *refinement* of $S_1, \ldots, S_m$ if every T_i is contained in some S_j—in fact, in a

---

[§] On the other hand, if we are given a maximal-block representation of S, it is relatively easy to tell whether a given point P lies in S or in $\bar{S}$—e.g., given the MAT, we check whether P is at a distance from each block's center not exceeding its radius; but it is harder to tell this from a border representation, where we would need to check whether P is inside some outer border of S but not inside a hole border of the same component of S.

unique one, since the S's are disjoint. (*Corollary*: Any S_j is a union of T_i's.) The same definition is used in the more general case where T is a subset of S, and $S_1, \ldots, S_m$ and $T_1, \ldots, T_n$ are partitions of S and T, respectively. As an important example, it is easy to show that if $T \subseteq S$, and $S_1, \ldots, S_m$ and $T_1, \ldots, T_n$ are the components of S and T, respectively, then $T_1, \ldots, T_n$ is a refinement of $S_1, \ldots, S_m$. Thus, e.g., if we threshold Σ at s and at t, where $s \leqslant t$, and let S, T denote the sets of above-threshold points, every component of T is contained in a unique component of S.

Suppose that we have a sequence of partitions $T_{h1}, \ldots, T_{hn_h}, h = 1, 2, \ldots, k$, each of which is a refinement of the preceding one, and where the first partition is the trivial one consisting of Σ itself. Thus each T_{hi} is contained in a unique $T_{h-1,j}$, Let us define a directed graph whose nodes are the T_{hi}'s, and where each node is joined to the node of the preceding partition which contains it. Readily, this graph is a tree rooted at $T_1 = \Sigma$. This *refinement tree* was introduced in [29] for the case where the partitions are the components of above-threshold points at a series of thresholds $0 = t_1 \leqslant \cdots \leqslant t_k$ (note that since $t_1 = 0$, the first partition consists of just Σ).

11.3.2 Size

In this section we consider size properties of a set S—area, perimeter, extent in a given direction, etc. Shape properties such as straightness, convexity, elongatedness, etc., will be considered in the next two sections.

a. Area

The *area* of S is simply the number of points of S. The areas of the sets in any partition $S_1, \ldots, S_m$ of Σ can be counted in a single scan of Σ, using one counter for each S_i. On a cellular array computer, marks corresponding to the points in each S_i can be shifted, e.g., leftward and upward, and counted at the upper left corner of Σ; the time required is on the order of the diameter of Σ. The areas of the connected components of S, or of any partition, can be measured in the process of labeling them (Sections 11.3.1a and 11.3.1b); each label occurrence adds 1 to the appropriate counter, and if two labels are found to be equivalent, the contents of their counters are combined.

Area computation from various other representations is also easy. Given the run length representation of S, we simply add the lengths of all the runs of 1's to obtain the area of S; given the quadtree representation, we add the areas of all the black leaf nodes (i.e., we add 4^{k-h} for each black leaf node at level h of the tree). On the other hand, it is not straightforward to obtain the area of S from its MAT representation, since there may be multiple overlaps of blocks.

Given a border representation of S, we get its area by adding the areas enclosed by all the outer borders and subtracting the areas enclosed by all the hole borders. The area enclosed by a given border can be obtained using standard integration formulas. For example, suppose we are given the crack code $\varepsilon_1 \cdots \varepsilon_r$, where each ε_i is 0, 1, 2, or 3. If we take the starting point (x_0, y_0) $= (0, 0)$ at the origin, then the y-coordinate at the end of the kth segment is $y_k = \sum_{i=1}^{k} \Delta y_i$, where $\Delta y_i = 1$ if $\varepsilon_i = 1$, 0 if $\varepsilon_i = 0$ or 2, and -1 if $\varepsilon_i = 3$. Let $\Delta x_i = 1$ if $\varepsilon_i = 2$, 0 if $\varepsilon_i = 1$ or 3, and -1 if $\varepsilon_i = 0$. Then it can be verified that the enclosed area is $\sum_{i=1}^{n} y_{i-1} \Delta x_i$. The analogous formula for a chain code is left as an exercise for the reader [19].

The area of S and the areas of other sets that can be derived from S (e.g., its connected components, its points at a given distance from $\bar{S}$, its convex hull, etc.) are all useful descriptive properties of S. Area-based criteria can also be used to derive new sets from a given S; e.g., we can discard all connected components of less than a given area—or, less trivially, if they are adjacent to only one other set in the given partition, we can merge them with that set. It should be realized that for some partitions, most of the connected components will be very tiny; for example, it has been found empirically [43] that the number of components of constant gray level having area N is proportional to $1/N^4$.

In the foregoing, we have treated the points of S as unit squares. Alternatively, we can regard S as a set of lattice points, and define the area of S as the sum of the areas of the polygons obtained by joining the outer border lattice points of each component of S, less the sum of the areas of the polygons obtained by joining the hole border lattice points. *Pick's theorem* states that if a simply connected S has b border points and i interior points, then its area is $\frac{1}{2}b + i - 1$ (rather than $b + i$, when we treat the lattice points as unit squares). More generally [69], if S has n holes, its area is $\frac{1}{2}b + i + n - 1$.

b. Perimeter and arc length

The *arc length* of a digital arc or curve is obtained from its chain code representation by counting horizontal and vertical moves as 1 and diagonal moves as $\sqrt{2}$. This gives reasonable values for an 8-arc or 8-curve, but it gives values that are somewhat too high in the 4-case, where a diagonal move is represented by a horizontal plus a vertical move and so is treated as having length 2.

The *perimeter* of S can be defined in a number of different ways. Some possible definitions are

(1) The sum of the lengths of the crack codes of all the borders of S.
(2) The sum of the arc lengths of all these borders, regarded as 8-curves.
(3) The sum of the areas of the borders of S.

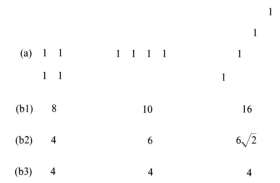

Fig. 18   Digital perimeter. (a) Three S's. (b1)–(b3) Their perimeters according to definitions (1)–(3).

Simple examples of the results obtained using these definitions are shown in Fig. 18.

It is straightforward to define algorithms for computing perimeter, as measured in any of these ways, from the array representation of S. We can easily compute the contribution of each border point and sum these contributions during a scan of S; or we can compute the contributions in parallel on a cellular array computer, and sum them by shifting and adding. Each border representation of S (crack code, chain code, $\chi_{S'}$) yields one of these measures directly; the problem of obtaining each of them from the other two representations (e.g., chain code length or border area from the crack codes) is left as an exercise for the reader.

Perimeter is easily measured from the run length representation of S; e.g., the crack code length is the sum of the number of run ends and the lengths of the subruns that are not overlapped by runs on the row above or on the row below. To compute crack code length from the quadtree representation, we visit each leaf node, and check the colors of the nodes whose blocks are adjacent to its block on two sides, say bottom and right, to determine which of these adjacencies contributes to the perimeter; the time required for this is proportional to the number of nodes times the tree height [68]. It is not straightforward to compute perimeter from the MAT representation; multiple overlaps of blocks make it complicated to determine how much each block contributes to the perimeter.

It should be pointed out that the perimeter of a digitized region often grows exponentially as the digitization becomes finer [35]; the area, on the other hand, tends to a finite limit. On the accuracy of estimating the lengths of straight lines and curves from digital representations see [18, 50].

If we regard S as a set of lattice points, its perimeter is the sum of the perimeters of the polygons formed by joining its border points. A formula for the perimeter computed in this way is obtained as follows [69]: Let $P_0, \ldots, P_7$ be the eight neighbors of P, numbered counterclockwise starting from the right-hand neighbor. Let

$$b_1(P) = \sum_{k=0}^{3} [(P \wedge P_{2k}) - (P_{2k} \wedge (P_{2k-1} \vee P_{2k-2}) \wedge (P_{2k+1} \vee P_{2k+2}))]$$

$$b_2(P) = \sum_{k=0}^{3} [(P \wedge P_{2k+1}) - (P_{2k} \wedge P_{2k+1} \wedge P_{2k+2})]$$

where the subscripts are all modulo 8. Then the perimeter of S is

$$\frac{1}{2} \sum_{P \in S} [b_1(P) + b_2(P)\sqrt{2}]$$

c.   *Extent and cross sections*

The *height* of S is the vertical distance between its highest and lowest points, and similarly its *width* is the horizontal distance between its leftmost and rightmost points. [Width should not be confused with *thickness*, which remains essentially the same when S is rotated or bent; see subsection (d).] More generally, the *extent* of S in a given direction θ is the distance between its extreme points as measured parallel to θ. Equivalently: If we drop a perpendicular from each point S onto a line l of slope θ, the extent of S in direction θ is the distance along l from the foot of the first perpendicular to the foot of the last one. In other words, the extent of S is the length of its *projection* on a line of slope θ. Alternatively, if we bring together a pair of parallel lines of slope $\theta + \pi/2$ until they hit S from opposite sides, the distance between them is the extent of S in direction θ.

Measuring the height or width of S is straightforward; measuring the extent in an arbitrary direction θ is more complicated, since it is harder to identify the extremal points. A brute-force solution is to rotate S by $-\theta$ and measure the width of the rotated S. In the next paragraph we indicate how to measure width from a given representation of S.

Given the array representation χ_S, we scan it to find the leftmost and rightmost columns containing 1's; the difference between the coordinates of these columns is the width. (A counting algorithm can be devised to compute this difference on a cellular array computer in time proportional to the picture diameter.) Given the run length representation, we scan it to find the coordinates of the leftmost and rightmost ends of runs of 1's; similarly, given the quadtree representation, we traverse the tree to find the leftmost and rightmost blocks of 1's. Given a border representation, we can scan the codes of the outer borders to find the leftmost and rightmost border points of S

(which are evidently also the leftmost and rightmost points of S). For example, given a crack code, we compute $x_k = \sum_{i=1}^{k} \Delta x_i$ for each k [see (a) above], and keep track of its greatest positive and negative values over all k; these indicate how far to the right and left of the starting point the given border extends.

The extent of S in a given direction is an orientation-sensitive property of S. Properties of this type are useful primarily when the orientation is known. Alternatively, we can *normalize* the orientation of S, e.g., by finding its greatest extent and rotating it to make it vertical, or by finding its smallest-area circumscribing rectangle (i.e., finding a pair of perpendicular directions in which the product of the extents of S is smallest) and rotating to make its long side vertical. Methods of normalizing the orientation of an arbitrary picture will be discussed in Section 12.1.1c; these methods can also be used for segmented pictures.

We obtain more detailed directional information about S if we examine its individual *cross sections* in a given direction (or along a given family of curves; but we will consider here for simplicity the cross sections defined by the rows of the picture). Each such cross section (of χ_S) consists of runs of 1's separated by runs of 0's. Various properties of these runs, as functions of row position, can provide useful descriptive properties of S; e.g., we can use the extent of the set of runs, their total length, the number of runs, etc. We can also use comparisons between runs on successive rows as a basis for segmenting S into pieces of simple shape [24]. In particular, we can break S into parts consisting of successions of single runs that do not change radically in length or position; whenever runs split or merge, or grow, shrink, or shift significantly, new pieces of S are defined. Thus each piece is a strip having approximately constant or slowly changing width. Of course, properties or segments derived from these runs are quite orientation-sensitive; they too should be applied only when the orientation is known, or after normalizing it.

d. Distances and thickness

The greatest extent of S in any direction is called its *diameter*. Readily, this is equal to the greatest distance between any two points of S. The extents of S in various directions, or the distances between points of S, are sometimes useful as descriptive properties.

Knowing the set of all its interpoint distances does not determine S up to congruence; for example [37],

$$
\begin{array}{ccccccc}
1 & 1 & 1 & & 1 & 1 & 1 \\
1 & & & \text{and} & 1 & & 1 \\
1 & & & & & &
\end{array}
$$

Fig. 19 Simplification by shrinking and reexpanding. (a) S; (b) $S^{(-1)}$; (c) $(S^{(-1)})^{(1)}$.

have the same set of distances (four 1's, two 2's, two $\sqrt{2}$'s, and two $\sqrt{5}$'s). On the other hand, if we have a one-to-one correspondence between S and T such that corresponding pairs of points have the same distance, S and T must be congruent; this generalizes the familiar theorem (corresponding to the case where S and T have three points each) that a triangle is determined up to congruence by specifying the lengths of its three sides.

One can also use the notion of distance to define new sets from a given S, e.g., the set of points of Σ at (or within) a given distance of S or of $\bar{S}$. We recall (Section 11.2.1c) that the k-step expansion $S^{(k)}$ of S is the set of points within distance k of S, while the k-step contraction $S^{(-k)}$ is the set of points further than k from $\bar{S}$. In the following paragraphs we show how combinations of expanding and shrinking can be used to extract new sets from S, and to "clean" it.

Note first that shrinking followed by expanding wipes out small parts of S; it can thus be used as a noise cleaning operation. A simple example is given in Fig. 19. We observe that this process also wipes out thin parts of S such as curves; it should not be used if such parts are significant. In fact, we can define the *thickness* of S as twice the number of shrinking steps required to wipe it out. Methods of detecting thin or elongated parts of S by shrinking, expanding, and comparing with the original S will be discussed in Section 11.3.4b.

We now show how expanding followed by shrinking can be used to identify isolated parts and clustered parts of S. Suppose that we expand S k times and then shrink the result k times, i.e., we compute $(S^{(k)})^{(-k)}$; we recall that this set always contains S. A small isolated piece of S (where "isolated" means "more than $2k$ away from the rest of S") simply expands and shrinks back, so that it gives rise to a small connected component of $(S^{(k)})^{(-k)}$. On the other hand, a cluster of pieces of S (less than $2k$ apart) "fuses" under the expansion, and only shrinks back at its edges, so that it yields a large component of $(S^{(k)})^{(-k)}$, as illustrated in Fig. 20. Thus to detect clusters or isolated pieces of S, we compute $(S^{(k)})^{(-k)}$ (for various k's) and examine its connected

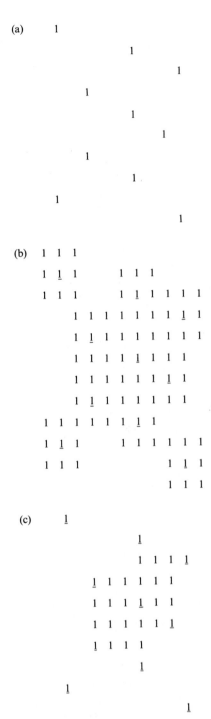

Fig. 20 Cluster detection by expanding and reshrinking. (a) S; (b) $S^{(1)}$; (c) $(S^{(1)})^{(-1)}$. Eight-neighbor expansion and shrinking were used. In (b) and (c) the original points of S are underlined.

components; if the area of a component is large relative to k^2, it must have arisen from a cluster of pieces of S, while if it is small, it must have arisen from an isolated piece.

11.3.3   Angle

In this section we discuss properties involving slope and curvature, as measured for curves or borders. Most of this material assumes that the given curve or border is specified by a chain code. We also briefly discuss relationships of relative position (e.g., "to the left of") between objects.

a.   Slope and curvature

The slope of a chain code at any point is a multiple of 45°, and the slope of a crack code is always a multiple of 90°. In order to measure a more continuous range of slopes, we must use some type of smoothing. For example, we can define the *left* and *right k-slopes* at a point P as the slopes of the lines joining P to the points k steps away along the curve on each side. Alternatively, we might define the k-slope at P as the average of the unit slopes (i.e., codes) at a sequence of k points centered at P.

Curvature is ordinarily defined as rate of change of slope. Here again, if we use differences of successive unit slopes, we always have a multiple of 45° for chain code, and a multiple of 90° for crack code. To obtain a more continuous range of values, we must again use smoothing. For example, we can define the *k-curvature* of P as the difference between its left and right k-slopes. (We would not want to use the average of the unit curvatures (i.e., differences of consecutive codes) at a sequence of k points centered at P, since curvatures should be cumulated rather than averaged.)

The k-slopes (and analogously for the k-curvature) can take on angular values whose tangents are rational numbers with denominator $\leqslant k$. Of course, k-slope is not defined if P is less than k away from an end of the arc. We have not specified here how to choose k; its choice depends on the particular application. Ordinarily, k should not be too large a fraction of the total arc length or perimeter. (Consider what happens in the case of a closed curve when k is half the perimeter or more!)

As a simple example, consider the arc in Fig. 21a and assume it to be continued indefinitely in both directions. Here the right 1-slope is 0° at some points and 45° at others, but the right 3-slope is $\tan^{-1}(\frac{1}{3})$ at every point (this is because the arc is periodic with period 3). For $k > 3$, the right k-slopes are not all equal, but as k increases, they all approach $\tan^{-1}(\frac{1}{3})$. Similarly, the k-curvatures fluctuate for small k (except that they are all zero for $k = 3$), and approach zero as k gets large.

For a border, say of a region of 1's, we can also measure curvature at a point P from the array representation by counting the number of 1's in a neighborhood of P. If this is about half the neighborhood size, the border is relatively straight at P; if it is much higher or much lower, the border has a sharp convex or concave curvature at P. The neighborhood size used determines the amount of smoothing in this definition.

b. Straightness

It is of interest to characterize those digital arcs which could arise from the digitization of a straight edge or line segment. We give a characterization here in terms of chain code; the crack code case is left to the reader. As examples, Figs. 21a and 21b are digital straight line segments, but Figs. 21c–21e are not.

The following conditions on the chain code are necessary for straightness [53]:

(1)   At most two slopes occur in the chain, and if there are two, they differ by 45°. This is violated by Fig. 21c.

(2)   At least one of the two slopes occurs in runs of length 1. This is violated by Fig. 21d.

(3)   The other slope occurs in runs of at most two lengths (except possibly at the ends of the arc, where the runs may be truncated), and if there are two lengths, they differ by 1. This is violated by Fig. 21e.

These conditions are not sufficient. In fact, it turns out that at least one of the two run lengths occurs in runs of length 1; the other occurs in runs of at

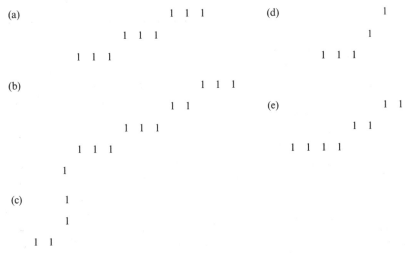

Fig. 21 Digital straightness. (a) and (b) are digital straight-line segments, but (c)–(e) are not.

most two lengths (except at the ends) which differ by 1; and so on. These conditions ensure that the runs of length 1 are spaced as evenly as possible in the chain. In any case, for many purposes one would not bother to check these conditions exactly; it would suffice to check that the smoothed slope is approximately constant, or that the fit of the arc to a real straight line is good.

Note that there may be more than one straight-line segment between two points, if the direction from one to the other is not a multiple of 45°. As a very simple example, consider

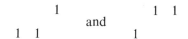

c. Curve segmentation

There are many ways of segmenting a curve into parts. Some of these are analogous to techniques for picture segmentation (Chapter 10), with slope (or smoothed slope) playing the role of gray level. Others do not depend on slope; for example, we can segment a curve at local or global extremum points, e.g., leftmost, rightmost, uppermost, or lowermost. It is assumed here that segmentation at junctions, if any, has already been done, so that we are dealing with a simple arc, curve or border.

The *slope histogram* of a curve tells us how often each slope occurs on the curve. Of course, it does not tell us how these slopes are arranged along the curve; a squiggle consisting of equal numbers of horizontal and vertical segments may have the same slope histogram as a square. However, for noncomplex curves it is reasonable to assume that a peak on the slope histogram provides information about overall orientation. It should also be pointed out that if the curve is the border of a convex object, its slope histogram does determine it, since in this case the sequence of slopes around the curve must be monotonic. Figures 22a–22c show a closed curve, its chain code, and its histogram of 1-slopes; the two peaks, 180° apart, correspond to the predominant orientation of the curve. The slope histogram is sometimes called the "directionality spectrum."

Information about the "wiggliness" of a curve can be obtained from its *curvature histogram*. For a smooth curve, the curvatures will be concentrated near 0, whereas for a wiggly curve they will be more spread out. This is analogous to the use of histograms of gray level difference magnitudes to describe the coarseness or busyness of a texture; see Section 12.1.5c. The 1-curvature histogram for Fig. 22a is shown in Fig. 22d; since the curve is smooth, the absolute values are nearly all ≤ 1. One can use the curvature histogram to determine a threshold for segmenting a curve into straight

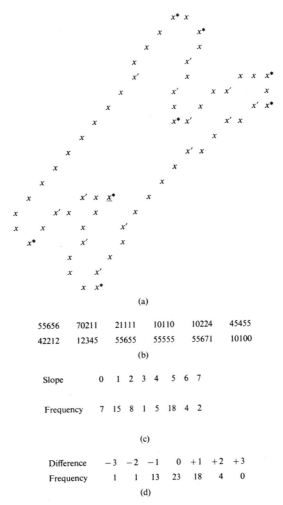

(a)

55656	70211	21111	10110	10224	45455
42212	12345	55655	55555	55671	10100

(b)

Slope	0	1	2	3	4	5	6	7
Frequency	7	15	8	1	5	18	4	2

(c)

Difference	−3	−2	−1	0	+1	+2	+3
Frequency	1	1	13	23	18	4	0

(d)

Fig. 22 Digital curve segmentation. (a) Curve; starting point underlined, curvature maxima starred, inflections primed. (b) Chain code. (c) Histogram of 1-slopes. (d) Histogram of 1-curvatures.

and curved parts; and similarly, one can use the slope histogram as a guide in segmenting a curve into relatively straight parts ("strokes") in various directions.

The analog of edges for curves are *corners*; these are points where the slope changes abruptly, i.e., where the absolute curvature is high. Thus the simplest types of "corners" are maxima of the (smoothed) curvature [57]. The maxima of 1-curvature in Fig. 22a are starred; generally speaking, they represent a reasonable set of corners. This simple definition, however, does not distinguish very well between smooth and sharp turns, e.g., between

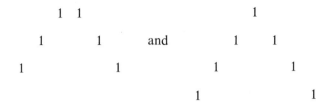

as long as the total amount of rotation is the same. One way to make this distinction is to compare the k-curvatures for several values of k; if small k's give the same value as large ones at the maximum, the corner must be sharp [20, 60].

Inflections are points where the sign of the curvature changes; they separate the curve into convex and concave parts. The 1-inflections in Fig. 22a are primed.

Various types of *local features* on a curve can be detected by drawing chords. The (maximum) distance between a chord and its arc is high (relative to the chord length) when the arc is sharply curved and low when it is relatively straight; but it might also be wiggly or S-curved. Another measure is the ratio of arc length to chord length; this is high for a sharply curved or wiggly arc, and low for a relatively straight one. Both of these measures are especially high for "spurs" or "spikes" on the curve. Of course, the sizes of the features that are detected in this way depend on the lengths of the arcs that are used (i.e., on k). (Note that this is a length-based, rather than slope- or curvature-based, curve segmentation technique.)

More generally, arbitrary curvature sequences can be detected on a curve by a *matching* process. For example, we can compute a sum of absolute slope differences between corresponding points, just as in picture matching, to obtain measures of the mismatch of a "template" with a curve in various positions. Much of the discussion in Chapter 9 about picture matching applies to curve matching as well, with appropriate modifications.

The curve segmentation methods mentioned up to now are all "parallel," i.e., they are applied independently at all positions along the curve. One can also use sequential methods (Section 10.4) analogous to tracking or region growing, e.g., keep extending an arc as long as it remains a relatively good fit to a straight line. Splitting techniques can also be used—for example, if a chord is not a good fit to its arc, subdivide the arc (e.g., at the point farthest from the chord), draw the chords of the two arcs, and repeat the process for each of them. Of course, splitting and merging can be combined; for a detailed discussion of this approach to the approximation of curves see [46],

Chapters 2 and 7, as well as [9]. When constructing polygonal approximations to a curve in this way, a good place to put the initial polygon vertices (i.e.,
the initial segmentation points) is at the curvature maxima, since the curve is
turning rapidly at these points and cannot be fit well there by a single line
segment.

Iterative "relaxation" methods (Section 10.5) can also be used in curve
segmentation. For example, suppose that at each point we estimate initial
corner and no corner ($=$ straight) probabilities at each point, based on the
k-curvature for some k. With each corner probability we associate the amount
of turn (e.g., the k-curvature itself, which is the difference between the left and
right k-slopes), and with each straight probability we associate the average
of these two k-slopes. Straights then reinforce nearby straights to the extent
that their slopes agree; a corner reinforces nearby straights on each side, to
the extent that its slope on that side agrees with the slope of the straight, and
it competes with nearby corners. When this reinforcement process is iterated,
the corner probabilities become high at sharp curvature maxima, and the
straight probabilities become high elsewhere [12].

d. Curve equations and transforms

In the real plane, various types of equations can be used to define curves.
Parametric equations specify the coordinates of the points on the curve as
functions of a parameter, i.e., $x = f(t)$, $y = g(t)$. The *slope intrinsic equation*
specifies slope as a function of arc length along the curve; this determines
the curve once a starting point is given. Similarly, the *curvature intrinsic
equation* specifies curvature as a function of arc length; this determines the
curve, given a starting point and initial slope. (The parameter t in the parametric equations might also be arc length.) A curve can sometimes be defined by specifying one coordinate as a function of the other, e.g., $y = f(x)$
or $r = f(\theta)$; but this is only possible when the function is single-valued.

All of these types of equations can be used for digital curves. Here the arc
length and the coordinates are all discrete-valued. The functions can be
specified by listing their values, or we can specify them analytically and
obtain the discrete values by quantization. Evidently, the chain code is a
digital version of the slope intrinsic equation, since it specifies slope as a
function of position along the curve; note, however, that the positions used
are not evenly spaced, since diagonal steps are $\sqrt{2}$ times as long as horizontal or vertical steps. Similarly, the difference chain code (Section 11.1.3)
is a digital version of the curvature intrinsic equation.

Given a string of numbers specifying a curve, we can compute a one-
dimensional discrete transform of the string, e.g., its discrete Fourier transform. (Note that for closed curves, the string can be regarded as periodic, which

simplifies its Fourier analysis.) In the case of parametric equations, we can still use a one-dimensional transform if we combine the two real coordinates into a single complex coordinate, i.e., $x + iy = f(t) + ig(t)$.

The transform can provide useful descriptive information about the curve. For example, a wiggly curve should have stronger high-frequency content than a smooth curve; compare the use of Fourier-based features to measure texture coarseness/busyness in Section 12.1.5d. Symmetries in the curve should be detectable as peaks in its transform, and the lower-frequency transform values should be good descriptors of the gross shape of the curve. If we derive these features from the magnitudes of the Fourier coefficients (i.e., from the power spectrum), they should be essentially rotation-invariant, since except for quantization effects, rotation of the curve corresponds to a cyclic shift of the string of values that specify the curve. We can get scale invariance by using ratios of transform values.

One can also "filter" a curve by operating on the transform and then inverse transforming, e.g., smooth the curve by suppressing high frequencies from the transform. Note, however, that the inverse transform may not be a simple closed curve unless the changes made to the transform are small.

Exercise 14. Prove that a crack code defines a closed curve iff $n_0 = n_2$ and $n_1 = n_3$, where n_i is the number of i's in the code. Similarly, prove that a chain code defines a closed curve iff $n_7 + n_0 + n_1 = n_3 + n_4 + n_5$ and $n_1 + n_2 + n_3 = n_5 + n_6 + n_7$. ∎

e. Relative position

In describing a picture, we often want to make statements about the relative positions of objects in the picture, e.g., "A is to the left of B," "A is above B," "A is near B," or "C is between A and B." Note that except for "near" (and "far"), these are relations of relative bearing, i.e., they involve the direction of A as seen from B, or the relative directions of A and B as seen from C.

For point objects, it is not hard to give fuzzy definitions for these relations, e.g., "A is to the left of B" has value 1 for the 180° direction (where B is at the origin), and drops off to zero in some appropriate way as we go from 180° to $\pm 90°$. For extended objects, however, defining these relations becomes quite complicated, as Fig. 23 shows. If we required that every point of A be to the left of every point of B, we exclude all but cases (a) and (b); but excluding case (e) seems unreasonable. If we require only that every point of A be to the left of some point of B, we include all but case (f); but including (d) seems unreasonable. If we require that A's centroid be to the left of B's centroid, we include all but case (d); but including (f) seems unreasonable. One proposed definition [76] requires that two conditions be satisfied: A's centroid must

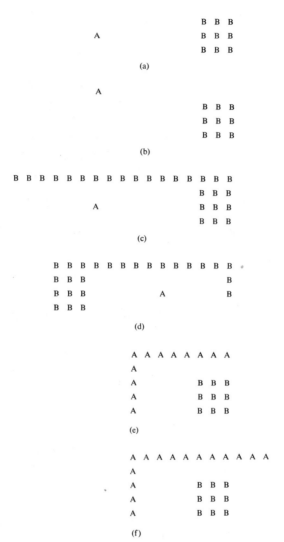

Fig. 23 The difficulty of defining "to the left of." In which of these cases is the object composed of *A*'s to the left of the object composed of *B*'s?

be to the left of *B*'s leftmost point, and *A*'s rightmost point must be to the left of *B*'s rightmost point. This definition excludes (c), (d), and (f), which is defensible. However, a purely coordinate-based rule of this type will not be adequate in all cases; our judgements about "to the left of" depend on our three-dimensional interpretation of the given picture, and may even depend on the meanings of the objects that appear in the picture.

11.3.4   Shape

Border features such as corners provide information about shape at a local level. In this section we discuss global shape properties such as complexity, elongatedness, and convexity. Other measures which provide global information about shape, such as moments and Fourier coefficients, will be discussed in Chapter 12, since they are applicable to unsegmented pictures. In this section, S is usually a connected set.

a.   Complexity

Human judgments of shape complexity depend on several factors. Of course, topological factors play an important role; the numbers of components and holes in S affect its judged complexity. We will assume in the following paragraphs that S is simply connected, i.e., is connected and has no holes, so that it has only a single border.

The *wiggliness* or *jaggedness* of S's border is an important complexity factor. This can be measured by the total absolute curvature summed over the border [77]. Alternatively, we might simply count the curvature maxima ("corners"), perhaps weighted by their sharpnesses.

Another frequently proposed complexity measure is p^2/A, *(perimeter)*^{2}/ *area.* (p is squared in this expression to make the ratio independent of size; when we magnify S by the factor m, p is multiplied by m and A is multiplied by m^2.) In the real plane, the "isoperimetric inequality" states that $p^2/A \geqslant 4\pi$ for any shape, with equality holding iff the shape is a circle; the ratio increases when the shape becomes elongated or irregular, or if its border becomes wiggly. In a digital picture, it turns out [52] that, depending on how perimeter is measured (see Section 11.3.2b), p^2/A is smaller for certain octagons than for digitized circles. However, we can still use it as a (rough) measure of complexity.[§]

The complexity of S also depends on how much *information* is required to specify S. Thus the presence of *equal parts, periodicities,* or *symmetries* in S reduce its complexity. As indicated in Section 3.5, there is a tendency to perceive certain ambiguous pictures in such a way that their descriptions are simplified. For example, Fig. 3.18a is easy to see as a three-dimensional cube, since this interpretation makes the lines and angles all equal, while Figs. 3.18b is easier to see as two-dimensional. The three-dimensional interpretation of Fig. 3.18b is also improbable, since it requires that two corners of the cube be exactly in line with the eye.

[§] Of course, if we want to measure *circularity*, we can compute, e.g., the standard deviation of the distances of the border points of S from its centroid [25]; this is small iff S is approximately circular.

The measurement of curvature, perimeter, and area have already been discussed. Periodicity or symmetry of (parts of) a picture can be detected using (approximate) matching techniques (Chapter 9); in the case of symmetry, we must rotate by 180° (for symmetry relative to a point) or reflect (for symmetry relative to a line) before matching. It is easiest to perceive symmetry of a picture about a vertical or horizontal axis; in these cases, symmetry of a shape is relatively easy to detect using any of the standard representations—array, run length, MAT, quadtree [2], or border—or from approximations [10]. The details will not be given here. On the use of moments of odd order as asymmetry measures see Section 12.1.3.

b. Elongatedness

As in Section 11.2.1c, let $S^{(k)}$ denote the result of expanding S k times, and $S^{(-k)}$ the result of shrinking it k times. Suppose that $S^{(-k)}$ is empty, i.e., S vanishes when we shrink it k times. Thus every point of S is within distance k of $\bar{S}$, so we can say that the *thickness* t of S is at most $2k$, as in Section 11.3.2d.

Suppose that the area A of S is large relative to k^2, say $A \geq 10k^2$. Then we can call S elongated, since its intrinsic "length" ($= A/t$) $\geq 10k^2/2k = 5k$ is at least $2\frac{1}{2}$ times its thickness. In general, we can define the *elongatedness* of a simply connected S as A/t^2, where A is the area of S and t is twice the number of shrinking steps required to make S disappear.

Note that this measure of elongatedness is unreliable for small values of t; for example,

$$
\begin{matrix} 1 & 1 \\ 1 & 1 \end{matrix} \qquad \text{and} \qquad 1 \quad 1 \quad 1 \quad 1
$$

both have $A = 4$ and $t = 2$, but we would only call the second one elongated. We can use this measure even when S has holes, but its reasonableness is less obvious; we might call a ring elongated, but what about a sieve? If S is noisy (i.e., has pinholes), it should be cleaned (e.g., by expanding and reshrinking; see Section 11.3.2d) before applying the methods of this subsection.

Somewhat more information about the elongatedness of S can be obtained by studying how S's area changes as we shrink it, rather than merely using A and t. In particular, for an everywhere elongated S the rate of decrease of area should be relatively constant, since the areas of the borders of S, $S^{(-1)}$, $S^{(-2)}$, ... are nearly the same; whereas for a compact S, the rate of decrease should steadily decline, since the border areas decrease.

Overall measures of the elongatedness of S are only of limited usefulness, since S may be partly elongated and partly not, and the elongated parts may have different thicknesses (e.g., S consists of streams, rivers, and lakes). We

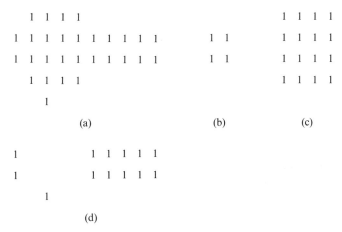

Fig. 24   Elongated part detection. (a) S; (b) $S^{(-1)}$; (c) $(S^{(-1)})^{(1)}$; (d) $S - (S^{(-1)})^{(1)}$. The 10-point component is large, hence is elongated.

will now show how shrinking and reexpanding can be used to detect elongated parts of S.

Consider $(S^{(-k)})^{(k)}$, which is the result of shrinking S k times and then re-expanding it k times. As mentioned in Section 11.2.1c, this is always contained in S. Let C be any connected component of the difference set $S - (S^{(-k)})^{(k)}$. Since C vanishes under k steps of shrinking, its thickness is at most $2k$; thus if its area is large relative to k^2, it is elongated. By doing this analysis for various values of k, we can detect elongated parts of S that have various thicknesses. A simple example, using only $k = 1$, is shown in Fig. 24.

The methods described in this subsection are designed for use with the array representation of S. They are especially efficient on a cellular array computer, but can also be implemented on a conventional computer, as discussed in Section 11.2.1c. Elongatedness is relatively easy to detect using maximal-block representations, e.g., S must be elongated if the number of MAT blocks is high and their radii are small, or the number of black quadtree leaves is large and their sizes are small. It is harder to detect from a border representation.

c.   Convexity

In this and the next two subsections we discuss properties of S that are defined in terms of the intersections of straight lines with S.

In the real plane, S is called *convex* if any straight line meets S at most once, i.e., in only one run of points. Evidently, a convex set must be connected and can have no holes; and an arc can be convex only if it is a straight-line segment.

It is easily seen that each of the following properties is equivalent to convexity:

(1) For any points P, Q of S, the straight-line segment from P to Q lies in S.
(2) For any points $P = (x, y)$, $Q = (u, v)$ of S, the midpoint $((x + u)/2, (y + v)/2)$ of P and Q lies in S.

In fact, it suffices, in these definitions, to assume that P and Q are border points of S.

We can use analogous definitions for digital pictures, but we must be careful to allow for quantization effects. The digital straight-line segment from P to Q is not always uniquely defined (see Section 11.3.3b); we might require, e.g., that at least one of the possible segments lie entirely in S. The midpoint of P and Q may not have integer coordinates; here we might require that for at least one rounding of each half-integer coordinate, either up or down, the result lies in S.

It should be pointed out that even if S is digitally convex [e.g., it has the digital version of property (2)], t may not be the digitization of any convex object. (Conversely, however, if S is such a digitization it can be shown [26] that S does have the digital property (2).) Note that any S can be the digitization of a concave object, e.g., one having tiny concavities that are missed by the sampling process.

There are various ways of subdividing a given S into convex pieces; for a detailed treatment of this subject see [46], Chapter 9. A useful heuristic is to make cuts in S that join deepest points of concavities (e.g., concave corners on the border of S).

Convexity is most readily detected from the border representation of S; in fact, we can define S to be convex if the curvature of its border never changes sign. (Compare the remarks in Section 11.3.3c.) It is much harder to detect convexity from the maximal block representations. Some methods of determining convexity from the array representation are described in the next subsection.

d. The convex hull

In the real plane, there is a smallest convex set S_H containing any given set S. (*Proof*: Readily, any intersection of convex sets is convex; in particular, the intersection of all the convex sets containing S is convex.) S_H is called the *convex hull* of S; it is schematically illustrated in Fig. 25. It can be shown that S_H is the intersection of all the half-planes containing S; and if S is connected, S_H is the union of all the line segments whose end points are in S.

Clearly S is convex iff $S_H = S$. Thus one way to decide whether S is convex is to construct its convex hull and see whether it properly contains S. In any

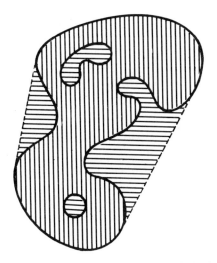

Fig. 25 A set (vertical shading) and its convex hull (includes the horizontally shaded parts).

case, we can define the *concavities* of S as the connected components of the difference set $S_H - S$.

The half-plane definition leads to the following construction for S_H [62], which can also be used to construct (and define) a convex hull in the digital case. Let P_1 be the leftmost of the uppermost points of S, and let L_1 be the horizontal line through P_1. Rotate L_1 counterclockwise about P_1 until it hits S; call the resulting rotated line L_2, and let P_2 be the point of S farthest from P_1 along L_2. Rotate L_2 counterclockwise about P_2 until it hits S; let L_3 be this rotated line, and let P_3 be the point farthest from P_2 along L_3. It is not hard to show that when this process is repeated, we eventually have $P_n = P_1$ and $L_n = L_1$. The polygon whose vertices are $P_1, \ldots, P_{n-1}$ is then S_H. Note that the L's bound half-planes that just contain S. Another characterization of the P's is as follows: For any border points P, Q of S, let $\bar{S}_{PQ}$ be the part of $\bar{S}$ surrounded by S and by the line segment PQ. Then $\bar{S}_{PQ}$ is maximal (i.e., is not contained in any other $\bar{S}_{P'Q'}$) iff P and Q are two consecutive P_i's (modulo $n - 1$).

The union of line segments definition leads [4] to a digital convex hull construction (and definition) that is more appropriate for a cellular array computer. Let S_θ denote the result of "smearing" S in direction θ, i.e., it is the union of all possible shifts of S by amounts $0, 1, 2, \ldots$ in direction θ; we need not use shifts greater than the diameter of S. Then $S_\theta \cap S_{\theta+\pi}$ is the set of points that have points of S on both sides of them in direction θ, i.e., these are just the points that are on line segments of slope θ between two points of S.

Thus $\bigcup_\theta (S_\theta \cap S_{\theta+\pi}) = S_H$, if S is connected. We cannot use just a few directions θ in this construction; for example,

$$
\begin{array}{cccccc}
1 & & & & & \\
& & & & & \\
1 & 1 & & & & \\
& & & & & \\
1 & 1 & 1 & 1 & 1 & 1
\end{array}
$$

contains every line segment between a pair of its points whose slope is a multiple of $45°$, but it is not convex.

Certain simple properties of S_H can be estimated without actually constructing it. For example, it can be shown [45], using methods of *integral geometry*, that the expected number of times that a random line meets S is proportional to the perimeter of S_H; thus we might estimate this perimeter by drawing a large number of random lines and counting intersections. Incidentally, the expected length of intersection of a random line with S (i.e., the sum of the lengths of the runs in which the line meets S) is proportional to the area of S.

e.   Generalizations of convexity

If S is in a known orientation, or if its orientation has been normalized, the intersections with S of lines in specific directions may provide useful descriptive properties of S. For example, we call S *row convex* if every row of the picture meets it only once; *column convex* is defined analogously. The S shown in the preceding subsection is both row and column convex, as well as diagonally convex for both diagonal directions.

These concepts can also be used to define new subsets in terms of S. For example, the "row convex hull" of a connected S might be defined as the union of all horizontal line segments whose end points are in S, i.e., as $S_0 \cap S_\pi$; and similarly for other directions. Analogously, the *shadow* of S from direction θ is the set of points from which a half-line in direction θ meets S; these are the points that would be in shadow if S were illuminated by a parallel beam of light in direction $\theta + \pi$. [22]. More generally, one can define sets of points from which the half-line in direction θ meets S a certain number of times, or in runs of a certain total length, etc.

Similar concepts can be defined if we use families of lines emanating from a given point, rather than in a given direction. In the real plane, for any point P of S, we call S *starshaped from* P if every line through P meets S exactly once. [Equivalently: (1) for all points Q of S, the straight-line segment from P to Q lies in S; (2) for all such Q, the midpoint of P and Q lies in S.] Readily, a starshaped S must be connected and can have no holes, but it need not be convex (a star is starshaped from its center). Evidently S is convex iff it is starshaped from every one of its points. In general, the set of points of S from which all of S is visible is called the *kernel* of S; note that it may be empty.

If S is starshaped from P, every point of S, and in particular every point of the border of S, is visible from P; thus this border has a single-valued polar equation $r = f(\theta)$ when we take the origin at P. If S is convex, this is true for any $P \in S$. For arbitrary sets, the visible part of S may vary from point to point. If [11] we associate the area (or some other property) of this visible part with each point of S, we can segment S into parts by classifying the points on the basis of these values; this is analogous to segmenting S on the basis of the distances of its points from $\bar{S}$.

11.4 BIBLIOGRAPHICAL NOTES

Only a few general references are mentioned here. Selected references on particular methods were given in the text; no attempt has been made to give a complete bibliography.

The MAT representation is due to Blum [6]; on its theory in the digital case see Rosenfeld and Pfaltz [58] and Mott-Smith [40]. The quadtree representation was introduced by Klinger [30, 31]; details of the algorithms for converting between quadtrees and other representations, and for computing geometric properties from quadtrees, can be found in a series of papers by Samet *et al.* [14, 16, 63–68, 72] (see also Alexandridis and Klinger [2] and Hunter and Steiglitz [27, 28]). Chain codes and their generalizations have been extensively studied by Freeman, who reviews them in [19]. On the representation of borders by sequences of circular arcs defined by successive triples of border points see Shapiro and Lipkin [71].

The theory of digital connectedness was developed by Rosenfeld, and is reviewed in [55]. On the theory of digital distance see Rosenfeld and Pfaltz [59]; on early work using shrinking and expanding for shape analysis see Moore [39].

Digital convexity has been extensively studied by Sklansky (e.g., [73]; see also [74] on a parallel approach to filling concavities by iteratively adding "concavity points" to S.) Another approach to convex hull construction based on a succession of inscribed polygons is described in [21]. Many fast algorithms have been proposed for solving various geometric problems involving intersections (e.g., constructing the convex hull of a set of points by intersecting a set of half-planes) and distances (e.g., finding pairs of points that are nearest neighbors).

Shape analysis techniques are reviewed by Pavlidis [47, 48]; for an extensive review of shape perception see Zusne [78].

REFERENCES

1. N. Ahuja, L. S. Davis, D. L. Milgram, and A. Rosenfeld, Piecewise approximation of pictures using maximal neighborhoods, *IEEE Trans. Comput.* **27**, 1978, 375–379.
2. N. Alexandridis and A. Klinger, Picture decomposition, tree data-structures, and identifying directional symmetries as node combinations, *Comput. Graphics Image Processing* **8**, 1978, 43–77.
3. C. Arcelli, A condition for digital points removal, *Signal Processing* **1**, 1979, 283–285.
4. C. Arcelli and S. Levialdi, Concavity extraction by parallel processing, *IEEE Trans. Systems Man Cybernet.* **1**, 1971, 394–396.
5. D. H. Ballard, Strip trees: a hierarchical representation for map features, *Proc. IEEE Conf. on Pattern Recognition and Image Processing* August 1979, 278–285.
6. H. Blum, A transformation for extracting new descriptors of shape, *in* "Models for the Perception of Speech and Visual Form" (W. Wathen-Dunn, ed.), pp. 362–380. MIT Press, Cambridge, Massachusetts, 1967.
7. H. Blum and R. Nagel, Shape description using weighted symmetric axis features, *Pattern Recognition* **10**, 1978, 167–180.
8. R. L. T. Cederberg, Chain-link coding and segmentation for raster scan devices, *Comput. Graphics Image Processing* **10**, 1979, 224–234.
9. L. S. Davis, Understanding shape: angles and sides, *IEEE Trans. Comput.* **26**, 1977, 236–242.
10. L. S. Davis, Understanding shape: symmetry, *IEEE Trans. Systems Man Cybernet.* **7**, 1977, 204–212.
11. L. S. Davis and M. L. Benedikt, Computational models of space: isovists and isovist fields, *Comput. Graphics Image Processing* **11**, 1979, 49–72.
12. L. S. Davis and A. Rosenfeld, Curve segmentation by relaxation labeling, *IEEE Trans. Comput.* **26**, 1977, 1053–1057.
13. I. De Lotto, Un inseguitore di contorno, *Alta Frequenza* **32**, 1963, 703–705.
14. C. R. Dyer, Computing the Euler number of an image from its quadtree, *Comput. Graphics Image Processing* **13**, 1980, 270–276.
15. C. R. Dyer and A. Rosenfeld, Thinning algorithms for grayscale pictures, *IEEE Trans. Pattern Anal. Machine Intelligence* **1**, 1979, 88–89.
16. C. R. Dyer, A. Rosenfeld, and H. Samet, Region representation: boundary codes from quadtrees. *Comm. ACM* **23**, 1980, 171–179.
17. R. B. Eberlein, An iterative gradient edge detection algorithm, *Comput. Graphics Image Processing* **5**, 1976, 245–253.
18. T. J. Ellis, D. Proffitt, D. Rosen, and W. Rutkowski, Measurement of the lengths of digitized curved lines. *Comput. Graphics Image Processing* **10**, 1979, 333–347.
19. H. Freeman, Computer processing of line-drawing images, *Comput. Surveys* **6**, 1974, 57–97.
20. H. Freeman and L. S. Davis, A corner-finding algorithm for chain-coded curves, *IEEE Trans. Comput.* **26**, 1977, 297–303.
21. H. Freeman and R. Shapira, Determining the minimum-area encasing rectangle for an arbitrary closed curve, *Comm. ACM* **18**, 1975, 409–413.
22. H. A. Glucksman, A parapropagation pattern classifier, *IEEE Trans. Electron. Comput.* **14**, 1965, 434–443.
23. S. B. Gray, Local properties of binary images in two dimensions, *IEEE Trans. Comput.* **20**, 1971, 551–561.
24. R. L. Grimsdale, F. H. Sumner, C. J. Tunis, and T. Kilburn, A system for the automatic recognition of patterns, *Proc. IEEE* **106B**, 1959, 210–221.

25. R. M. Haralick, A measure for cirularity of digital figures, *IEEE Trans. Systems Man Cybernet.* **4**, 1974, 394–396.
26. L. Hodes, Discrete approximation of continuous convex blobs, *SIAM J. Appl. Math.* **19**, 1970, 477–485.
27. G. M. Hunter and K. Steiglitz, Operations on images using quad trees, *IEEE Trans. Pattern Anal. Machine Intelligence* **1**, 1979, 145–153.
28. G. M. Hunter and K. Steiglitz, Linear transformation of pictures represented by quad trees, *Comput. Graphics Image Processing* **10**, 1979, 289–296.
29. R. A. Kirsch, Resynthesis of biological images from tree-structured decomposition data, *in* "Graphic Languages" (F. Nake and A. Rosenfeld, eds.), pp. 1–19. North-Holland Publ., Amsterdam, 1972.
30. A. Klinger, Data structures and pattern recognition, *Proc. Internat. Joint Conf. Pattern Recognition, 1st* 1973, 497–498.
31. A. Klinger and C. R. Dyer, Experiments on picture representation using regular decomposition, *Comput. Graphics Image Processing* **5**, 1976, 68–105.
32. R. S. Ledley, J. Jacobsen, and M. Belson, BUGSYS: a programming system for picture processing—not for debugging, *Comm. ACM* **9**, 1966, 79–84.
33. G. Levi and U. Montanari, A grey-weighted skeleton, *Informat. Control* **17**, 1970, 62–91.
34. S. Levialdi, On shrinking binary picture patterns, *Comm. ACM* **15**, 1972, 7–10.
35. B. B. Mandelbrot, "Fractals: Form, Chance, and Dimension." Freeman, San Francisco, California, 1977.
36. S. J. Mason and J. K. Clemens, Character recognition in an experimental reading machine for the blind, *in* "Recognizing Patterns" (P. A. Kolers and M. Eden, eds.), pp. 156–167. MIT Press, Cambridge, Massachusetts, 1968.
37. M. L. Minsky and S. Papert, "Perceptrons—An Introduction to Computational Geometry." MIT Press, Cambridge, Massachusetts, 1969.
38. U. Montanari, Continuous skeletons from digitized images, *J. ACM* **16**, 1969, 534–549.
39. G. A. Moore, Automatic scanning and computer processes for the quantitative analysis of micrographs and equivalent subjects, *in* "Pictorial Pattern Recognition" (G. C. Cheng *et al.*, eds.), pp. 275–326. Thompson, Washington, D.C., 1968.
40. J. C. Mott-Smith, Medial axis transformations, *in* Picture Processing and Psychopictorics" (B. S. Lipkin and A. Rosenfeld, eds.), pp. 267–283. Academic Press, New York, 1970.
41. Y. Nakagawa and A. Rosenfeld, A note on the use of local min and max operations in digital picture processing. *IEEE Trans. Systems Man Cybernet.* **8**, 1978, 623–635.
42. R. Nevatia and T. O. Binford, Description and recognition of curved objects, *Artificial Intelligence* **8**, 1977, 77–98.
43. S. Nishikawa, R. J. Massa, and J. C. Mott-Smith, Area properties of television pictures, *IEEE Trans. Informat. Theory* **11**, 1965, 348–352.
44. S. Nordbeck and B. Rystedt, Computer cartography—point-in-polygon programs, *BIT* **7**, 1967, 30–64.
45. A. B. J. Novikoff, Integral geometry as a tool in pattern perception, *in* "Principles of Self-Organization" (H. von Foerster and G. W. Zopf, eds.), pp. 347–368. Pergamon, Oxford, 1962.
46. T. Pavlidis, "Structural Pattern Recognition." Springer, New York, 1977.
47. T. Pavlidis, A review of algorithms for shape analysis, *Comput. Graphics Image Processing* **7**, 1978, 243–258.
48. T. Pavlidis, Algorithms for shape analysis of contours and waveforms, *Proc. Internat. Joint Conf. on Pattern Recognition, 4th* 1978, 70–85.
49. J. L. Pfaltz and A. Rosenfeld, Computer representation of planar regions by their skeletons, *Comm. ACM* **10**, 1967, 119–122, 125.

50.   D. Proffitt and D. Rosen, Metrication errors and coding efficiency of chain-encoding schemes for the representation of lines and edges, *Comput. Graphics Image Processing* **10**, 1979, 318–332.

51.   C. V. Kameswara Rao, B. Prasada, and K. R. Sarma, A parallel shrinking algorithm for binary patterns, *Comput. Graphics Image Processing* **5**, 1976, 265–270.

52.   A. Rosenfeld, Compact figures in digital pictures, *IEEE Trans. Systems Man Cybernet.* **4**, 1974, 211–223.

53.   A. Rosenfeld, Digital straight line segments, *IEEE Trans. Comput.* **23**, 1974, 1264–1269.

54.   A. Rosenfeld, A characterization of parallel thinning algorithms, *Informat. Control* **29**, 1975, 286–291.

55.   A. Rosenfeld, "Picture Languages: Formal Models for Picture Recognition," Chapter 2. Academic Press, New York, 1979.

56.   A. Rosenfeld and L. S. Davis, A note on thinning, *IEEE Trans. Systems Man Cybernet.* **6**, 1976, 226–228.

57.   A. Rosenfeld and E. Johnston, Angle detection on digital curves, *IEEE Trans. Comput.* **22**, 1973, 875–878.

58.   A. Rosenfeld and J. L. Pfaltz, Sequential operations in digital picture processing, *J. ACM* **13**, 1966, 471–494.

59.   A. Rosenfeld and J. L. Pfaltz, Distance functions on digital pictures, *Pattern Recognition* **1**, 1968, 33–61.

60.   A. Rosenfeld and J. Weszka, An improved method of angle detection on digital curves, *IEEE Trans. Comput.* **24**, 1975, 940–941.

61.   D. Rutovitz, Data structures for operations on digital images, *in* "Pictorial Pattern Recognition" (G. C. Cheng *et al.*, eds.), pp. 105–133. Thompson, Washington, D.C., 1968.

62.   D. Rutovitz, An algorithm for in-line generation of a convex cover, *Comput. Graphics Image Processing* **4**, 1975, 74–78.

63.   H. Samet, Region representation: quadtree-to-raster conversion, Computer Science Center TR-768, Univ. of Maryland, College Park, Maryland, June 1979.

64.   H. Samet, Region representation: quadtrees from boundary codes, *Comm. ACM* **23**, 1980, 163–170.

65.   H. Samet, Region representation: quadtrees from binary arrays, *Comput. Graphics Image Processing* **13**, 1980, 88–93.

66.   H. Samet, Connected component labelling using quadtrees, *J. ACM* **28**, 1981, 487–501.

67.   H. Samet, An algorithm for converting rasters to quadtrees, *IEEE Trans. Pattern Anal. Machine Intelligence* **3**, 1981, 93–95.

68.   H. Samet, Computing perimeters of images represented by quadtrees, *IEEE Trans. Pattern Anal. Machine Intelligence* **3**, 1981, 683–687.

69.   P. V. Sankar and E. V. Krishnamurthy, On the compactness of subsets of digital pictures, *Comput. Graphics Image Processing* **8**, 1978, 136–143.

70.   J. Serra, Theoretical basis of the Leitz texture analysis system, *Leitz Sci. Tech. Informat. Suppl.* **1** (4), 1974, 125–136.

71.   B. Shapiro and L. Lipkin, The circle transform, an articulable shape descriptor, *Comput. Biomed. Res.* **10**, 1977, 511–528.

72.   M. Shneier, Calculations of geometric properties using quadtrees, *Comput. Graphics Image Processing* **16**, 1981, 296–302.

73.   J. Sklansky, Recognition of convex blobs, *Pattern Recognition* **2**, 1970, 3–10.

74.   J. Sklansky, L. P. Cordella, and S. Levialdi, Parallel detection of concavities in cellular blobs, *IEEE Trans. Comput.* **25**, 1976, 187–196.

75.   J. P. Strong III and A. Rosenfeld, A region coloring technique for scene analysis, *Comm. ACM* **16**, 1973, 237–246.

76. P. H. Winston, Learning structural descriptions from examples, *in* "The Psychology of Computer Vision" (P. H. Winston, ed.), pp. 157–209. McGraw-Hill, New York, 1975.
77. I. T. Young, J. E. Walker, and J. E. Bowie, An analysis technique for biological shape, *Informat. Control* **25**, 1974, 357–370.
78. L. Zusne, "Visual Perception of Form." Academic Press, New York, 1970.

Chapter 12

Description

Picture descriptions generally specify properties of parts of the picture and relationships among these parts. Thus such descriptions are often represented by relational structures such as graphs in which the nodes correspond to the parts; each node is labeled with its associated property values; and the arcs correspond to relations between parts, labeled with the associated relation values if the relations are quantitative. In the special case where the description does not refer to parts of the picture, it consists simply of a list of property values, defined for the picture as a whole.

Geometric properties of picture subsets were treated in Section 11.3; such properties depend only on the set of points in the given subset, but not on the gray levels of these points. In Section 12.1 we deal with properties of pictures and picture subsets that do depend on gray level. Section 12.1.1 discusses some general classes of picture properties—in particular, linear and transformation-invariant properties. Sections 12.1.2–12.1.5 deal with specific types of properties, including local properties and property complexity, template properties such as moments, properties of projections and transforms, and statistical and textural properties.

Section 12.2 deals with descriptions of pictures and with models for classes of pictures. Section 12.2.1 discusses descriptions at the array, region, and relational structure levels; Sections 12.2.2 and 12.2.3 discuss "declarative" and "procedural" (in particular: syntactic) models at each of these levels.

Section 12.2.4 treats the problem of constructing models appropriate to a given picture (or scene) analysis task, and of establishing correspondences between these models and descriptions of the picture, so that the picture can be interpreted in terms of the models.

12.1   PROPERTIES

12.1.1   Types of Properties

Let $\mathcal{P}$ be a function that takes pictures into numbers[§]; the result of applying $\mathcal{P}$ to the picture f is denoted by $\mathcal{P}[f]$. $\mathcal{P}$ is called a *property*, and $\mathcal{P}[f]$ is called the value of $\mathcal{P}$ for f. Properties can be real or complex valued (e.g., the value of f's Fourier transform at a particular spatial frequency, or the average gray level of f) or integer valued (e.g., the gray level of f at a particular point, or its highest gray level).

Let $\mathcal{Q}$ be a proposition about pictures; thus $\mathcal{Q}$ is either true or false for a given picture f. $\mathcal{Q}$ is called a *predicate*, and $\mathcal{Q}[f]$ is its value for f. It is convenient to regard a predicate as a special type of property that takes on only the values 1 and 0, representing "true" and "false," respectively.

The value of a property $\mathcal{P}$ may depend only on the values of f at a given set S of points, which is the same for all f. For any $\mathcal{P}$, there is evidently a smallest such set $S_{\mathcal{P}}$; it is called the *set of support of* $\mathcal{P}$. *Local* properties, which are properties having small sets of support, will be discussed in Section 12.1.2.

On the other hand, the value of $\mathcal{P}$ may depend only on some set T defined in terms of f (e.g., the set of points where f has gray level above some threshold), but not on the values of f at these points (except insofar as these values define T itself). Here T can be any set defined by segmenting f using any of the methods in Chapter 10. In this case, we can regard $\mathcal{P}$ as a property of T. Such $\mathcal{P}$'s are called *geometric* properties. Many examples of such properties were given in Section 11.3. If $\mathcal{P}$ also depends on the values of f at the points of T—e.g., the average gray level of f on T—it is no longer called geometric. This section deals primarily with nongeometric properties.

It is often desirable to perform operations on f before computing property values. For example, in order to compute geometric properties, we first segment f. Other types of properties are computed on transforms or projections of f; see Section 12.1.4. Still others are computed on "normalized" versions of f, to insure that they are invariant under certain types of transformations of f; see Subsections (b) and (c) below. If we "preprocess" f, e.g., by

[§] Our use of $\mathcal{P}$ here should not be confused with its use in Chapters 2 and 5 to denote the probability of an event.

noise cleaning, before property computation, we may obtain more reliable property values.

In the following subsections we discuss general classes of picture properties, including linear (a) and transformation-invariant [(b) and (c)] properties. Sections 12.1.2–12.1.5 will treat various specific types of properties.

a. Linear properties

$\mathscr{P}$ is called *linear* if $\mathscr{P}[af + bg] = a\mathscr{P}[f] + b\mathscr{P}[g]$ for all pictures f, g, and all constants a, b. We saw in Section 2.1.2 that if $\mathcal{O}$ is an operation that takes pictures into pictures, and $\mathcal{O}$ is linear and shift invariant, then it is a convolution operation—i.e., there exists a function h such that

$$\mathcal{O}[f] = \int\!\!\!\int_{-\infty}^{\infty} h(x - \alpha, y - \beta) f(\alpha, \beta)\, d\alpha\, d\beta$$

for all f, so that $\mathcal{O}[f] = h * f$ is the convolution of h with f. In this subsection we show, analogously, that if $\mathscr{P}$ is a linear picture property, then there exists a function h such that

$$\mathscr{P}[f] = \int\!\!\!\int_{-\infty}^{\infty} h(\alpha, \beta) f(\alpha, \beta)\, d\alpha\, d\beta$$

for all f, provided $\mathscr{P}$ is " bounded " in a certain sense. Note that $\mathscr{P}$ need not be shift-invariant. Invariant properties will be discussed in the next subsection.

We first prove this result in the "continuous" case, where it is known as the *Riesz representation theorem*. Let $\mathscr{L}_2$ be the set of functions f such that $\int\!\!\int f^2 < \infty$.

Theorem.   Let $\mathscr{P}$ be any real-valued, linear function defined on $\mathscr{L}_2$, and suppose that there exists a real number M such that $(\mathscr{P}[f])^2 \leqslant M(\int\!\!\int f^2)$ for all f in $\mathscr{L}_2$. Then there exists an h in $\mathscr{L}_2$ such that $\mathscr{P}[f] = \int\!\!\int hf$ for all f in $\mathscr{L}_2$.

Proof: This is trivial if $\mathscr{P}[f] = 0$ for all f (just take $h = 0$). Otherwise, it can be shown that there exists an h' in $\mathscr{L}_2$ such that

$$(\mathscr{P}[h'])^2 \Big/ \int\!\!\int h'^2 = \max\left[(\mathscr{P}[f])^2 \Big/ \int\!\!\int f^2 \right]$$

We first show that if g in $\mathscr{L}_2$ is such that $\mathscr{P}[g] = 0$, then $\int\!\!\int h'g = 0$. Indeed, for any real λ we have

$$(\mathscr{P}[h'])^2 \Big/ \int\!\!\int h'^2 \geqslant (\mathscr{P}[h' - \lambda g])^2 \Big/ \int\!\!\int (h' - \lambda g)^2$$

Since $\mathscr{P}$ is linear and $\mathscr{P}[g] = 0$, we have $\mathscr{P}[h' - \lambda g] = \mathscr{P}[h'] \neq 0$; thus we can cancel $(\mathscr{P}[h'])^2$ from both sides and invert the fractions to obtain

$$\iint h'^2 \leqslant \iint (h' - \lambda g)^2 = \iint h'^2 - 2\lambda \iint h'g + \lambda^2 \iint g^2$$

Thus canceling $\iint h'^2$ gives

$$0 \leqslant -2\lambda \iint h'g + \lambda^2 \iint g^2$$

If $\iint g^2 = 0$, we have $g = 0$, so that $\iint h'g = 0$ trivially. Otherwise, we can set $\lambda = \iint h'g / \iint g^2$ to obtain $0 \leqslant -(\iint h'g)^2 / \iint g^2$, which implies $\iint h'g = 0$.

Finally, for any f in $\mathscr{L}_2$ we have

$$f = [f - (\mathscr{P}[f]/\mathscr{P}[h'])h'] + (\mathscr{P}[f]/\mathscr{P}[h'])h'$$

If we call the bracketed term g, we see that $\mathscr{P}[g] = 0$, so that by the preceding paragraph, $\iint h'g = 0$. Let $h = (\mathscr{P}[h']/\iint h'^2)h'$; thus $\iint hg = 0$ too. It follows that

$$\iint hf = \iint hg + (\mathscr{P}[f]/\mathscr{P}[h']) \iint hh'$$

$$= (\mathscr{P}[f]/\mathscr{P}[h'])(\mathscr{P}[h'] / \iint h'^2) \iint h'^2 = \mathscr{P}[f] \quad \blacksquare$$

The proof for digital pictures is exactly the same, with sums replacing integrals. In this case $\sum \sum f^2$ is automatically finite, since it is a finite sum of finite values. Moreover, since there are only finitely many possible digital pictures f, $\mathscr{P}[f]$ has a largest possible value; hence we can always find the M required by the theorem. (*Proof*: Since $\mathscr{P}$ is linear, we have $\mathscr{P}[f] = \mathscr{P}[f+0] = \mathscr{P}[f] + \mathscr{P}[0]$ for all f; hence $\mathscr{P}[g] = 0$ when $g \equiv 0$. Thus $(\mathscr{P}[f])^2 \leqslant M \sum \sum f^2$ is automatically satisfied when $f \equiv 0$. Let m be the smallest possible nonzero value of $\sum \sum f^2$, and let M' be the largest possible value of $\mathscr{P}[f]$; then for $M = M'/m$ we evidently have $\mathscr{P}[f] \leqslant M \sum \sum f^2$ for all f.) The existence of h' in the first part of the proof is also obvious, since $(\mathscr{P}[f])^2 / \sum \sum f^2$ has only finitely many possible values.

This theorem shows that any linear picture property is of the form $\sum \sum hf$, i.e., its value for a given picture f is simply a linear combination of the gray levels of f. A variety of useful linear picture properties will be described in Section 12.1.3.

b. Invariant properties

Let $\mathcal{O}$ be an operation that takes pictures into pictures. We say that the property $\mathscr{P}$ is *invariant* under $\mathcal{O}$ if $\mathscr{P}[\mathcal{O}[f]] = \mathscr{P}[f]$ for all f. More generally, let $\mathcal{S}$ be a set of such operations; we say that $\mathcal{P}$ is invariant under $\mathcal{S}$ if $\mathcal{P}[\mathcal{O}[f]] = \mathcal{P}[f]$ for all f and all $\mathcal{O}$ in $\mathcal{S}$.

We often are interested in picture descriptions that remain the same when certain types of operations are performed on the pictures. For example, we may want our descriptions to be insensitive to gray scale operations such as overall changes in lightness or contrast, or to geometric operations such as translation, rotation, or magnification. If so, our descriptions should make use of properties that are invariant under these operations. In this subsection we discuss general methods of defining invariant properties. Many examples of such properties will be given in the following sections.

It is easy to define picture properties that are invariant under changes in lightness or contrast, i.e., under shifting or stretching of the gray scale. Differences between gray levels (or between linear combinations of gray levels) are invariant under gray scale shifting, i.e.,

$$\sum a_i(z_i + b) - \sum a_i(w_i + b) = \sum a_i z_i - \sum a_i w_i$$

Similarly, ratios of gray levels are invariant under gray scale stretching or shrinking, i.e., $cz/cw = z/w$. To obtain invariance under both shifting and stretching, we can use ratios of differences, i.e.,

$$[c(z + b) - c(w + b)]/[c(z' + b) - c(w' + b)] = (z - w)/(z' - w')$$

There are several ways of defining picture properties that are invariant under geometrical operations. The gray level histogram of a picture is invariant under any one-to-one geometrical operation on the picture; hence so are properties derived from the histogram, e.g., statistical properties (Section 12.1.5). Invariant properties based on moments will be discussed in Section 12.1.3; invariant properties derived from transforms of the picture will be treated in Section 12.1.4.

Most of the geometric properties of picture subsets discussed in Section 11.3 are invariant under particular classes of geometrical operations. (In fact, the various branches of geometry can be regarded as the study of properties that remain invariant under various types of geometric operations.) For example:

(1) Extent and slope properties (e.g., the slope intrinsic equation) are invariant under translation.

(2) Area, perimeter, distance, thickness, and curvature properties (e.g., the curvature intrinsic equation) are also invariant under rotation.

(3) Ratios of sizes, angles, and shape properties such as complexity, elongatedness, and convexity, are also invariant under magnification.

(4) The cross ratio (Section 9.1.3c) is invariant under projection (e.g., under the imaging transformation).

(5)   Topological properties such as number of components are invariant under arbitrary "rubber-sheet" geometrical distortions.

Of course, for digital pictures, most of these statements are only approximately correct, since a picture must be redigitized after a geometric transformation is applied to it, as we saw in Section 9.3.

Let $\mathscr{G}$ be a *group* of operations, i.e., the identity operation is in $\mathscr{G}$, the composite of two operations of $\mathscr{G}$ is in $\mathscr{G}$, and any operation in $\mathscr{G}$ has an inverse in $\mathscr{G}$. For example, the gray scale shifts or stretches, the translations, the rotations, and the magnifications are all groups of operations. (We ignore here the fact that for digital pictures, there can be only finitely many different shifts, stretches, translations, or magnifications, and we also ignore the need to redigitize.) Then it is not hard to see that for any $\mathcal{O}$ in $\mathscr{G}$, the set of composites $\{\mathcal{O} \circ \mathcal{O}' | \mathcal{O}' \in \mathscr{G}\}$ is just $\mathscr{G}$ itself. This implies that if we start with any $\mathcal{O}[f]$, and apply every $\mathcal{O}'$ to it, we get the same set of pictures, independent of $\mathcal{O}$. Hence if we measure any property $\mathscr{P}$ for the set of pictures

$$\{\mathcal{O}'[\mathcal{O}[f]] | \mathcal{O}' \in \mathscr{G}\}$$

we get the same set of property values, independent of $\mathcal{O}$. Thus any property that depends only on this set of values, such as its average or maximum, will be the same no matter which $\mathcal{O}[f]$ we started with. As a simple example, when we cross-correlate a template with a picture, we are measuring its match with every translate of the picture, and the maximum of these matches is a translation-invariant property of the picture. Alternatively, rather than measuring $\mathscr{P}$ for every transformed version of $\mathcal{O}[f]$, we may be able to apply a set of transformed $\mathscr{P}$'s to the original $\mathcal{O}[f]$; this will still yield the same set of property values, independent of $\mathcal{O}$. For example, in cross-correlating a template with f, matching the template with every translate of f is the same as matching every translate of the template with f.

c.   Normalization

An important approach to defining properties that are invariant under a set $\mathscr{S}$ of operations is to *normalize* the given picture f with respect to $\mathscr{S}$—in other words, to construct a picture $\mathscr{N}[f]$ such that $\mathscr{N}[\mathcal{O}[f]] = \mathscr{N}[f]$ for all $\mathcal{O}$ in $\mathscr{S}$. Any property of $\mathscr{N}[f]$ can then be regarded as an invariant property of f.

A general approach to defining $\mathscr{N}[f]$ when $\mathscr{S} = \mathscr{G}$ is a group is as follows: Let $\mathscr{P}_1, \ldots, \mathscr{P}_k$ be a set of properties such that, for any f, there is a unique $\mathcal{O}'$ in $\mathscr{G}$ such that $\mathscr{P}_1[\mathcal{O}'[f]], \ldots, \mathscr{P}_k[\mathcal{O}'[f]]$ have specified values $w_1, \ldots, w_k$. By the remarks at the end of Subsection (b), if we start with any $\mathcal{O}[f]$ rather than with f, there is a unique $\mathcal{O}''$ in $\mathscr{G}$ (namely, the composite of $\mathcal{O}'$ and $\mathcal{O}^{-1}$, where $\mathcal{O}^{-1}$ is the inverse of $\mathcal{O}$) such that $\mathcal{O}''[\mathcal{O}[f]] = \mathcal{O}'[f]$. Thus we can take $\mathcal{O}'[f]$ as our $\mathcal{N}[f]$.

For example, let $\mathscr{G}$ be the group of gray scale shifts. Readily any f has a unique shifted version $\mathcal{N}[f]$ that has a given mean gray level (e.g., 0), and evidently we get the same $\mathcal{N}[f]$ no matter what shifted version of f we start with. Similarly, let $\mathscr{G}$ be the group of gray scale shifts and stretches; then we can take $\mathcal{N}[f]$ to be the unique shifted and stretched version of f that has a given mean and standard deviation of gray level (e.g., 0 and 1).[§] Let $\mathscr{G}$ be the group of translations; then we can take $\mathcal{N}[f]$ to be the unique translated version of f that has its centroid at the origin (see Section 12.1.3), and similarly for other groups of geometric operations.

Another approach to normalization is to use transforms of f as $\mathcal{N}[f]$'s. For example, the autocorrelation R_f and the power spectrum $|F|^2$ remain the same when f is translated (see Section 2.1.4), so that they can be used as normalized versions of f with respect to translation. A number of other transform methods of normalization will be described in Section 12.1.4.

Some methods of normalizing a segmented picture were mentioned in Section 11.3.2c. For example, given a subset S of the picture, we can normalize with respect to translation by shifting S, say, to the upper left corner of the picture (i.e., until it touches the top row and left column); with respect to magnification, by rescaling S to make its height and width equal to standard values; and with respect to rotation, by rotating S so that its greatest extent is in the vertical direction, or so that its minimum-area circumscribing rectangle is upright. Analogously, we can standardize the shape of S by constructing, say, a best-fitting quadrilateral to S and then transforming coordinates to make this quadrilateral into a square [20]; this approach can be used to counteract the effects of various types of geometrical distortions.

12.1.2   Local Properties and Property Complexity

a.   Local properties

As indicated earlier, a property $\mathscr{P}$ is called *local* if it has a small set of support, i.e., if its value depends only on the values of f at a small set $S_\mathscr{P}$ of points. Here "small" can be interpreted in two ways:

(a) $|S_\mathscr{P}|$ is small, i.e., $S_\mathscr{P}$ consists of only a few points;

(b) $S_\mathscr{P}$ has small diameter, i.e., its points all lie within a small neighborhood.

[§] More realistically, let $\mathcal{N}[f]$ be the result of flattening f's histogram, as described in Section 6.2.4; then $\mathcal{N}[f]$ is approximately invariant under monotonic transformations of f's gray scale, provided the number of points having any given gray level does not get too large.

Clearly (b) implies (a), but the converse is not true, since the points of S can be extremely far apart. The extreme case of a local property is a *point* property, i.e., the value of $\mathscr{P}$ depends only on the value of f at a single point; evidently such a property is local in both senses.

Exercise 1. If $f_1, \ldots, f_n$ are distinct pictures, show that there exist points $(x_1, y_1), \ldots, (x_{n-1}, y_{n-1})$ such that no two of the n pictures have the same values at all $n - 1$ of the points. Thus any set of n distinct pictures can be distinguished using at most $n - 1$ point properties (see [25] for references).

∎

Individual local properties are useful primarily for pictures that have been normalized with respect to geometric transformations. For example, if the picture shows a character in a standard position, orientation, and scale, gray level measurements at specified places, or streak detector outputs in specified positions and orientations, can be used to check for the presence of specific strokes in the character. As another example, if we normalize a picture with respect to translation by computing its autocorrelation or Fourier power spectrum (Section 12.1.4), then values at specific points give information about the picture's texture (see Section 12.1.5). If we do not normalize, then when we compute a local property in a given position on the picture, we do not know where this information is coming from in the scene.

Sets of local property values can provide useful information even if the picture has not been normalized. For example, suppose that a binary picture χ_S contains only 2×2 patterns of the forms

$$
\begin{array}{cc}
1 \ 1 \\
1 \ 1
\end{array}
\qquad
\begin{array}{cc}
1 \ 1 \\
0 \ 0
\end{array}
\qquad
\begin{array}{cc}
1 \ 0 \\
0 \ 0
\end{array}
\quad \text{and} \quad
\begin{array}{cc}
0 \ 0 \\
0 \ 0
\end{array}
$$

and their rotations by multiples of $90°$. Then [18] the set S of 1's in the picture must consist of a collection of disjoint upright rectangles (how else could all the other 2×2 patterns be absent?). In fact, if the patterns

$$
\begin{array}{cc}
0 \ 0 \\
0 \ 1
\end{array}
\qquad
\begin{array}{cc}
0 \ 0 \\
1 \ 0
\end{array}
\qquad
\begin{array}{cc}
0 \ 1 \\
0 \ 0
\end{array}
\quad \text{and} \quad
\begin{array}{cc}
1 \ 0 \\
0 \ 0
\end{array}
$$

occur exactly once each, S must be a single rectangle.

We know that histograms of local property values (gray levels, slopes, etc.) are useful for segmentation purposes. In Section 12.1.5 we will discuss statistical properties (i.e., statistics of local properties) and their role in texture description. Some early pattern recognition systems, e.g., those known as *perceptrons*, made use of sets of randomly defined local properties.

b.   Predicate length [14]

The complexity of a property $\mathscr{P}$ can be defined as the length of a program, written in some standard way, that computes $\mathscr{P}$. For example, the complexity of a predicate $\mathscr{P}$ can be defined as the length of the description of an automaton that accepts a picture f iff it satisfies $\mathscr{P}$, or of a grammar that generates f iff it satisfies $\mathscr{P}$. Alternatively, we can define the *space complexity* of $\mathscr{P}$ as the amount of tape that an automaton requires for computing $\mathscr{P}$, and its *time complexity* as the amount of time required. The time complexities of various picture properties for various types of automata are discussed in [27].

Another way of defining the "length" of a predicate is to express it as a Boolean function. Let $P_1, \ldots, P_N$ be the points of the binary picture χ_S having set of 1's S; then predicates of χ_S are Boolean functions of the Boolean variables $P_1, \ldots, P_N$. For example, the predicate "$|S| = 1$" ("there is only one 1 in χ_S") corresponds to the function

$$\bigvee_{i=1}^{N} (\bar{P}_1 \wedge \cdots \wedge \bar{P}_{i-1} \wedge P_i \wedge \bar{P}_{i+1} \wedge \cdots \wedge \bar{P}_N)$$

where the overbars denote logical negation; evidently this function has value 1 iff exactly one of the P's is 1.

We can define the *length* of a predicate $\mathscr{P}$ as the length (=number of variables) of the shortest Boolean function expression for $\mathscr{P}$. Thus our example shows that the length of "$|S| = 1$" is at most N^2. In fact, it can be shown that the length of this predicate cannot be a linear function of N. Similarly, it can be shown that "S is convex" and "S is connected" cannot have lengths that are linear in N. On the other hand, "$|S| \geqslant 1$" has length N, since it can be expressed as the Boolean function $\bigvee_{i=1}^{N} P_i$.

If $\mathscr{P}$ has a small set of support, its length is also small; for example, the length of "$P_j = 1$" is 1, since it corresponds to the Boolean function P_j. Evidently, $\mathscr{P}$ has length 0 iff it is a constant, i.e., is true or false for every χ_S. If $\mathscr{P}$ and $\mathscr{Q}$ have lengths m and n, respectively, then logical combinations such as $\mathscr{P} \vee \mathscr{Q}, \mathscr{P} \wedge \mathscr{Q}$, etc, have lengths at most $m + n$; and if $\mathscr{P}$ has length m, so has its negation $\bar{\mathscr{P}}$. All of the familiar properties seem to have lengths that are at most quadratic functions of N.

c.   Predicate order [18]

Another definition of the complexity of a predicate is based on expressing the predicate by thresholding a linear combination of real-valued properties. Let $\mathscr{S}$ be a set of properties $\mathscr{P}_i$, and define the predicate $\mathscr{Q}$ by

$$\mathscr{Q}[f] = 1 \qquad \text{if } \sum_{\mathscr{S}} a_i \mathscr{P}_i[f] \geqslant t$$

$$= 0 \qquad \text{otherwise}$$

where the coefficients a_i and threshold t are real numbers. Such a $\mathcal{2}$ is said to be linear threshold with respect to $\mathcal{S}$ [for brevity: $LT(\mathcal{S})$]. We say that $\mathcal{2}$ has order k if $\mathcal{2}$ is $LT(\mathcal{S})$ and $|S_{\mathcal{P}_i}| \leq k$ for all $\mathcal{P}_i$ in $\mathcal{S}$, where k is as small as possible.

Evidently $\mathcal{2}$ has order 0 iff it is constant (i.e., 0 or 1 for all f). If $f = \chi_S$, then for any set T of picture points, the predicate "$T \subseteq S$" has order 1. (Such a predicate is called a *mask*.) Indeed, let $T = \{Q_1, \ldots, Q_M\}$, and let $\mathcal{P}_i$ be the predicate "$Q_i \in S$"; then $T \subseteq S$ iff $\sum_{i=1}^M \mathcal{P}_i \geq M$. It can be shown that the only translation-invariant predicates of order 1 are those of the form "$|S| \leq k$" (or $<$, $>$, or $\geq$).

Any predicate $\mathcal{2}$ is LT (the set of all masks). Indeed, $\mathcal{2}$ can be expressed as a Boolean function of $P_1, \ldots, P_N$ having normal form

$$\bigvee_{i=1}^{K} (P_{i1} \wedge \cdots \wedge P_{iN})$$

where each P_{ij} is either P_j or $\bar{P}_j$. For any given χ_S, at most one of these K conjunctions can be 1; hence their disjunction is the same as their sum. Moreover, each conjunction is equivalent to a product, if we replace $\bar{P}_j$'s by $(1 - P_j)$'s. If we multiply out these products and group like terms in their sum, we obtain a linear combination of products of P_j's with integer coefficients; but a product of P_j's is a mask, since it is 1 iff all its terms are in S.

For any points P, Q, let R be a midpoint of P and Q, i.e., if $P = (x, y)$ and $Q = (u, v)$, then $R = ((x + y)/2, (u + v)/2)$, where the arguments are rounded to integer values if necessary. Then "S is convex" (using the midpoint definition of Section 11.3.4c) is equivalent to $\sum_{P, Q, R} -PQ\bar{R} \geq 0$, so that it has order at most 3.

Since the genus of S can be computed by counting the numbers of various 2×2 patterns in χ_S (see Section 11.3.1d), we see that, e.g., "S has genus ≥ 0" has order at most 4. In fact, it can be shown that except for functions of the genus, any predicate invariant under connectedness-preserving transformations of the picture has unbounded order, i.e., order that grows with the picture size. In particular, "S is connected" has unbounded order.

Many other simple picture predicates do not have bounded order; "$|S|$ is odd" is an example. In fact, simple logical combinations of low-order predicates need not have low order; for example, for any k, there exist predicates $\mathcal{P}$ and $\mathcal{2}$ of order 1 such that $\mathcal{P} \wedge \mathcal{2}$ and $\mathcal{P} \vee \mathcal{2}$ have order $> k$. Moreover, even if a predicate is linear threshold with respect to some given set of properties $\mathcal{S}$, the coefficients in the linear combination may span an impractically large range. For example, when we express "$|S|$ is odd" as a linear threshold function with respect to the set of all masks, it turns out that the ratio of the largest to the smallest coefficient grows exponentially with the picture size.

d. Predicate diameter [18], rank [1], etc. [2]

We say that $\mathscr{Q}$ has *diameter d* if it is linear threshold with respect to a set of properties whose sets of support all have diameters $\leq d$, where d is as small as possible. Here again, many familiar predicates, such as "S is connected," turn out to have diameters that grow with the picture size.

More generally, we can consider predicates that are arbitrary functions of a set of linear threshold functions defined on f. In other words, let

$$\mathscr{Q}_i[f] = 1 \quad \text{if } \sum_x \sum_y a^{(i)}_{x,y} q_i(x, y) \geq t_i$$

$$= 0 \quad \text{otherwise}$$

where $q_i(x, y)$ is a predicate whose value depends only on $f(x, y)$, and let $\mathscr{Q}$ be any predicate having the values of the $\mathscr{Q}_i$'s as arguments. Note that if $f = \chi_S$, the sum in each $\mathscr{Q}_i$ is just

$$\sum_{\substack{x \\ (x,y)\in S}} \sum_y a^{(i)}_{x,y}$$

We say that $\mathscr{Q}$ has *rank r* if it can be expressed in this form using r $\mathscr{Q}_i$'s, where r is as small as possible. Here again, predicates such as "$|S|$ is odd" and "S is connected" do not have bounded rank, and neither does any predicate invariant under connectedness-preserving transformations unless it is a function of the genus.

Other measures of predicate complexity involve the average number of sets of support (of individual predicates) that contain a given point of the picture, or the size of the union of these sets. For the details of these definitions, see [2]; connectedness turns out to be highly complex in these terms too.

12.1.3 Linear Properties; Moments

We saw in Section 12.1.1a that for any linear property $\mathscr{P}$, there exists a real-valued array h such that $\mathscr{P}[f] = \sum \sum hf$ for all f. Many different types of h's can be used to define linear properties. We can regard h as a template, and $\sum \sum hf$ as a measure of its match with f (see Section 9.4.1). Such template match properties are useful primarily in cases where f is normalized with respect to geometric transformations. For example, one can use templates of strokes, or even of entire characters, for character recognition purposes.

An important class of linear properties makes use of h's that are digitized versions of standard mathematical functions. For example, if h is a sinusoid, $\sum \sum hf$ is a Fourier coefficient of f; and similarly for the coefficients in other expansions of f. Note that the coefficients obtained from a set of mutually orthogonal h's constitute a set of uncorrelated properties of f.

a. Moments

In the remainder of this section we discuss *moment* properties, which are obtained using h's of the form $x^i y^j$. The (i, j) *moment* of f is defined by

$$m_{ij} \equiv \sum_x \sum_y x^i y^j f(x, y)$$

(in the continuous case, $\sum \sum$ becomes $\iint dx\, dy$). The first few moments of the picture

$$
\begin{array}{ccc}
2 & 1 & 1 \\
3 & 1 & 0 \\
3 & 2 & 1
\end{array}
$$

are as follows, if we take the origin at the pixel in the lower left-hand corner of the picture:

i	j	m_{ij}
0	0	14
1	0	8
0	1	12
2	0	12
1	1	7
0	2	20

Moments can be given a physical interpretation by regarding gray level as mass, i.e., regarding f as composed of a set of point masses located at the points (x, y). Thus m_{00} is the total mass of f, and m_{02} and m_{20} are the moments of inertia of f around the x and y axes, respectively. The moment of inertia of f around the origin $m_0 \equiv \sum \sum (x^2 + y^2) f(x, y) = m_{20} + m_{02}$. It is easily verified that m_0 is invariant under rotation of f about the origin (see Section 10.2.1a). Moreover, if f is rescaled, say by the factor c, it is not hard to see that m_0 is multiplied by c^4. Thus we can normalize f with respect to magnification by rescaling it to give m_0 a specified value (see Section 12.1.1c). Alternatively, a ratio of two moments that have the same value of $i + j$, e.g., m_{01}/m_{10}, is invariant under magnification.

If we substitute $-x$ for x in the definition of m_{ij}, we obtain

$$\sum \sum (-x)^i y^j f(-x, y) = (-1)^i \sum \sum x^i y^j f(-x, y)$$

so that if f is symmetric about the y-axis [i.e., $f(-x, y) = f(x, y)$ for all x, y], we have $m_{ij} = (-1)^i m_{ij}$. Thus if i is odd, m_{ij} must be zero. Similarly, if f is symmetric about the x-axis and j is odd, $m_{ij} = 0$; and if f is symmetric about the origin $[f(-x, -y) = f(x, y)$ for all x, $y]$, and $i + j$ is odd, $m_{ij} = 0$.

Moments for which i, j, or $i + j$ is odd can thus be used as measures of asymmetry about the y-axis, x-axis, and origin, respectively.

Exercise 2. Prove that if f is symmetric about the line $y = x$ [i.e., if $f(x, y) = f(y, x)$ for all x, y], then $m_{ij} = m_{ji}$ for all i, j. ∎

If f is a binary-valued picture, say with S as its set of 1's, the moments of f provide useful information about the spatial arrangement of the points of S. To compute moments from the binary array representation χ_S of S, we simply sum the $x^i y^j$ values for all (x, y) in S. To compute them from the run length representation of S, we compute them for each run and sum the results; for example, the (i, j) moment of the run whose end points are (x', y) and (x'', y) is $y^j \sum_{x = x'}^{x''} x^i$. Similarly, they can be computed from the quadtree representation of S by computing them for each black leaf node, based on its position in the tree, and summing the results; see [72] in Chapter 11 for the details. They are not easy to compute from the MAT representation, since the blocks overlap. They can be computed from the crack or chain code representations of the borders of S in much the same way that area is computed from these representations (see Section 11.3.2a). As an example, for each horizontal crack c_k, let S_k be the vertical rectangle of width 1 extending from the bottom of the picture to c_k; then S is the union of the S_k's for which c_k is an upper boundary of S, minus the union of those for which c_k is a lower boundary. The coordinates of c_k determine the moments of S_k, just as in the case of runs; to compute the (i, j) moment of S, we add the (i, j) moments of all the upper-boundary S_k's, and subtract the sum of the (i, j) moments of all lower-boundary S_k's. For further details in the case of chain code see [19] in Chapter 11.

b.   The centroid; central moments

The *centroid* of f is the point $(\bar{x}, \bar{y})$ defined by

$$\bar{x} = m_{10}/m_{00}, \qquad \bar{y} = m_{01}/m_{00}$$

Thus the centroid of the 3×3 picture shown earlier is $(\frac{4}{7}, \frac{6}{7})$. It is easily verified that if f is shifted, its centroid shifts by the same amount. [*Proof:* If we shift f by (α, β), the origin is now at $(-\alpha, -\beta)$, and the new coordinates of (x, y) are $(x + \alpha, y + \beta)$. Hence $\sum \sum (x + \alpha) f(x, y)/\sum \sum f(x, y) = m_{10}/m_{00} + \alpha = \bar{x} + \alpha$, and similarly for $\bar{y}$.] Thus if we take the origin at the centroid of f, we have normalized f with respect to translation. Note that since the centroid does not have integer coordinates, if we take it at the origin we should redigitize f; alternatively, we can normalize f by taking the origin at the integer-coordinate point closest to the centroid. [Analogous remarks apply in the case of normalizing with respect to magnification in Subsection (a).]

When we take the origin at the centroid, moments computed with respect to this origin are called *central moments*, and will be denoted by $\bar{m}_{ij}$. Evidently $\bar{m}_{00} = m_{00}$, and it can be verified that $\bar{m}_{10} = \bar{m}_{01} = 0$. [*Proof:* Take $(\alpha, \beta) = (-\bar{x}, -\bar{y})$ in the preceding paragraph to obtain $\bar{m}_{10} = \sum\sum (x - \bar{x}) f(x, y) = m_{10} - \bar{x} m_{00} = 0$, and similarly for $\bar{m}_{01}$.]

c. The principal axis

The moment of inertia of f about the line $(y - \beta) \cos \theta = (x - \alpha) \sin \theta$, which is the line through (α, β) with slope θ, is

$$\sum\sum [(x - \alpha) \sin \theta - (y - \beta) \cos \theta]^2 f(x, y)$$

We can find the α, β, and θ for which this is a minimum by differentiating it with respect to α or β and equating the result to zero; this yields

$$\sum\sum [(x - \alpha) \sin \theta - (y - \beta) \cos \theta] f(x, y) = 0$$

or

$$m_{10} \sin \theta - \alpha m_{00} \sin \theta - m_{01} \cos \theta + \beta m_{00} \cos \theta = 0$$

Dividing through by m_{00} gives

$$(\bar{x} - \alpha) \sin \theta - (\bar{y} - \beta) \cos \theta = 0$$

Thus the moment of inertia around the minimum-inertia line must be

$$\sum\sum [(x - \bar{x}) \sin \theta - (y - \bar{y}) \cos \theta]^2 f(x, y)$$

which is the moment of inertia about the line of slope θ through $(\bar{x}, \bar{y})$. In other words, the minimum-inertia line passes through the centroid of f. This line is called the *principal axis* of f.

To find the slope of the principal axis, take the origin at the centroid; then the moment of inertia of f about the line $y = x \tan \theta$ is

$$\sum\sum (x \sin \theta - y \cos \theta)^2 f(x,y)$$
$$= \bar{m}_{20} \sin^2 \theta - 2\bar{m}_{11} \sin \theta \cos \theta + \bar{m}_{02} \cos^2 \theta$$

Differentiating this with respect to θ and equating to zero gives

$$2\bar{m}_{20} \sin \theta \cos \theta - 2\bar{m}_{11}(\cos^2 \theta - \sin^2 \theta) - 2\bar{m}_{02} \cos \theta \sin \theta = 0$$

or

$$\bar{m}_{20} \sin 2\theta - 2\bar{m}_{11} \cos 2\theta - \bar{m}_{02} \sin 2\theta = 0$$

so that $\tan 2\theta = 2\bar{m}_{11}/(\bar{m}_{20} - \bar{m}_{02})$. Since $\tan 2\theta = 2 \tan \theta/(1 - \tan^2 \theta)$, we can obtain $\tan \theta$ as a root of the quadratic equation

$$\tan^2\theta + \frac{\bar{m}_{20} - \bar{m}_{02}}{\bar{m}_{11}} \tan \theta - 1 = 0$$

Exercise 3. Show that the last equation above is equivalent to

$$(\bar{m}_{11}\tan\theta + \bar{m}_{20})^2 - (\bar{m}_{20} + \bar{m}_{02})(\bar{m}_{11}\tan\theta + \bar{m}_{20}) + (\bar{m}_{20}\bar{m}_{02} - \bar{m}_{11}^2) = 0$$

This implies that $m_{11}\tan\theta + m_{20}$ is an eigenvalue of the matrix

$$\begin{pmatrix} \bar{m}_{20} & \bar{m}_{11} \\ \bar{m}_{11} & \bar{m}_{02} \end{pmatrix}$$

Show that the principal axis is in the direction of the eigenvector corresponding to the larger eigenvalue of this matrix. ∎

A standard method of normalizing f with respect to rotation is to rotate it so that its principal axis has some standard orientation, say vertical. (Here again, this involves redigitization.) More generally, f can be normalized with respect to various types of geometrical distortions by transforming it so as to give standard values to various combinations of its moments.

The principal axis of f can be regarded as a line that "best fits" f. More generally [21], one can find higher-order curves that "best fit" f in various senses. For example, given a general quadratic curve

$$q(x, y) \equiv ax^2 + bxy + cy^2 + ux + vy + w = 0$$

we can attempt to find the values of the coefficients a, b, c, u, v, w such that

$$\sum\sum (q(x, y))^2 f(x, y)$$

is a minimum. The curve having these coefficients would be a sort of "quadratic principal axis" for f. Given f's best-fitting quadratic curve q_0, one can attempt to perform "shape normalization" on f by transforming coordinates so that q_0 becomes some standard type of curve—for example, if q_0 is an ellipse, we could transform to make it a circle.

12.1.4 Properties of Projections and Transforms

a. Projections

The vector of column sums $\sum_y f(x, y)$ is called the *x-projection* of the digital picture f, and the vector of row sums $\sum_x f(x, y)$ is called the *y-projection* of f. For example, if f is

$$
\begin{array}{ccccc}
0 & 1 & 1 & 2 & 2 \\
1 & 1 & 2 & 3 & 2 \\
1 & 3 & 2 & 2 & 3
\end{array}
$$

its x-projection is (2, 5, 5, 7, 7) and its y-projection is (11, 9, 6).

More generally, we can define the projection of f on a line of slope θ by summing the gray levels of f along the family of lines perpendicular to θ. Note that these sums do not all have the same number of terms, since some of the lines cross f near its center, while others cross it near its corners. As an example, the projection of the 3×5 f shown above on a $45°$ line is $(1, 4, 3, 5, 7, 4, 2)$. For slopes that are not multiples of $45°$, care must be taken in defining the families of lines, since there may be many digital line segments through a given point that have a given slope; see Section 11.3.3b.

Still more generally, we can define a "projection" of f by summing its gray levels along any one-parameter family of curves, e.g., the lines through a given point, the circles or squares centered at a given point, etc. Note that these sums may be redundant; e.g., lines through a given point have points in common near the given point if their slopes are nearly the same.

Projections are sometimes useful in detecting objects in a picture. For example, suppose f contains a compact region of high gray level on a low-level background; then the x- and y-projections of f will contain peaks near the x- and y-positions of the region, as shown in Fig. 1a. On reconstruction of a picture from a set of its projections see Chapter 8. If f contains a set of high-level points that lie along a curve belonging to a given family, then the projection of f defined by summing along the curves of that family will have a peak corresponding to the high-level curve. Figure 1b illustrates the detection of high-level vertical and near-vertical lines in this way using the x-projection of f (i.e., summing along the vertical lines). This method of curve detection is related to the Hough transform approach discussed in Section 10.3.3.

Projections also provide useful sets of picture properties, particularly for pictures that are normalized with respect to rotation. The x- (y-)projection of f describes how the gray levels of f are distributed in the x- (y-)direction; thus projections can be used to distinguish between f's for which these distributions differ. (As an example, consider the x- and y-projections of a set of digitized characters, e.g., 0, 1, 2, ...) The distributions can be described, e.g., by their moments; note, in fact, that the moments m_{i0} of f depend only on the x-projection of f, and the m_{0j}'s depend only on the y-projection.

A projection is derived from the set of cross sections of f along a given family of curves by summing the gray levels along each cross section. We can define a "generalized projection" by computing any other desired property for each cross section, e.g., the number of points of the cross section having gray levels above a given threshold, the length of the longest run of such points, etc. On the measurement of properties of a binary-valued picture from its cross sections see Section 11.3.2c.

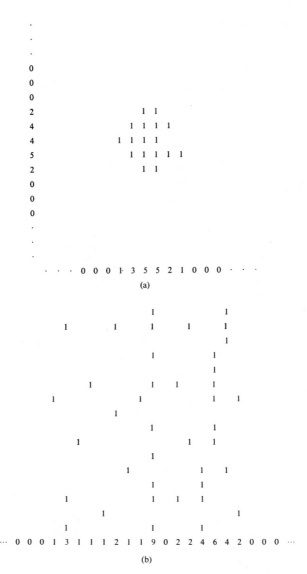

Fig. 1 Object detection using projections. In (a), the *x*- and *y*-projections show peaks at the position of the blob; in (b), the *x*-projection shows peaks at the positions of the broken near-vertical lines.

b. *The autocorrelation and power spectrum*

As mentioned earlier, the Fourier power spectrum $|F|^2$ of f remains the same when f is shifted (cyclically shifted, in the digital case). In other words, $|F|^2$ is invariant under translation of f, so that any properties derived from it are translation-invariant properties of f. Note that $|F|^2$ is also invariant under rotation of f by 180°. The same remarks apply to the (cyclic) auto-correlation R_f of f; in fact, R_f and $|F|^2$ are the Fourier transforms of one another.

It should be pointed out that R_f (or $|F|^2$) does not determine f up to a translation or 180° rotation. As a simple example,

$$3 \quad 10 \quad 3 \qquad \text{and} \qquad 1 \quad 6 \quad 9$$

(on backgrounds of 0's) both have autocorrelation

$$9 \quad 60 \quad 118 \quad 60 \quad 9$$

(on a background of 0's). [These examples have the fewest possible nonzero gray levels; if f has only two (or fewer) nonzero gray levels, R_f does determine it up to translation and 180° rotation.] On the other hand, it can be shown that the second-order autocorrelation of f, defined by

$$S(a, b, c, d) \equiv \sum \sum f(x, y) f(x + a, y + b) f(x + c, y + d)$$

determines f up to translation. On the determination of functions by their autocorrelations see the references cited in [24].

As we shall see in Section 12.1.5d, simple properties of $|F|^2$ or R_f, such as their values in specified regions, provide useful descriptors of the texture of f. Cross sections of R_f (or $|F|^2$) also provide interesting properties of f. For example, note that the values of R_f along a circle C of radius r centered at the origin measure the degree to which f matches itself when "nutated" around itself in an "orbit" of radius r. If f is binary, say $f = \chi_S$, then $R_f(a, b)$ measures the probability that $(x + a, y + b)$ is in S given that (x, y) is in S—i.e., if we drop a line segment l of length $\sqrt{a^2 + b^2}$ and slope $\tan^{-1}(b/a)$ on χ_S in a random position, this is the probability that if one end of l is in S, so is the other end. Thus the integral of R_f along C measures the probability that if we drop a line segment l of length r on χ_S in a random position and orientation, and one end of l is in S, so is the other end; see [24] for references.

Translation invariance can also be achieved using the one-dimensional autocorrelation or power spectrum of a projection of f. Indeed, the projection f_θ of f on a line in direction θ is invariant under translation of f in the direction $\theta + \pi/2$. Moreover, R_{f_θ} and $|F_\theta|^2$ are also invariant under translation of f in direction θ, hence in any direction. This can also be seen by noting that

$|F_0|^2$ is the cross section of $|F|^2$ along the line through the origin with slope $\theta + \pi/2$; see Sections 6.1.1, 7.1.2, and 8.2.

c. Generalized "autocorrelations" and spectra (see [24])

Let the polar coordinate "autocorrelation" of a continuous f be defined by

$$P_f(s, \varphi) \equiv \int_0^{2\pi} \int_0^{\infty} f(r, \theta) f(sr, \theta + \varphi) r \, dr \, d\theta$$

This is evidently invariant under rotation of f about the origin. Moreover, it changes by a constant factor if f is magnified; hence ratios of its values are invariant under magnification of f. If we compute it for R_f rather than for f, it is also invariant under translation of f. Unfortunately, this polar coordinate approach is not readily applicable to digital pictures; in particular, polar-coordinate digitization involves unequally spaced sampling of f.

Even more simply, since $f(r, \theta)$ or $R_f(r, \theta)$ is a periodic function of θ, we can expand it in a Fourier series, and the sum of the squares of the coefficients of this series is invariant under rotation of f about the origin. Magnification invariance can again be obtained by using ratios. On this "polar power spectrum" approach compare the methods of obtaining invariant shape descriptors using transforms of curve equations, as described in Section 11.3.3d.

Another polar-coordinate approach to rotation and magnification invariance is as follows: Scan f along a spiral centered at the origin, where the turns of the spiral are spaced logarithmically, and sample f at equally spaced angular positions around the spiral. Let $\hat{f}$ be a rectangular array each row of which contains the samples from a given turn of the spiral. Then if f is rotated, $\hat{f}$ shifts (cyclically) horizontally; and if f is magnified, $\hat{f}$ shifts vertically, since magnification becomes a shift on the logarithmic vertical scale. These shifts can be eliminated by taking the autocorrelation or power spectrum of $\hat{f}$. Of course, this description is not exact; in particular, it breaks down near the top and bottom of $\hat{f}$, since we can only use finitely many turns of the spiral. A similar method can be applied to R_f or $|F|^2$ rather than f; note that they rotate when f rotates, and (de)magnify when f is magnified (see Section 2.1.4).

Let $\mathcal{G} = \{\mathcal{O}_1, \ldots, \mathcal{O}_n\}$ be a group of linear picture operations, and let $\bar{f}$ be the average of $\mathcal{O}_1[f], \ldots, \mathcal{O}_n[f]$. As pointed out in Section 12.1.1b, for any $\mathcal{O}$ in $\mathcal{G}$ we have $\{\mathcal{O}_1[\mathcal{O}[f]], \ldots, \mathcal{O}_n[\mathcal{O}[f]]\} = \{\mathcal{O}_1[f], \ldots, \mathcal{O}_n[f]\}$; hence $\overline{\mathcal{O}[f]} = \bar{f}$ for all $\mathcal{O}$—in other words, $\bar{f}$ is the same whether we start with f or with any of the $\mathcal{O}[f]$'s. Thus $\bar{f}$ is a normalized version of f with respect to $\mathcal{G}$. In particular, $\mathcal{G}$ can be any group of geometric operations, since all such operations are linear. As an example, $\bar{f}(r) \equiv \int_0^{2\pi} f(r, \theta + \varphi) d\varphi$ is invariant under rotation of f about the origin; this $\bar{f}$ is what f would look like if it were rotating very rapidly about the origin.

12.1.5 Statistical Properties: Texture

This section discusses statistical picture properties, and in particular, properties that can be used to describe the "visual texture" of a picture, or better, of a statistically homogeneous region in a picture. We will not attempt to define conditions under which a region would be called uniformly textured. Such regions are often described as consisting of large numbers of small uniform patches, or "primitive elements," arranged according to "placement rules," where the patch shapes and positions are governed by random variables.

a. Gray level statistics

The histogram $p_f(z)$ of a digital picture f tells us how often each gray level occurs in f; it provides an estimate of the gray level probability density in the ensemble of pictures of which f is a sample. If there are k possible gray levels, $z_1, \ldots, z_k$, p_f is a k-element vector. Statistics computed from p_f give us general information about this gray level population. For example:

(1) The *mean* gray level of f, $\mu_f \equiv (1/N) \sum z p_f(z)$, where $N \equiv \sum p_f(z)$ is the number of points in f, is a measure of the overall lightness/darkness of f. The *median* gray level, i.e., the gray level m_f such that (about) half the points of f are lighter than m_f and half are darker, is another such measure.

(2) The ·gray level *variance* of f, $\sigma_f{}^2 \equiv (1/N) \sum (z - \mu_f)^2 p_f(z)$, and the *standard deviation* σ_f, are measures of the overall contrast of f[§]; if they are small, the gray levels of f are all close to the mean, while if they are large, f has a large range of gray levels. Another such measure is the *interquartile range* r_f, which is defined as follows: Let m_{1f} be the gray level such that $\frac{1}{4}$ of the pixels of f are lighter than m_{1f} and $\frac{3}{4}$ are darker; let m_{3f} be defined analogously, with the $\frac{1}{4}$ and $\frac{3}{4}$ interchanged; then $r_f \equiv |m_{1f} - m_{3f}|$. Other percentiles can be used here in place of the quartiles m_{1f} and m_{3f}.

b. Second-order gray level statistics

Statistics computed from the histogram p_f are of only limited value in describing f, since p_f remains the same no matter how the points of f are permuted—for example, p_f is the same when f is half black and half white, when f is a checkerboard, or when f consists of salt-and-pepper noise. More insight into the nature of f is obtained by studying how often the possible pairs of gray levels occurs in given relative positions.

Let $\delta \equiv (\Delta x, \Delta y)$ be a displacement, and let M_δ be the $k \times k$ matrix whose (i, j) element is the number of times that a point having gray level z_i occurs in

[§] Note that if we define one-dimensional moments by $m_i \equiv \sum z^i p_f(z)$, we have $N = m_0$ and $\mu_f = m_1/m_0$ (so that μ_f is the centroid of p_f); moreover, if we define central moments $\bar{m}_i \equiv \sum (z - \mu_f)^i p_f(z)$ by taking μ_f as the origin, we have $\sigma_f^2 = \bar{m}_2/\bar{m}_0$.

position δ relative to a point having gray level z_j, $1 \leqslant i, j \leqslant k$. For example, if f is

$$
\begin{array}{cccc}
1 & 1 & 2 & 2 \\
0 & 2 & 2 & 1 \\
0 & 0 & 2 & 1 \\
1 & 0 & 0 & 1
\end{array}
$$

and δ is $(1, 0)$, then M_δ is

$$
\begin{array}{ccc}
2 & 1 & 2 \\
1 & 1 & 1 \\
0 & 2 & 2
\end{array}
$$

Note that the size of M_δ depends only on the number of gray levels, not on the size of f. Elements near the main diagonal of M_δ correspond to pairs of gray levels that are nearly equal, while elements far from the diagonal correspond to pairs that are very unequal.

Let N_δ be the number of point pairs in f in relative position δ; this is less than the total number of points in f, since if (x, y) is near the border of f, $(x + \Delta x, y + \Delta y)$ may lie outside f. Then in the matrix $P_\delta \equiv M_\delta / N_\delta$ (i.e., if we divide each element of M_δ by N_δ), the (i, j) element is an estimate of the joint probability that a pair of points in relative position δ will have the pair of gray levels (z_i, z_j). P_δ is called a gray level *co-occurrence matrix* for f.

The matrices P_δ, for various δ's, provide useful information about the spatial distribution of gray levels in f. For example, suppose that f is composed of patches of approximately constant gray level of a certain size s. If the length of δ is small relative to s, then the high-valued entries in P will be concentrated near its main diagonal, since a pair of points δ apart will often have nearly the same gray level. On the other hand, if δ is long relative to s, the entries in P will be more spread out. If f consists of elongated streaks oriented in a given direction, the spread of values in P_δ will depend on both the length and slope of δ. If directionality is not important, we can use matrices $\bar{P}$ that are averages of P_δ's (or matrices $\bar{M}$ that are sums of M_δ's) for sets of displacements of a given size in various directions. For example, if f is the 4×4 picture shown above, and we use the displacements $(1, 0)$, $(0, 1)$, $(-1, 0)$, and $(0, -1)$, then the combined matrix $\bar{M}$ is

$$
\begin{array}{ccc}
8 & 4 & 4 \\
4 & 6 & 5 \\
4 & 5 & 8
\end{array}
$$

(Note that $\bar{M}$ is symmetric, since the set of directions used is symmetric.)

Figure 2 shows a set of pictures (Fig. 2a) and some of their co-occurrence matrices. These are 8×8 matrices, corresponding to 8-level requantization of the pictures. For the matrices in Figs. 2b–e, the set of displacements is $\{\pm(4, 0), \pm(0, 4), \pm(4, 4), \pm(4, -4)\}$. We see that for the "coarse" pictures (the second and third ones in Fig. 2a), the values are more concentrated near the diagonals of the matrices (Figs. 2c and 2d), whereas for the "busy" pictures (the first and fourth in Fig. 2a) the values are more spread out (Figs. 2b and 2e). Figure 2f shows the matrix for the fourth picture when we use displacements $\{\pm(1, 0), \pm(0, 1), \pm(1, 1), \pm(1, -1)\}$; here the diagonal concentration is noticeably greater than in Fig. 2e. Finally, Figs. 2g and 2h show the matrices for the third picture using only the displacements $\pm(4, 4)$ and $\pm(4, -4)$, respectively; the diagonal concentration is greater in the first of these, corresponding to the fact that the picture contains streaks oriented at about $45°$.

In principle, a large set of P_δ matrices is needed to completely specify the second-order gray level statistics of f. In practice, however, matrices corresponding to large displacements are not necessary. As δ becomes long, the pairs of gray levels separated by δ become uncorrelated, and $P_\delta(i, j)$ approaches the probability that a pair of randomly chosen points of f have gray levels z_i and z_j. Thus for practical purposes we need not use δ's having lengths greater than the distance over which f's gray levels remain correlated, or greater than the size of the "patches" of which f is composed. In fact, the most important P_δ's are usually those for which δ has length 1. Historically, gray level transition probabilities $p(z_j|z_i)$ have been used to characterize textures; $p(z_j|z_i)$ is the probability that a point has level z_j given that the preceding point (with respect to a scan of the picture) has level z_i. Note that the joint probability $p(z_i, z_j)$, which is equal to $p(z_i)p(z_j|z_i)$, is just the (i, j) element of P_δ for $\delta = (1, 0)$. Other investigators have characterized textures by fitting a time series model to the sequence of gray levels, and using the parameters of this model as texture descriptors; this approach will not be discussed here in detail.

Haralick (see [13]) has suggested a number of statistics that can be used to describe a given co-occurrence matrix P_δ. Four of these are:

(1) "Contrast," $\sum_i \sum_j (i - j)^2 P_\delta(i, j)$; this is the moment of inertia of P_δ about its main diagonal. Evidently, it is low when the diagonal concentration of P_δ is high, and vice versa.

(2) "Inverse difference moment," $\sum_i \sum_j P_\delta(i, j)/[1 + (i - j)^2]$; this is high when the diagonal concentration is high.

(3) "Angular second moment," $\sum_i \sum_j P_\delta^2(i, j)$; this is lowest when the $P_\delta(i, j)$'s are all equal, and high when they are very unequal, so that in particular it tends to be high when the diagonal concentration is high.

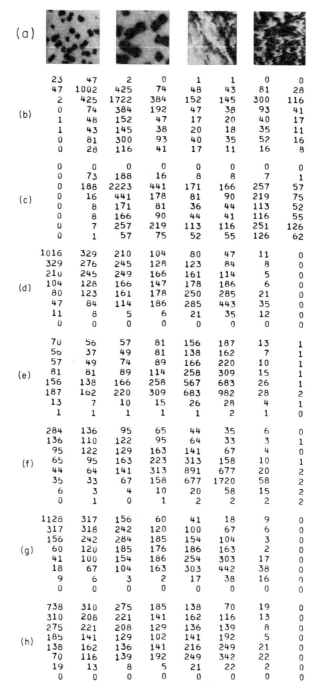

(a)

(b)

23	47	2	0	1	1	0	0
47	1002	425	74	48	43	81	28
2	425	1722	384	152	145	300	116
0	74	384	192	47	38	93	41
1	48	152	47	17	20	40	17
1	43	145	38	20	18	35	11
0	81	300	93	40	35	52	16
0	28	116	41	17	11	16	8

(c)

0	0	0	0	0	0	0	0
0	73	188	16	8	8	7	1
0	188	2223	441	171	166	257	57
0	16	441	178	81	90	219	75
0	8	171	81	36	44	113	52
0	8	166	90	44	41	116	55
0	7	257	219	113	116	251	126
0	1	57	75	52	55	126	62

(d)

1016	329	210	104	80	47	11	0
329	276	245	128	123	84	8	0
210	245	249	166	161	114	5	0
104	128	166	147	178	186	6	0
80	123	161	178	250	285	21	0
47	84	114	186	285	443	35	0
11	8	5	6	21	35	12	0
0	0	0	0	0	0	0	0

(e)

70	56	57	81	156	187	13	1
56	37	49	81	138	162	7	1
57	49	74	89	166	220	10	1
81	81	89	114	258	309	15	1
156	138	166	258	567	683	26	1
187	162	220	309	683	982	28	2
13	7	10	15	26	28	4	1
1	1	1	1	1	2	1	0

(f)

284	136	95	65	44	35	6	0
136	110	122	95	64	33	3	1
95	122	129	163	141	67	4	0
65	95	163	223	313	158	10	1
44	64	141	313	891	677	20	2
35	33	67	158	677	1720	58	2
6	3	4	10	20	58	15	2
0	1	0	1	2	2	2	2

(g)

1128	317	156	60	41	18	9	0
317	318	242	120	100	67	6	0
156	242	284	185	154	104	3	0
60	120	185	176	186	163	2	0
41	100	154	186	254	303	17	0
18	67	104	163	303	442	38	0
9	6	3	2	17	38	16	0
0	0	0	0	0	0	0	0

(h)

738	310	275	185	138	70	19	0
310	208	221	141	162	116	13	0
275	221	208	129	136	139	8	0
185	141	129	102	141	192	5	0
138	162	136	141	216	249	21	0
70	116	139	192	249	342	22	0
19	13	8	5	21	22	2	0
0	0	0	0	0	0	0	0

Fig. 2 A set of pictures (a) and some of their gray level co-occurrence matrices, based on subdivision of the gray scale into eight ranges. In parts (b)–(e), each matrix entry (i, j) represents the number of co-occurrences of a gray level in range i with a gray level in range j at any of the displacements $(0, \pm 4)$, $(\pm 4, 0)$, $(\pm 4, \pm 4)$, and $(\pm 4, \mp 4)$. Part (f) shows the matrix for the last picture only, for displacements $(0, \pm 1)$, $(\pm 1, 0)$, $(\pm 1, \pm 1)$, and $(\pm 1, \mp 1)$. Parts (g)–(h) show matrices for the next-to-last picture only, for displacements $(\pm 4, \pm 4)$ and $(\pm 4, \mp 4)$, respectively.

(4) "Entropy," $-\sum_i \sum_j P_\delta(i, j) \log P_\delta(i, j)$; this is highest when the $P_\delta(i, j)$'s are all equal, and hence is low when the diagonal concentration is high.

It should be pointed out that the arrangements of values in the co-occurrence matrices depend not only on the coarseness or busyness of the given picture, but also on its lightness and contrast. For example, if we stretch the gray scale of a picture, the entries in the matrices will spread away from the diagonal, since the pairs of gray levels will be farther apart. Features (1) and (2) defined above will be especially sensitive to such changes [this is why feature (1) is called "contrast"], while features (3) and (4) will be less so. To avoid confusing the effects of the first and second order statistics of the picture, it is common practice to normalize its gray scale (e.g., by histogram flattening; see Section 6.2.4) before computing the matrices, so that the first-order statistics have standard values.

We can define co-occurrence matrices that may be more sensitive to the spatial structure of the given texture by using only selected pairs of points in constructing the matrices, rather than using all possible pairs having a given relative position. For example, suppose that we consider only point pairs (Q, R) in which Q is on an edge (e.g., is at a local maximum of the gradient magnitude), and R is a given distance δ away from Q in the gradient direction. In the matrix P_δ' defined in this way, diagonal concentration still corresponds to coarseness, since if δ is small relative to the texture patch size, R should be interior to the patch on the edge of which Q lies. However, P_δ' may be more sensitive to coarseness changes than the P_δ matrices were, since P_δ' is not influenced by point pairs that are both interior to patches.

c. Local property statistics

Another way of obtaining information about the spatial arrangement of the gray levels in f is to compute statistics of various local property values f' measured at the points of f.

As an illustration of how local properties can be used for texture description, let $\delta \equiv (\Delta x, \Delta y)$ be a displacement, let

$$f_\delta(x, y) \equiv f(x, y) - f(x + \Delta x, y + \Delta y)$$

and let p_δ be the histogram of f_δ. Suppose that f is composed of patches of size s. If δ is short relative to s, the high entries in p_δ will be concentrated near 0, since pairs of points δ apart will usually have small differences in value; but if δ is large, the entries in p_δ will be more spread out. [Note, in fact, that $p_\delta(z)$ is the sum of the entries in the matrix M_δ along the line parallel to its main diagonal for which $i - j = z$.] Thus the concentration of p_δ near 0 is a measure of the "coarseness" of f relative to δ, or equivalently, the spread of p_δ away from 0 is a measure of the "busyness" of f. Here again, these properties may depend on direction. Similar remarks apply if we use absolute rather than signed differences; this simply folds p_δ over on itself at the origin. Absolute difference values and their histograms for $\delta = (1, 0)$ and $(0, 1)$ are shown in Fig. 3 for the same pictures as in Fig. 2.

The gray level (absolute) difference histograms p_δ are not affected by shifting the gray scale (as co-occurrence matrices are), but they are affected by stretching it; thus they too should be used in conjunction with gray scale normalization. Various statistics can be used to describe p_δ, including its mean $[(1/N) \sum z p_\delta(z)$, if we use absolute differences], its second moment $[\sum z^2 p_\delta(z)$; this is proportional to the "contrast" statistic for the corresponding cooccurrence matrix], its entropy $[-\sum p_\delta(z) \log p_\delta(z)]$, and so on.

A wide variety of local properties f' can be used in place of f_δ for texture description. For example, we can use combinations of differences, such as the gradient (magnitude) or Laplacian; matches to local templates, such as spot, line, corner, or line end detectors; and so on. f' can be a predicate, e.g., 1 if an above-threshold difference is present and 0 otherwise; in this case, the histogram consists of only two values, and its mean tells us how many edges (or spots, lines, etc.) are present in f per unit area. Another possibility is to count local gray level maxima and minima in f; evidently, both the number of edges and the number of extrema per unit area are measures of "busyness." More generally, we can count occurrences of arbitrary local patterns of values in f.

We can use second-order as well as first-order local property statistics as texture descriptors, by constructing co-occurrence matrices of the values of f' in given relative positions. If desired, we can use only selected pairs of points in constructing the matrices, e.g., pairs of extrema or pairs of above-threshold edge points, and we can use displacements at each point that depend on f', e.g., displacements in the gradient direction, as at the end of Subsection (b). For a general discussion of this approach see [7].

Rather than using a set of local properties, e.g., differences computed for a set of displacements, we can use a single property and measure it for pictures derived from the original one by a set of local operations. For example, suppose that we use a sequence of local min (or max) operations, and at each step, measure the average gray level; the rate at which this decreases (or increases) is a measure of the coarseness of the high-valued (low-valued) patches in f. For a binary-valued f, the analogous idea is to shrink (or expand) the 1's in f repeatedly, and at each step, count the number of 1's. This approach, using generalized shrinking and expanding operations (see Section 11.2.1c), has been extensively used for texture analysis in microscopy.

d. Autocorrelation and power spectrum

In the previous two subsections we saw how various statistics of the co-occurrence matrix, difference histogram, etc., for a given displacement $\delta \equiv (\Delta x, \Delta y)$ provide useful information about a texture. Thus the set of values of a given statistic α as a function of δ (in particular, for relatively short δ's) can be used as a texture descriptor.

For example, let $f_\delta'' \equiv f(x, y)f(x + \Delta x, y + \Delta y)$, and let α be the mean of f_δ''; then α as a function of δ is just the *autocorrelation* R_f, i.e., the expected value of the product of the gray levels of a pair of points δ apart. By the Cauchy–Schwartz inequality, this takes on its maximum value for $\delta = (0, 0)$. [*Proof*: We know that

$$\frac{\sum f(x, y)f(x + \Delta x, y + \Delta y)}{[\sum f(x, y)^2 \sum f(x + \Delta x, y + \Delta y)^2]^{1/2}} \leqslant 1$$

but the two factors in the denominator are the same, so that the denominator is equal to $\sum f(x, y)^2$, which is $R_f(0, 0)$.] The rate at which R_f falls off as δ moves away from $(0, 0)$ is a measure of the coarseness of f; the falloff is slower for a coarse texture, and faster for a busy one.

Similarly, let $f_\delta^2 = [f(x, y) - f(x + \Delta x, y + \Delta y)]^2$, and let $\alpha \equiv v$ be its mean, i.e., the expected squared gray level difference at two points δ apart; this descriptor is sometimes called the *variogram* of f. [Compare the use of the mean of the absolute difference histogram as a texture descriptor in Subsection (c).] The rate at which its value rises as δ moves away from $(0, 0)$ is a measure of the coarseness of f; the rise is slow for a coarse texture and fast for a busy one. Note that

$$\begin{aligned} v(\delta) &= E\{[f(x, y) - f(x + \Delta x, y + \Delta y)]^2\} \\ &= E\{f^2(x, y)\} + E\{f^2(x + \Delta x, y + \Delta y)\} \\ &\quad - 2E\{f(x, y)f(x + \Delta x, y + \Delta y)\} \end{aligned}$$

Here the first two terms are just $R_f(0, 0)$ and the third is $-2R_f(\delta)$, so that $v(\delta) = 2(R_f(0, 0) - R_f(\delta))$. If f is isotropic, the values of R_f and v depend only on the length of δ, not on its direction, so that they become functions of a single variable.

A texture can be modeled as a correlated random field, e.g., as an array of independent identically distributed random variables, to which a filtering operator has been applied. This model suggests that a texture can be described by its autocorrelation and by the probability density of the original random variables; the latter can be approximated by a histogram after a "whitening" operation has been applied to decorrelate the texture. If we use the gradient or Laplacian as an approximate whitening operation, the histogram is just a histogram of difference values, as in the preceding subsection [23].

As we saw in Section 2.1.4, the Fourier power spectrum $|F|^2$ and the autocorrelation R_f are Fourier transforms of each other. Thus $|F|^2$ can also be used as a texture descriptor. The rate at which $|F|^2$ falls off as the spatial frequency (u, v) moves away from $(0, 0)$ is again a measure of the coarseness of f; the falloff is faster for a coarse texture and slower for a busy one, since fine detail gives rise to more power at high spatial frequencies. Samples of $|F|^2$ taken over rings centered at $(0, 0)$, or over sectors emanating from $(0, 0)$ (to detect directional biases), have often been used as texture features. Figure 4 shows the log scaled power spectra of the pictures used in Figs. 2 and 3, together with plots of their summed values over the rings and sectors.

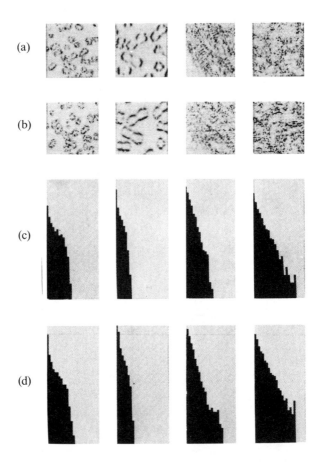

Fig. 3 Absolute gray level differences in the horizontal and vertical directions [(a) and (b)] and their histograms [(c) and (d); log scaled] for the pictures in Fig. 2.

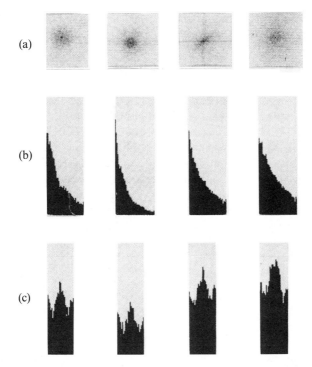

Fig. 4 Log power spectra (a), sums over rings centered at (0, 0) (b), and sums over sectors emanating from (0, 0) (c), for the pictures in Fig. 2.

Other transforms of f can also be used as a source of texture features. In practice, features based on $|F|^2$ (or R_f) seem to be somewhat less effective for texture discrimination than features based on second-order or local property statistics. At the same time, computation of $|F|^2$ is more costly than computation of a few statistical features (recall that they usually need only be computed for a few δ's), unless we compute it optically (see Section 4.2.3).

e. Region-based descriptions

The texture descriptors considered so far are derived from local or point pair properties. We conclude by briefly discussing texture description in terms of homogeneous patches or "primitive" regions. Several types of texture models are based on such decompositions into regions. For example, textures can be generated by using a random geometric process to tessellate the plane into cells, or to drop objects onto the plane, and then selecting gray levels (or gray level probability densities) for the cells or objects in accordance with some probability law [30].

If we can explicitly extract a reasonable set of primitives from f, we can describe the texture of f using statistics of properties of these primitives—e.g., the mean or standard deviation of their average gray level, area, perimeter, orientation (of principal axis), eccentricity, etc. Second-order statistics can also be used—i.e., we can construct matrices for pairs of values of the area (etc.) at pairs of neighboring primitives (perhaps in directions defined by each primitive's orientation [17]). Of course, this approach depends on being able to extract a good set of primitives from f at a reasonable computational cost. A related, but much simpler, idea is to extract maximal homogeneous blocks (e.g., runs of constant gray level in various directions) from f, and describe f in terms of (first- or second-order) statistics of the block sizes (e.g., run lengths).

In general, the description of textures in terms of primitives may be hierarchical; the primitives may be composed of subprimitives, etc., or they may be arranged into groupings which in turn form larger groupings, etc. This makes it possible to define placement rules for the primitives in the form of stochastic grammars. Texture analysis can thus be carried out, in principle, by parsing with respect to a set of such grammars. Picture grammars will be discussed further in Section 12.2.3.

12.2   MODELS

12.2.1   Descriptions and Models

Up to now we have discussed methods of representing and describing a specific, given picture. In this section we discuss picture *models*, which can be thought of as descriptions of classes of pictures. In Sections 12.2.2 and 12.2.3 we define many types of models at various levels of description. Section 12.2.4 considers how to construct models appropriate to a given picture analysis task, and how to establish correspondences between descriptions and models, so that a picture can be interpreted in terms of the models.

a.   Levels of description

It is convenient to distinguish three levels of description and modeling: the *array* level, the *region* or *geometric* level, and the *relational* level. These levels are discussed in the following paragraphs.

Pictures are initially specified as arrays of numerical values, representing gray levels, color components, or intensities in given spectral bands. Various types of derived information, obtained by processing these input arrays, are also commonly represented in array form. These include:

(1) Results of applying operations or transforms to the input, e.g., computing its gradient magnitude at each point, taking its Fourier transform, etc.

(2) "Symbolic" or "overlay" arrays obtained by classifying or partition-ing the input and assigning symbols to the classes or parts, e.g., the results of thresholding, connected component labeling, etc.

(3) Arrays representing pointwise properties of the original scene that are not sensed directly, e.g., range, surface slope, illumination, gloss, etc.

Any of these types of arrays, including the original picture itself, can be regarded as a description of the picture. Of course, such descriptions are usually too detailed to be the final goal of the description process, but as the above examples suggest, they play important roles in the early stages of that process. As we shall see in the next two sections, many types of models can be defined at the array level.

A second level of description treats a picture as being composed of regions, and specifies the geometry of these regions, but does not give a point by point description of their gray levels. The region descriptions can be of many types, as indicated in Chapter 11. For example, curves or region borders can be specified by chain (or crack) codes, or by piecewise approximation by simple functions. Solid regions can be specified as unions of maximal blocks, e.g., runs or maximal "disks" (as in the MAT); these may be restricted to have standard positions and sizes, as in quadtree representations. If the block centers lie on a set of curves, the region is a union of "generalized ribbons," defined by specifying the curves and their associated "width functions." [Of course, regions can also be represented by overlay arrays (e.g., 1's at points of the region, 0's elsewhere), but as we saw in Chapter 11, nonarray representations are more commonly used, because they are more compact when the regions are simple.] Classes of regions can be modeled in terms of any of these representations, as will be seen below.

A third level of description deals with properties of and relationships among picture parts. The parts themselves may be geometrically specified as regions, or they may only be characterized by a set of geometric properties. The description may also specify the positions of the parts and nongeometric properties of the parts (e.g., it may specify an array model for each part, or may only give a set of gray level dependent properties of the part), as well as relationships among the parts. Such a description can generally be repre-sented by a graph structure in which the parts correspond to nodes, labeled with lists of property values (and with pointers to descriptions at lower levels, if appropriate), and the arcs are labeled with lists of relationship values. Models based on such graphlike representations will be discussed below. Note that if a description does not refer to parts of the picture, it reduces to a list of nongeometric properties of the picture as a whole.

Other levels of picture description are possible; for a review of the types of data structures used in picture analysis and description see [31]. Moreover, some aspects of a description may not directly relate to picture parts at all, e.g., statements about the intentions of the people who appear in the picture. We shall consider here only descriptions that do explicitly refer to picture parts or to the picture as a whole.

b. Types of models

We shall consider two general types of picture models, which we call *declarative* and *procedural*.

A declarative model consists of a set of constraints on the properties of or relationships among the parts (pixels, regions, etc.) at the given level of description. The pictures whose descriptions satisfy these constraints constitute the class defined by the model. We will give many examples of such models, on all levels, in Section 12.2.2.

A procedural model, in general, is any process that generates or recognizes pictures; the class that it defines consists of the pictures that it generates or accepts. Since this type of definition can be quite arbitrary, we consider here only a special class of procedural models that have been extensively studied: *grammatical* or *syntactic* models. Various types of grammars will be defined in Section 12.2.3, including string, array, cycle, plex, tree, and graph grammars, as well as grammars in which the symbols have numerical parameters associated with them.

Picture models are often fuzzy or probabilistic; rather than sharply defining a class of pictures, they define membership functions or probability densities on the space of all pictures. In particular, declarative models often involve probabilistic constraints; while in a syntactic model, rule application may be controlled probabilistically. Many of the examples of models given in the next two sections will be probabilistic or fuzzy.

12.2.2 Declarative Models

This section discusses picture models defined by constraints on properties or relationships at a given level of description. In Subsections (a)–(d) we give general examples of such models at the array, region, and relational levels. Subsection (e) briefly discusses models for classes of scenes and their relationship to picture models.

a. Arrays

At the pixel level, declarative models are specified by constraints on properties of the pixels. Constraints on relationships are not appropriate at this level, since the pixels have a fixed arrangement; however, one can impose constraints on the properties of pixels that are related in specified ways. The constraints may be deterministic (e.g., we can specify which gray levels are allowed to occur in which positions), but we will primarily consider probabilistic constraints here.

The weakest constraints on pixel properties are those which do not depend on the positions of the pixels. Such constraints impose restrictions on the pixel population without regard to its spatial arrangement; they can thus be regarded as constraints on the probability density of pixel values (e.g., gray levels or spectral signatures). For example, a model of this type might specify that the density has high variance, is bimodal, or is (approximately) a mixture of two Gaussian densities. We saw in Section 10.1 how such models provide a basis for segmenting a picture by classifying its pixels. Since the histogram (or scatter plot) of the pixel values in a picture is an estimate of their probability density, we can decide how well a given picture fits a model of this type by analyzing its histogram.

More generally, we can define models in terms of the probability densities of various arrays derived from the given picture. For example, one can define array models for Fourier transforms or power spectra of pictures, e.g., by specifying which spatial frequencies are present, or by prescribing the rate of falloff from the origin. As we have seen in earlier chapters, models for the autocorrelations of pictures, based on exponential falloff, are quite commonly used.

Another possibility is to define models for arrays of local property values such as local average gray level, absolute gray level difference, etc., derived from the picture. Note that these densities depend on the arrangement of the pixels, at least locally. We saw in Section 10.1 how local property values, possibly in conjunction with gray level, can be used in segmentation by pixel classification, and in Section 12.1.5c how statistics of local property values are useful as a basis for defining texture descriptors; all of these applications are implicitly based on assumptions about populations of local property values.

A more general class of models is defined by constraining the joint probability density of pixel values (or local property values) for pairs (or *n*-tuples) of pixels in given relative positions. Evidently, these second- and higher-order probability density models constrain the spatial arrangement of the pixels that have various values. As in the first-order case, these models can range from highly specific to highly general; the density can be completely specified, or it can be required to have some given property. We saw in Sections 12.1.5b and 12.1.5c how second-order statistics of gray levels or local property values, measured for small displacements, can be useful as texture descriptors; the value of such descriptors in characterizing textures depends on assumptions about the second-order probability densities of the textures. Note that when we model pictures in terms of autocorrelation, we are using a second-order statistic, the expected value of the product of two gray levels in a given relative position, measured for all displacements.

Another method of constraining pairs or n-tuples of pixel values is based on conditional, rather than joint, probabilities. The probability density for a given pixel is assumed to depend, at least in part, on those of one or more "preceding" pixels (e.g., relative to a row-by-row scan of the picture). This approach is used in *time series* models, in which each gray level (or property value) z_i, regarded as a random variable, is assumed to depend on a set of preceding z_{i-j}'s, and/or on "noise" terms v_{i-j}, which are a set of independently distributed random variables. For example, in *autoregressive models*, z_i is assumed to be a linear combination of preceding z's and the current v_i; while in *moving average* models, z_i is a linear combination of the current and preceding v's. Considerable use has been made of such models in texture modelling. A problem with this approach is its one-sidedness (i.e., dependence of a pixel on a set of "predecessors" rather than on all its neighbors); unfortunately, symmetrical constraints seem to be mathematically much less tractable than one-sided constraints.

The most general type of model at the array level is a *random field*, in which the probability density of values of a pixel can depend on the pixel's position. If the dependence on position is very strong, such a model can be regarded as a noisy template; this interpretation is appropriate, for example, if each pixel's values are normally distributed, with mean depending on position. At the other extreme, if all pixel values are identically and independently distributed, we have a pure noise model. To introduce some spatial structure into such a model without making it templatelike, we can assume that all pixel values are identically, but not independently, distributed; and we can further assume that joint densities of pixel values (of various orders) depend only on relative position, as discussed above. Such homogeneous random field models are commonly used in image processing, as we saw in earlier chapters. Another possibility is to define a model by applying a filtering operation (which may or may not be linear or space-invariant) to a noise model; as indicated in Section 12.1.5d, textures can be modeled in this way too.

b. Regions

Models at the region level are concerned with the shapes of regions in the picture. They do not deal with the gray level population of each region;

this can be characterized by an array model of any of the types considered in Subsection (a), restricted to the given region. They also do not deal with the positions of the regions, nor with relationships among them; models on this level will be discussed in the next subsection.

In principle, declarative models for classes of region shapes can be defined in terms of any of the region representations of Chapter 11. For example, we can define a class of simply connected regions by specifying a family of border codes, or, more generally, a class of intrinsic or parametric equations, or a class of transforms of such equations. Some region shape properties are easy to constrain in this way, e.g., we can make the shapes smooth by constraining the rate of change of curvature, or convex by requiring that the curvature never changes sign; but it is harder to control properties such as elongatedness in this representation. Similarly, we can define a class of regions in terms of generalized ribbons, using equational representations of the MA arcs and width functions; here we can impose smoothness by making the arcs smooth and constraining the rate of change of width, and elongatedness by making the widths small relative to the arc lengths, but it is harder to control properties such as convexity.

One can also define region models in terms of constraints on the population of slopes, curvatures, or widths. For example, first-order slope statistics (the "directionality spectrum") can be used to characterize the overall orientation of a region border, while second-order slope statistics (for given displacements along the curve), or first- (or higher-) order curvature statistics, can be used to characterize its smoothness or wiggliness. Such models should be useful in the design of border following or curve tracking algorithms. Similarly, the "size spectrum" (of widths, or maximal block sizes), or higher-order width statistics, should be useful in designing region growing algorithms.

Conditional probability constraints on slopes, widths, etc., can also be used to define models. For example, one can define "random walk" or time series models for border codes (i.e., for sequences of slopes or curvatures) or for the width functions of generalized ribbons. If a region meets each row of the picture in at most one run, time series models can be used for the sequences of run lengths and run end positions.

The models considered so far involve single regions that do not interact with other regions. Such models are applicable to pictures containing isolated objects on a background. More generally, however, regions may interact in various ways, and this imposes constraints on their shapes. The following are a few of the many possibilities:

(1) Regions can be built up by superposition or concatenation of other regions. The use of grammars to builds up regions or objects out of parts will be discussed in Section 12.2.3c. As an extreme example, regions can be

generated by "bombing" processes such as those studied in random geometry, or by random packing processes.

(2)   Regions can be defined by partitioning the picture. Partitions can be constructed using a variety of random geometric processes—e.g., start with a spatial distribution of points, and allow them to expand until they collide; or define the borders of the parts by a set of curves, specified by a distribution of points in some curve parameter space. Some limited use has been made of such processes in texture modeling.

When regions interact, one must also specify what happens in their zone of interaction. For example, when regions are superimposed, do they mix or occlude? Are interregion borders sharp or blurred?

c.   Relations

At the relational level we deal with picture modeling in terms of parts, their properties, and their interrelationships. The parts may be geometrically specified by models at the region level, and may have their gray level populations characterized by array level models; but at the relational level we treat the parts as "atomic" objects.

In general, a relational model may specify or constrain the positions of the objects within the picture. For example, it might specify, for each position, the probability that an object of a given type occurs (e.g., has its centroid) at that position. This type of specification is basically a random field in which the values of the random variables are object occurrences. As in the array case, such a model can range from templatelike (i.e., very strong dependence on position) to random (i.e., position-independent); compare the "bombing models" mentioned in Subsection (b). As an intermediate possibility, the relative positions of objects can be specified or constrained; in particular, we can constrain the relative positions of nearest-neighbor pairs of objects, e.g., we can require them to touch, or conversely, forbid them to occur closer than a given distance from one another. Note that for objects of arbitrary shapes these types of constraints may be quite difficult to formulate. Grammars that generate arrangements of objects satisfying positional constraints will be discussed in the next section.

Relational models can specify or constrain all types of object properties. These properties can be geometric—for example, size, orientation (e.g., of the principal axis), elongatedness, compactness, convexity, etc., as well as qualitative size or shape descriptors (tall, small, etc.), or the property of having a specific shape (square, triangular, etc.). They can also be non-geometric, e.g., properties of the gray level population within the object, or qualitative descriptors of lightness, color, or texture. Note that some of these properties are numerical-valued, while others are predicates, which may have very fuzzy definitions. The values of these properties can be constrained in various ways, individually or jointly. The constraints may be independent of position, i.e., they may specify only the population of property values or statistics of this population. They may depend on relative position, e.g., they may specify second-order statistics of property values measured for neighboring pairs of objects. In the most general case, they may depend on absolute position, as specified by a random field. We saw in Section 12.1.5e how first-order statistics of properties of texture primitives, or second-order statistics of properties measured for neighboring pairs of primitives, are useful texture descriptors. By making the properties position-dependent, we can model abrupt or gradual changes in property values, corresponding, for example, to changes in the distance to the textured surface.

Relational models can also specify all types of relations among objects, such as whole/part relations, adjacency, surroundedness, betweenness, relative position, etc., as well as relations defined by the relative values of properties (larger than, darker than, etc.). Here again, these relations may be (fuzzy) predicates or numerical-valued, and can be position-dependent. Note that most of them are binary relations; the principal exception is the ternary relation "between."

As a very simple example of a model at the relational level, consider the following description of the "arch" shown in Fig. 5a:

(1) Objects A, B, C are upright rectangles;
(2) A and B are congruent;
(3) A and B are nonadjacent;
(4) C is above and adjacent to A and B.

Note that this is a very partial description; it defines a class of pictures, not just the single picture in Fig. 5a. The pictures it describes are all "archlike," but they may have unreasonable proportions, or may be asymmetric; for example, the three pictures in Fig. 5b all satisfy the definition. On the other hand, this description excludes arches that are not made up of three rectangles, such as those shown in Fig. 5c. This example illustrates the non-triviality of modeling even simple objects such as arches, even at the relatively abstract relational level. Note also that constraint (3) is a negative condition; such conditions are often very important in picture modeling.

If a description on the relational level does not refer to parts of the picture, it reduces to a list of properties of the picture as a whole. The description then defines a point in "property space," i.e., the point whose coordinates are the values. Correspondingly, models on this level can be defined by regions in property space, specifying allowable combinations of values; or by probability densities over the space, specifying the probability that a picture in the given class has each combination of values. In this case, deciding

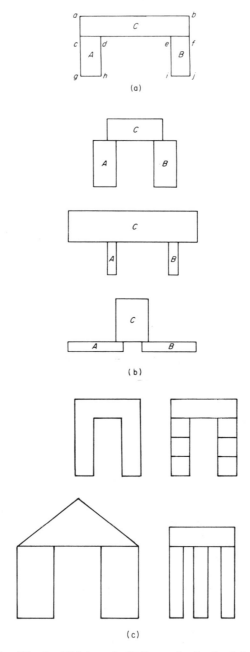

Fig. 5 Examples of "arches." (a) An arch. (b) Some other "arches" that fit the description given in the text. (c) Some configurations that do not fit the description.

whether a given picture "matches" a given model becomes a pattern classification problem (see Section 10.1).

d. Representation of relational descriptions

Descriptions and models at the relational level can be conveniently represented by labeled graphs in which

(1)   each node corresponds to an object, and is labeled with a list of property names and property values that hold for that object;
(2)   each arc between a pair of objects is labeled with a list of names and values of the relations that hold between that pair of objects.

Note that the arcs must be directed, since the relations may not be symmetric (e.g., we must be able to distinguish "A is to the left of B" from "B is to the left of A"); thus the graph should more properly be called a *digraph*. If desired, we can allow more than one arc between a given pair of nodes; such a graph is more properly called a *multigraph*. The graph corresponding to the description of Fig. 5a above [statements (1)–(4)] is shown in Fig. 6a. Note that ternary relations such as "between" cannot be easily represented in this way.

The graph structure just described is strongly object oriented; it is easy to extract from the graph the properties of a given object, or the relations into which it enters, but it is harder to determine which objects enjoy a given property, or enter into a given relation, without searching through the entire graph. Alternative types of graph structures can be devised in which objects, properties, and relations are all treated alike. For example, we can use a structure of the following type:

(1)   Nodes correspond to objects, to properties, or to relations.
(2)   If object A has property $\mathscr{P}$, we join node A to node $\mathscr{P}$ by an arc, and label the arc with the value of $\mathscr{P}$ for A.
(3)   If objects A and B are in relation $\mathscr{R}$, we create an "instance node" $\mathscr{R}_i$, joined to $\mathscr{R}$ by an arc labeled "instance of." We then join A and B to $\mathscr{R}_i$, and label the arc(s) with the value of $\mathscr{R}$ for (A, B).
(4)   Property or relation nodes can themselves be joined by arcs if they are themselves related, e.g., if they are opposite, if one is a special case of the other, etc.

This type of graph can also be used to represent certain ternary relations; e.g., to represent "C is between A and B," one could join A, B, and C to an instance node $\mathscr{R}_i$ of the "between" relation, but have the arc between C and $\mathscr{R}_i$ go in the opposite direction from the other two arcs. A graph of this type representing our description of Fig. 5a is shown, in part, in Fig. 6b.

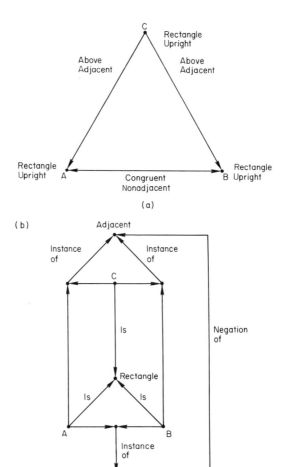

(a)

(b)

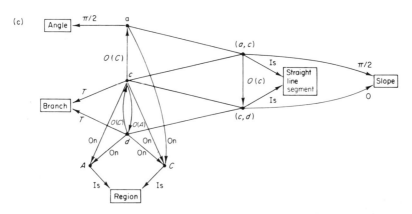

Fig. 6 Relational descriptions of an "arch." (a) Description in which objects are represented by nodes, properties by node labels, and relations by arc labels. (b) Partial description in which objects, properties, and relations are all represented by nodes. (c) Description at the line-drawing level. $O(c)$ is the clockwise order at vertex c; $O(C)$ is the clockwise order around the border of region C.

It is often desirable to use relational descriptions that are hierarchical, i.e., the parts are composed of subparts having given properties and relations, and so on. For example, we can model each rectangle in Fig. 5a as being composed of four straight line segments, meeting at four vertices (e.g., c, d, g, h), and satisfying the following (redundant) constraints:

(1)   Segments (c, g) and (d, h) are vertical and equal.

(2)   Segments (c, d) and (g, h) are horizontal and equal.

(3)   Segments (c, d) and (c, g) make a right angle at vertex c (and similarly at the other four vertices).

(4)   The cyclic sequence of vertices on the outer boundary of rectangle A, in clockwise order, is c, d, h, g.

As a more general illustration of relational description at the level of line or curve segments, vertices, and regions, consider a line drawing composed of curves that have been segmented at branch points, angles, and inflections, as in Section 11.3.3c; the border-based description of a picture segmented into regions might give rise to such a drawing. (In particular, this applies to the border description of a single simply connected region, if we segment its border at angles and inflections.) Descriptions of such drawings might include the following types of information:

(1)   properties of the arcs, such as straightness, (average) curvature, (average) slope, arc length, chord length, etc.;

(2)   properties of the vertices at which arcs begin and end—in particular, whether they are end points, angles (and if so, whether acute, obtuse, etc.), inflections, or branch points of various types;

(3)   incidence relations between arcs and vertices—which arcs emanate from which vertices, and which vertices lie on which arcs;

(4)   incidence relations between arcs (or vertices) and regions, e.g., which arcs lie on the borders of which regions;

(5)   order relations—in particular, the cyclic order in which the arcs are encountered as one moves around a given vertex (say clockwise), or in which vertices are encountered as one moves around a given border. Note that with respect to the direction of border following, angles can be identified as convex or concave.

For example, if we regard Fig. 5a as a line drawing, its description might contain the following information:

(1)   Vertices a, b, g, h, i, j are right angles; vertices c, d, e, f are "T"-type branch points.

(2)   There are arcs joining the pairs of vertices $(a, b), (g, h), (i, j), (c, d), (d, e),$ $(e, f), (a, c), (b, f), (c, g), (d, h), (e, i),$ and (f, j). They are all straight line segments; the first six are horizontal, the last six vertical.

(3)   The clockwise order of the line segments at vertex c is (c, a), (c, d), (c, g); and similarly for the other three branch points.

(4)   In clockwise order, the cyclic sequence of vertices (a, b, f, e, d, c) is on the outer boundary of a region, and all its angles are convex; similarly for other sequences that form region boundaries.

In this type of description, the arcs, vertices, and regions can be represented by the nodes of a labeled graph, and their incidence relations by arcs of the graph. Order relations at vertices and around region borders can be incorporated by "threading" directed cycles through the graph. (In the case of a single , simply connected region, the graph itself is essentially a cycle, with arcs and vertices alternating.) Part of such a graph, for Fig. 5a, is shown in Fig. 6c. One could also create a graph representation in which the properties and relations, as well as the entities, correspond to nodes (compare Figs. 6a and 6b); the details are left as an exercise to the reader. It should be mentioned that descriptions at this level are very similar to the types of descriptions of line drawings that are used in computer graphics.

Exercise 4.   Give a description of a printed capital "A" at the line drawing level. ∎

e.   Scene models

The types of descriptions and models considered so far deal with "objects" defined as subsets of a picture, rather than real objects in the original scene. A more natural approach, in many cases, is to model the scene as a three-dimensional arrangement of solid objects, and then map this model, by an imaging transformation (Section 9.1), into a two-dimensional arrangement of picture objects. Under the transformation, three-dimensional properties of and relationships among the solid objects map into two-dimensional properties of and relationships among their projections. In a scene model, we may want to refer to properties and relations that are physical, not merely geometrical; for example, we may want to say that one object is "supported by" (not merely "on top of"), or "leaning on" (not merely "touching"), another object, so that we can more easily make inferences about the structure of the scene.

Exercise 5.   What constraints must be satisfied by the sizes and positions of rectangles A, B, and C in Fig. 5a in order that C be stably supported by A and B? You may assume that these rectangles all have the same uniform density (i.e., mass per unit area). ∎

Scene models can be used not only at the relational level, but also at the array and regional levels. We can model a scene as a three-dimensional array of "voxels" (e.g., each characterized by its x-ray absorptivity), or we can use two-dimensional arrays to represent scene surface properties at the points that project into the image pixels (e.g., range, slant, illumination, reflectivity, gloss, etc., as mentioned in Section 12.2.1a). Realistic models for the pixel gray levels can be derived in this way, given models for the scene illumination and reflectivity.

Similarly, we can use three-dimensional geometrical representations, as in Section 11.1.6, to model solid objects. These models can then be mapped into two-dimensional array or region models by applying an imaging transformation.

12.2.3 Procedural Models: Grammars

As indicated in Section 12.2.1b, any procedure that generates or accepts pictures can be regarded as an implicit model for the class of pictures that it generates or accepts. This section deal with an important special class of procedural models, called *grammars*. Since these models are of a rather specialized nature, their definitions are complicated; in order to adequately define and illustrate them, this section is somewhat more detailed than Section 12.2.2.

Grammatical models can be used at all three levels of description: arrays, regions, and relations. As a preliminary to defining array grammars, Subsection (a) introduces string grammars, which can be thought of as defining classes of one-dimensional (i.e., $1 \times n$) pictures. Subsection (b) defines a special class of string grammars which have a natural generalization to arrays. Subsection (c) describes some methods of using grammars to define classes of region shapes, while Subsection (d) discusses grammars for relational structures. Finally, in Subsection (e) we treat grammars that make use of numerical parameters, and discuss their uses in picture modeling at the region and relational levels.

In the theory of formal languages it is shown that grammars of various types are equivalent to "acceptors" (i.e., automata) of various types—in other words, the classes (of strings, arrays, etc.) generated by grammars are the same as the classes accepted by automata. Acceptor models will not be treated here. For an introduction to array acceptors and their relationships to array grammars see [27]; on relational structure acceptors and their relationship to grammars see [6, 19, 26].

a. String grammars

A *string grammar G* consists of

(a) a set V_T, called the *terminal vocabulary* of G;
(b) a set V_N, called the *nonterminal vocabulary* of G, where $V_N \cap V_T = \varnothing$, and a distinguished element $S \in V_N$, called the *start symbol* of G;

(c)   a set P of pairs of strings (α, β), each composed of elements of $V = V_N \cup V_T$, where α is not the null string. The pairs in P are called the (re-writing) *rules* or *productions* of G; the pair (α, β) is usually written "$\alpha \to \beta$."

We say that the string τ is *directly derived* from the string σ in G (notation: $\sigma \Rightarrow \tau$) if there exists a rule $\alpha \to \beta$ of G such that α is a substring of σ, and τ is obtained from σ by replacing some occurrence of α (as a substring of σ) by β. We say that τ is *derived* from σ in G (notation: $\sigma \overset{*}{\Rightarrow} \tau$) if there exists a sequence of strings $\sigma = \sigma_0, \sigma_1, \ldots, \sigma_n = \tau$ such that σ_i is directly derived from σ_{i-1} in G, $1 \leqslant i \leqslant n$. The set of strings of elements of V_T (in brief: "terminal strings") that can be derived in G from the initial string consisting of the start symbol S is called the *language* of G.

The following examples may serve to illustrate how string grammars work. In all of them we have $V_T = \{a, b\}$, while the symbols in V_N are denoted by capital letters.

(1)   *Periodic strings with a specified period.* Let α be any specific string of a's and b's, and let G have the rules

$$S \to \alpha S, \qquad S \to \alpha\alpha$$

The first rule can be used any number of times to produce a string of the form $\alpha \cdots \alpha S$. As soon as the second rule is used, the process terminates, since there are no more S's. Thus the final result is a string of the form $\alpha \cdots \alpha$, i.e., a periodic string with period α, and consisting of at least two periods.

(2)   *Symmetric strings.* Let the rules of G be

$$S \to aSa, \qquad S \to bSb, \qquad S \to aa, \qquad S \to bb, \qquad S \to a, \qquad S \to b$$

When the first two rules are used in any sequence, they create a string of the form $\sigma S \sigma^R$, where σ^R is the reversal of σ, and σ is any string of a's and b's. When one of the last four rules is used, the process terminates, since no S's remain. Thus the language of this grammar is the set of all nonnull symmetric strings of a's and b's, i.e., strings of the form $\sigma\sigma^R$ or $\sigma a \sigma^R$ or $\sigma b \sigma^R$.

(3)   *Repeated strings.* We begin with the rules

$$S \to aa, \qquad S \to bb, \qquad S \to aCT, \qquad S \to bDT$$
$$T \to ACT, \qquad T \to BDT, \qquad T \to Aa, \qquad T \to Bb$$

These rules create a string of AC's and BD's, beginning with an aC or bD and ending with an Aa or Bb. (There may be no A's, B's, C's, or D's in the string, in which case it is just aa or bb).

$$CA \to AC, \qquad CB \to BC, \qquad DA \to AD, \qquad DB \to BD$$

These rules allow C's and D's to shift rightward relative to the A's and B's. Note that the order of the A's and B's, and of the C's and D's, does not change.

Thus the sequence of C's and D's, omitting its first term, is always the same as the sequence of A's and B's, omitting its last term, since AC's and BD's were originally created in pairs.

$$aA \to aa, \qquad aB \to ab, \qquad bA \to ba, \qquad bB \to bb$$
$$Ca \to aa, \qquad Da \to ba, \qquad Cb \to ab, \qquad Db \to bb$$

These rules change A's and B's to a's and b's when they have a's or b's on their left, and change C's and D's to a's and b's when they have a's or b's on their right. Thus we can obtain a string τ consisting entirely of terminal symbols (a's and b's) provided all the C's and D's have shifted to the right of all the A's and B's. Note that τ must be composed of two identical substrings (i.e., τ is of the form $\sigma\sigma$), since the sequence of A's and B's corresponds to the sequence of C's and D's. In other words, the language of this grammar is the set of all strings of the form $\sigma\sigma$, where σ is any nonnull string of a's and b's.

G is called *monotonic* if, for each rule $\alpha \to \beta$ of G, we have $|\alpha| \leqslant |\beta|$, where $|\sigma|$ denotes the length of σ. It is called *context-sensitive*, or "type 1," if for each rule $\alpha \to \beta$ of G there exist strings ξ, η, and ω, where ω is nonnull, and a symbol $A \in V_N$, such that $\alpha = \xi A \eta$ and $\beta = \xi \omega \eta$. In other words, in a context-sensitive grammar, each rule replaces a single symbol (A) by a nonnull string (ω) when it occurs in a specified context (ξ, η). Evidently context-sensitive implies monotonic. It can be shown, conversely, that for any monotonic grammar there exists a context-sensitive grammar having the same language. Examples (1) and (2) of grammars just given are context-sensitive, and (3) is monotonic. To make (3) context-sensitive, we can replace the rule $CA \to AC$ by the three rules $CA \to C'A$, $C'A \to C'C$, $C'C \to AC$, and similarly for the rules $CB \to BC$, $DA \to AD$, and $DB \to BD$.

G is called *context-free*, or "type 2," if its rules are all of the form $A \to \beta$, where $A \in V_N$ and β is nonnull. Evidently context-free implies context-sensitive. Examples (1) and (2) are both context-free. G is called *finite-state*, or "type 3," if its rules are all of the form $A \to \tau B$ or $A \to \tau$, where $A, B \in V_N$ and τ is a terminal string. Evidently finite-state implies context-free; example (1) is finite-state. A language is called context-sensitive, context-free, or finite-state if it is generated by a context-sensitive, context-free, or finite-state grammar, respectively.

Up to now we have considered a grammar as *generating* its language by repeated application of rules starting from S. Conversely, we can think of G as *accepting* or *parsing* the strings of its language by reducing them to S using repeated application of the rules of G in reverse. In other words, we say that G parses the terminal string τ if there exists a sequence of strings $\tau = \sigma_n, \sigma_{n-1}, \ldots, \sigma_0 = S$ such that σ_i is directly derivable from σ_{i-1} in G, $1 \leqslant i \leqslant n$. Although these definitions are equivalent, in practice parsing may be much harder than generation. A simple illustration is provided by example (2). If we are given a symmetric string σ of a's and b's, we can apply the reverse of one of the rules $S \to a$ or $S \to b$ to it in any position, and we can probably also apply the reverse of $S \to aa$ or $S \to bb$ to it in many positions as well; but if we pick the wrong rule or the wrong position (not at the center of σ), we will not be able to complete the parse.

At any given step in a derivation, more than one rule of G may be applicable, and a given rule may be applicable in more than one place (i.e., α may occur several times as a substring of σ). We have not specified here how to choose which rule to apply or in which place to apply it; the language of G is defined by allowing all possible choices. A *programmed grammar* specifies which rules are allowed at a given step, after a given rule has just been successfully applied or unsuccessfully tried.

To illustrate how programmed grammars work, consider the following set of rules:

(1) $S \to UV$
(2) $U \to aU$ (3) $U \to bU$ (4) $U \to a$ (5) $U \to b$
(6) $V \to aV$ (7) $V \to bV$ (8) $V \to a$ (9) $V \to b$

Initially we use rule (1), and after it, one of rules (2), (3), (4), or (5). Subsequently, the next rule to be applied is chosen according to the following table:

Rule just used	Next rule
(2)	(6)
(3)	(7)
(4)	(8)
(5)	(9)
(6) or (7)	(2), (3), (4), or (5)
(8) or (9)	None

It is easily seen that the language of this grammar is the set of all strings of the form $\sigma\sigma$, where σ is any nonnull string of a's and b's, i.e., the same language as in Example (3) above. Note that in the present grammar the rules are all context-free. Note also that in this example we have no need to specify which rules to use if a rule is tried and found not to apply.

Rather than applying a rule to one particular substring of the current string, we can define *parallel grammars* in which, when a rule $\alpha \to \beta$ is applied, we rewrite every instance of α in the current string as β. (This may cause problems if the instances of α can overlap!) For example, we can generate the set of strings $\sigma\sigma$ using the parallel context-free grammar whose rules are

$$S \to TT \qquad T \to aT, \qquad T \to bT, \qquad T \to a, \qquad T \to b$$

In a *stochastic grammar*, the rules to be used are chosen in accordance with a given probability density on P. This allows some degree of control over the distribution of sizes of the resulting strings, or over the proportions in which particular symbols occur.

Exercise 6.   Define grammars for the languages consisting of the set of all strings of a's whose lengths are (a) even, (b) powers of 2, (c) perfect squares.

b.   *Isometric grammars and array grammars*

In order to extend the definitions of a grammar from strings to arrays, we first introduce a modified form of the string definition. Let $\#$ be a special symbol in V_N, and let us require that for any rule $\alpha \to \beta$ in P:

(c_1) $|\alpha| = |\beta|$ (α and β have the same length).
(c_2) α does not consist entirely of $\#$'s.
(c_3)   Replacing α by β cannot disconnect or eliminate the non-$\#$'s. [Readily, this is equivalent to requiring that the non-$\#$'s in β exist and are connected, and if α has a non-$\#$ at its left (right) end, so has β.]

If G satisfies these conditions, it is called *isometric* (or sometimes *isotonic*). We define the language of G as the set of terminal strings τ such that the infinite string $\#^\infty\tau\#^\infty$ (i.e., τ embedded in an infinite string of $\#$'s) can be derived in G from the infinite initial string $\#^\infty S\#^\infty$.

It is not hard to show that the languages generated by isometric grammars are the same as those generated by ordinary grammars—in other words, for any grammar G, there exists an isometric grammar having the same language as G, and vice versa. This is easy to see for examples (1) and (3) in the preceding subsection. In (1), if $|\alpha| = k$, we simply replace the rules $S \to \alpha S$ and $S \to \alpha\alpha$ by the rules $S\#^k$ (i.e., S followed by $k\#$'s) $\to \alpha S$ and $S\#^{2k-1} \to \alpha\alpha$. We can make similar replacements for the first group of rules in (3) (e.g., $T \to ACT$ becomes $T\#\# \to ACT$), while the remaining rules are already isometric. The translation to isometric form is harder for example (2), where strings grow in the middle rather than at the end; here we must introduce rules that turn $\#$'s into $\natural$'s (say) at the right end of the string and shift the $\natural$'s leftward until they reach the S, so that we can rewrite $S\natural\natural$ as aSa or bSb, or $S\natural$ as aa or bb.

It can be shown that the languages generated by isometric grammars that never create $\#$'s are the same as the context-sensitive languages. Moreover, the languages generated by isometric grammars in which, for all rules $\alpha \to \beta$, α consists of a single non-$\#$ symbol in V_N together with $\#$'s, and β consists of non-$\#$'s, are the same as the finite-state languages. (Note that the rules here are analogous to context-free rules, not to finite-state rules!)

We are now ready to define grammars whose languages are sets of arrays, rather than sets of strings. An *array grammar* G is defined exactly as in the preceding subsection, except that α and β are connected (sub)arrays rather than strings. G is called *isometric* if, for any rule $\alpha \to \beta$:

(c$_1'$) α and β are geometrically identical.
(c$_2'$) α does not consist entirely of #'s.
(c$_3'$) Replacing α by β cannot disconnect or eliminate the non-#'s.

We define the language of G as the set of connected terminal arrays τ such that τ, embedded in an infinite array of #'s, can be derived in G from the infinite initial array consisting of a single S embedded in an infinite array of #'s.

Exercise 7. Show that (c$_3'$) holds if the following conditions are satisfied:

(1) If the non-#'s of α do not touch the border of α, then there must be non-#'s in β, and they must be connected.

(2) Otherwise, every connected component of non-#'s in β must contain the intersection of some component of non-#'s in α with the border of α; and conversely, every such intersection must be contained in some component of non-#'s in β. ∎

The reason for requiring array grammars to be isometric is that if we replaced the subarray α of σ by a subarray β which was not geometrically identical to α, we would have to shift parts of σ around to make room for β, and this would cause changes in the adjacency relationships of symbols in σ arbitrarily far away from α. In the one-dimensional case, this problem does not arise; even if β does not have the same length as α, the adjacency relationships in the rest of σ do not change when we replace α by β.

The following isometric array grammar generates upright rectangles of a's on a background of #'s:

$$S\# \to aS, \quad S \to A, \quad S \to a, \quad \begin{matrix} A \\ \# \end{matrix} \to \begin{matrix} a \\ B \end{matrix}$$

These rules generate a horizontal row of a's from left to right, possibly ending in an A; if so, the A turns into an a and creates a B below it.

$$\begin{matrix} a \\ \#B \end{matrix} \to \begin{matrix} a \\ Ba \end{matrix}, \quad \begin{matrix} \# \\ \#B \end{matrix} \to \begin{matrix} \# \\ \#a \end{matrix}, \quad \begin{matrix} \# \\ \#B \end{matrix} \to \begin{matrix} \# \\ \#C \end{matrix}, \quad \begin{matrix} C \\ \# \end{matrix} \to \begin{matrix} a \\ D \end{matrix}$$

The B moves leftward, leaving a trail of a's. When it reaches the end of the row of a's above it, it turns into either an a or a C; in the latter case, the C turns into an a and creates a D below it.

$$\begin{matrix} a \\ D\# \end{matrix} \to \begin{matrix} a \\ aD \end{matrix}, \quad \begin{matrix} \# \\ D\# \end{matrix} \to \begin{matrix} \# \\ a\# \end{matrix}, \quad \begin{matrix} \# \\ D\# \end{matrix} \to \begin{matrix} \# \\ A\# \end{matrix}$$

The D moves rightward, leaving a trail of a's. When it reaches the end of the row of a's above it, it turns into either an a or an A, in which case the previous processes can be repeated.

In this example the arrays generated by the grammar consist entirely of a single symbol a; but the grammar can easily be extended to generate patterns of symbols, or random symbols, representing gray levels. Thus stochastic array grammars can be used, at least in principle, as texture models.

The various types of grammars (monotonic, finite-state, etc.), as well as the concepts of stochastic, parallel, and programmed grammars, can all be defined for array grammars just as they were for string grammars. Similar remarks apply to the other classes of grammars introduced in the following subsections.

c. Grammars for regions

The array grammar example just given also indicates how array grammars can be used to model various classes of region shapes, defined by overlay arrays. As another simple example, the following array grammar generates all connected regions of a's on a background of #'s:

$$S\# \to SS, \quad \#S \to SS, \quad \begin{matrix} S & S \\ \# & S \end{matrix} \to \begin{matrix} S \\ S \end{matrix}, \quad \begin{matrix} \# & S \\ S & S \end{matrix} \to \begin{matrix} S \\ S \end{matrix}, \quad S \to a$$

In the following paragraphs we describe some other types of grammars that define classes of regions.

Classes of borders or curves can be modeled by *code grammars* whose terminal symbols are the unit moves in a chain or crack code. [Note that in the case of a border or closed curve, the objects derived by the grammar should be regarded as cycles rather than as strings.] It can be shown that most of the interesting code languages are context-sensitive—e.g., to get the codes of digitized straight line segments, context-sensitive rules are necessary to insure (near-) periodicity; to get the codes of (upright) rectangles or squares, context-sensitive rules are necessary to insure equality of the sides.

As an example, consider the language consisting of the crack codes of all upright rectangles, traversed clockwise starting from their upper left corners. These codes are of the form $0^n 3^m 2^n 1^m$, where $m, n \geqslant 1$. It is straightforward to define a grammar for this language analogous to that in example (3) of Subsection (a): a string of the form $(AB)^n (CD)^m$ is generated; B's are allowed to shift rightward through A's and C's, and C's leftward through D's and B's, until the string is rearranged as $A^n C^m B^n D^m$; and finally A's, C's, B's, and D's turn into 0's, 3's, 2's, and 1's, respectively, when they are in the right order. The details are left to the reader.

It is not as easy to define grammars for sequences of run lengths or end positions, or for width sequences along the MA arcs in a generalized ribbon representation. This is because the lengths, positions, and widths can have values that grow with the picture size, so that they cannot be represented by symbols in a vocabularly of fixed size, unlike the directions in a chain or crack code. One way of defining shape grammars for the run or MA representations is to extend the concept of a grammar by associating a numerical-valued property (e.g., length, position, or width) with each symbol; with each rule $\alpha \rightarrow \beta$, we must then associate a set of functions that compute the values for the symbols in β in terms of the values for α's symbols. Grammars of this type will be introduced in Subsection (e) below.

Grammars that build up objects or regions out of parts (in particular, line drawings out of lines, arcs and curves) can be defined in various ways. One possibility is to let the symbols represent parts or assemblages of parts ("plexes") and have a set of *attaching points* associated with each symbol. For example, the terminal symbols might represent arcs whose attaching points are their endpoints; or they might represent primitive regions with a specified set of (say) border points as attaching points. The attaching points need not belong to the plex, but can be in specified positions relative to it; this allows us to generate nonconnected plexes. In a rule $\alpha \rightarrow \beta$ of such a plex grammar, the "subplexes" α and β represent subassemblages of parts. When α is replaced by β in a plex σ, the rule must specify how β is to be attached to σ after α is removed.

As a simple example, consider the set of upright grids composed of rectangular boxes of a given size. To generate this language, we can use a grammar analogous to that for rectangular arrays in Subsection (b), but where each derivation step is interpreted in terms of adjoining line segments to those already constructed. Thus, initially we construct a single box with attaching points at its NE, SE, and SW corners. When generating the rest of the top row of the grid, we repeatedly adjoin ⊐'s, attaching the NW and SW corners of each new one to the NE and SE corners of the part already generated. When we generate the first box on the next row, we adjoin a ⊔ to the rightmost box on the preceding row, attaching its NW and NE corners to the SW and SE corners of that box. To generate the remainder of the new row, we repeatedly adjoin └'s, attaching their NW and SE corners to the appropriate corners on the row above and the new row, respectively; and so on. Here again, the details are left to the reader.

d. Grammars for relational structures

For picture modeling at the relational level, we can define grammars whose languages are sets of labeled graphs (sometimes called "webs"; we omit the word "labeled" from now on). Here, in any rule $\alpha \to \beta$, α and β are graphs; to apply this rule to a graph σ, we replace some occurrence of α, as a subgraph of σ, by β. The rule must also specify how β is to be attached ("embedded") to the neighborhood of α. The language of such a grammar can be defined as the set of graphs, having labels in the terminal vocabulary, that can be derived by successive rule applications, starting from a one-node graph labeled S.

To illustrate the concept of a *graph grammar*, we give four simple examples, corresponding to four basic types of directed graphs: strings, cycles, trees, and binary trees. In these examples, for simplicity, we use only a single terminal symbol; but they could easily be modified to allow more than one symbol.

(1)   Let the rules be

$$S \to a \underrightarrow{\quad} S, \qquad S \to a$$

where all nodes that were joined by arcs to the rewritten "S" node become joined to the "a" node. It is evident that the language of this grammar is the set of all directed strings whose nodes are all labeled "a," i.e., the set of all digraphs of the form

$$\underrightarrow{a} \ \underrightarrow{a} \ \underrightarrow{a} \quad \cdots \quad \underrightarrow{a}$$

(2)   Let the rules be

$$S \to \overset{T}{\underset{a \longrightarrow a}{\triangle}}, \qquad T \to a \underrightarrow{\quad} T, \qquad T \to a$$

where in the second rule, nodes that were joined to the "T" became joined to the "a," while if the "T" was joined to a node, the new "T" becomes joined to that node; and in the third rule, all arcs to or from the "T" are transferred to the "a." Readily, the language of this grammar is the set of all directed cycles labeled "a," i.e.,

(3)   Let the rules be

$$S \to S \underrightarrow{\quad} S, \qquad S \to a$$

where all arcs to or from the "S" become transferred to the left-hand "S" in the first rule, and to the "a" in the second rule. Here the language is the set of all directed trees labeled "a", rooted at the initial node.

  (4)   Let the rules be

$$S \to a \overset{\nearrow\, S}{\underset{\searrow\, S}{\phantom{<}}}, \qquad S \to a$$

where in both rules, the S's arcs are tranferred to the "a." Here the language is the set of all directed binary trees labeled "a," rooted at the initial node.

In these examples, only the nodes are labeled; but it is straightforward to define graph grammars in which both the nodes and the arcs are labeled. Grammars for multigraphs can also be defined; we omit the details here. The graphs can be undirected; this simplifies the definition of the neighborhood attachments, since we need not distinguish between incoming and outgoing arcs.

  An important special case of a graph grammar is a *tree grammar*; here, in the rule $\alpha \to \beta$, α and β are labeled (sub)trees, and we apply the rule to a tree σ by replacing some occurrence of α, as a subtree of σ, by β. Note that in this case it is obvious how we attach β to α's neighborhood; the parent of α's root simply becomes the parent of β's root. The language of a tree grammar is the set of terminally labeled trees that can be generated in this way from a one-node tree labeled S. Example (4) of a graph grammar can be regarded as a tree grammar, but example (3) cannot, since it allows a new outgoing arc to be added to a tree node without replacing the entire subtree at that node. Tree grammars have been extensively used by Fu and others in a variety of picture modeling applications.

e.   Parametrized grammars

  A powerful extension of the grammar concept is obtained by associating numerical-valued parameters with the symbols. This allows us, for example, to generate graphs whose nodes and arcs have numerical-valued labels, e.g., the symbol represents the name of a property or relation, and the parameter is its value. As another example, mentioned in Subsection (c), it allows us to define "ribbon grammars" in which a width value is attached to each arc point.

  In such a *parametrized grammar* (sometimes called an "attributed grammar"), we assign a set of initial parameter values to the initial symbol S; and we associate with each rule $\alpha \to \beta$ a set of functions that allow us to compute the values for the symbols in β from the values for the symbols in α. The functions can be probabilistic; this allows the final parameter values to be probabilistic, even if the symbols themselves are uniquely determined. If desired, more than one parameter can be associated with a given symbol.

As a simple example, suppose that the terminal symbol in a plex grammar represents a straight line segment, and that we associate with it two parameters representing its length and slope. A parametrized plex grammar that generates the set of all rectangles can be informally defined as follows:

(1) The initial S is rewritten as the plex A, B in which one end of A is attached to one end of B; the lengths of A and B are arbitrary nonnegative numbers; and the slopes of A and B differ by $90°$. (Here S itself needs no initial parameter values; the parameters for A and B are chosen randomly, perhaps in accordance with given probability densities, subject to the constraint that they are perpendicular.)

(2) A, B is rewritten as a, b, A, where the original A and B now have terminal labels and unchanged parameter values; the new A is attached to the other end of b and has the same slope and length as the old A.

(3) a, b, A is rewritten as a, b, a, b, where the new A has now become a terminal; the new b is attached to the free ends of both a's, and has the same slope and length as the other b.

Note that we could have generated a, b, a, b from S in a single step; we broke the derivation into three steps to make the steps simpler.

An important special class of parametrized grammars are *coordinate grammars*, in which position coordinates [(x, y), in two dimensions] are associated with each symbol. We can model spatial arrangements of points or objects using coordinate grammars in which, in any rule $\alpha \to \beta$, α and β are tuples of symbols, say $\alpha = (A_1, \ldots, A_m)$, $\beta = (B_1, \ldots, B_n)$, and we associate with the rule a $2n$-tuple of functions, each of $2m$ variables, that specify the coordinates of the B's in terms of those of the A's. (Note that plex rules can be represented in this way, by defining the new coordinates in such a way that the desired attachment points coincide.) The language is the set of patterns of terminal symbols derivable in this way from a single S, say at the origin. As an example, consider the language consisting of all 4-tuples of a's located at the corners of an upright rectangle; this can be generated using the following rules, where the subscripts denote the coordinates:

$$S_{0,0} \to A_{x,y} \qquad\qquad x, y \text{ arbitrary}$$

$$A_{x,y} \to (a_{x,y}, B_{x',y}) \qquad x' \text{ arbitrary}$$

$$B_{x',y} \to (a_{x',y}, C_{x',y'}) \qquad y' \text{ arbitrary}$$

$$C_{x',y'} \to (a_{x',y'}, a_{x,y'})$$

Here again, we have used a four-step derivation for reasons of simplicity.

12.2.4   Model Construction and Matching

In the preceding subsections we have defined various types of picture descriptions and models at the array, regional, and relational levels. In this final subsection we discuss how models are constructed and used in picture (or scene) analysis.

A picture description, even at the relational level, is usually not the desired goal of the picture analysis process, since it merely provides information about the relationships of parts of the picture, without providing any real-world identifications for the parts. On the other hand, if we are also given a relational model for a class of pictures, and we can establish a *correspondence* between the description and the model, we now have an *interpretation* of the picture in terms of the model. For example, if we are given a picture of an "arch" and a declarative model for the class of arches, as in Section 12.2.2c, we can set up a correspondence between the regions in the picture and the nodes in the model, so that these regions now have interpretations as functional parts of an arch. Similarly, if we have a syntactic model for the class of arches, and we use it to successfully parse the description, we have established a correspondence between the regions in the picture and the symbols of the grammar, which also provides interpretations for the regions.

In short, we can regard the goal of picture analysis as that of establishing correspondences between descriptions of a given picture and models for classes of pictures. Any such correspondence provides an interpretation of the picture in terms of the model.

Note that when we speak of a "correspondence," the implication is that we are dealing with descriptions and models on the relational level, where we must determine which picture parts correspond to which parts referred to in the model. At the array or region level, deciding that a (sub)picture or region "matches" a given declarative model is basically a pattern classification problem. In this section we will deal primarily with models on the relational level.

The formulation just given raises several basic questions. How do we construct models that characterize useful, realistic classes of pictures? Even if we have such a model, how do we find descriptions of a given picture that match it, in view of the fact that a picture can have many possible descriptions? Even if we are given both the description and the model, can we find a correspondence between them without checking an enormous number of possible matchings?

The problem of matching a given description to a given model is treated in Subsection (a). Subsection (d) discusses the more realistic problem of finding descriptions that match a given model. In Subsection (c) we discuss the difficulty of defining models for nontrivial classes of pictures, while

Subsection (b) deals with methods of inferring a model for a class of pictures from a set of descriptions of representative pictures that belong to that class.

a. Matching descriptions to models

Establishing a correspondence between a syntactic model and a description is done by parsing the description. As mentioned in Section 12.2.3a, parsing is not always a straightforward process, since there may be many possible sequences of (reverse) rule applications, only a few of which lead to a successful parse. There is an extensive theory of string parsing, particularly for context-free languages, some of which can be applied to picture grammars; but we will not pursue this subject here. Relevant material can be found in books on syntactic pattern recognition [8, 11]. In this subsection we discuss correspondences between descriptions and declarative models on the relational level.

Let us assume that descriptions and models are represented by labeled graphs, with nodes representing picture parts, arcs representing relations between parts, node labels representing property values, and arc labels representing relation values; see Sections 12.2.2c and 12.2.2d. Formally, a graph G consists of a set of nodes N_G and a set of arcs A_G, which are unordered pairs of distinct nodes (or ordered pairs, if G is directed). To define a labeled graph, we also need a set L_N of node labels and a set L_A of arc labels, as well as two functions $\varphi_G: N_G \to L_N$ and $\psi_G: A_G \to L_A$.

Let G and H be labeled graphs representing a description and a model, respectively, both having the same label sets L_N and L_A. By a (partial) correspondence between G and H we mean a set R of pairs of nodes (m, n), where $m \in N_G$ and $n \in N_H$, such that

(1) $(m, n) \in R$ implies $\varphi_G(m) = \varphi_H(n)$: corresponding nodes have the same labels.

(2) If (m_1, n_1) and (m_2, n_2) are in R, then $(m_1, m_2) \in A_G$ iff $(n_1, n_2) \in A_H$. In other words, if two nodes are joined by an arc in G, then the corresponding nodes are joined by an arc in H, and vice versa. Moreover, $\psi_G(m_1, m_2) = \psi_H(n_1, n_2)$: these corresponding arcs have the same labels.

More generally, we could allow the label sets of G and H to be different, and define a correspondence to involve mappings between these label sets, so that corresponding nodes or arcs have corresponding labels; but the simpler definition just given is sufficient for our purposes.

Note that a correspondence need not involve all the nodes of G or of H, and it may pair several nodes of G with the same node of H or vice versa. If R does involve every node of H, and it pairs only one node of G with any given node of H, then R is a function from H into G. Such a function that satisfies (1) and (2) is called a labeled graph *homomorphism*. If the function is also one-to-one, i.e., no two nodes of H are paired with the same node of G, it is called an *isomorphism*. We often want to find correspondences that are isomorphisms, i.e., to find subgraphs of the description graph G that are isomorphic to a given model graph H.

As a very simple illustration of these concepts, consider the relational model of an arch shown in Fig. 6a, and suppose that we are given the relational description of a picture shown in Fig. 7. To determine whether the picture contains an arch, we want to find isomorphisms between the arch model, call it H, and the picture description, call it G. Comparison of Figs. 6a and 7b shows that there does in fact exist such an isomorphism.

In general, finding correspondences or isomorphisms between two labeled graphs involves a combinatorial search process; in the worst case, to find a subgraph of G that is isomorphic to H, we may have to consider all possible subsets of N_G that are the same size as N_H. In practice, however, various types of heuristics can usually be used to shorten the search. Evidently, one need only consider node pairs (m, n) that have the same (or corresponding) labels; if the number of labels is large, this "node consistency" condition immediately eliminates most of the combinations. Further eliminations can be achieved by enforcing "arc consistency." Let $m_1, \ldots, m_r$ be the nodes of G that are joined to m by arcs (in short, the neighbors of m in G), and let $n_1, \ldots, n_s$ similarly be the neighbors of n in H. If m and n correspond, then for each n_j there must be an m_i having the same or corresponding node label, and such that the arcs (n, n_j) and (m, m_i) have the same or corresponding arc labels. If any one of these conditions fails to hold, the pair (m, n) can be rejected.

Further eliminations can be achieved using a discrete relaxation process (Section 10.5.4) that enforces both node and arc consistency. For each node m of G, let N_m be initially defined as the set of nodes of H that have the same label and could thus correspond to m. If m and n correspond, then for each n_j there must exist an m_i such that (n, n_j) and (m, m_i) have the same arc labels, and such that $n_j \in N_{m_i}$. If any of these conditions fails to hold, we eliminate n from N_m, just as in the preceding paragraph. Now, however, when we have done this for every m, the sets N_m have become smaller, so we can repeat the process and eliminate further pairs. The process usually stabilizes after only a few iterations, and the result is a relatively unambiguous set of pairs [16].

When the node or arc labels have numerical values, the notion of a correspondence should be defined more loosely, so as not to require that the labels exactly match. Given any pairing of nodes of G with nodes of H, we can define its match merit in terms of the smallness of the maximum difference, or the sum of squared or absolute differences, between the labels of corresponding nodes and arcs. Here again, in the worst case, finding the pairing

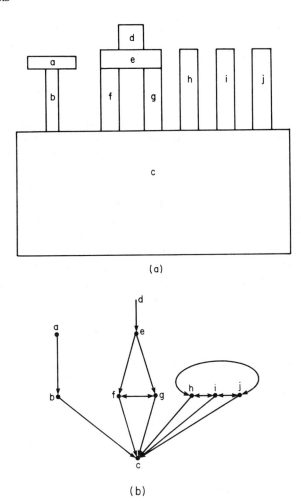

Fig. 7 A picture containing an arch (a), and its relational description (b), in which all nodes are labeled "upright rectangle"; all double-headed arcs are labeled "congruent, nonadjacent"; and all single-headed arcs are labeled "above, adjacent." The graph in Fig. 6a is isomorphic to the subgraph of Fig. 7b whose nodes are e, f, g.

that has highest merit requires combinatorial search. A fuzzy relaxation process (Section 10.5.3) can be used to simplify the search by proceeding along the following lines: Let $\mu(m, n)$ be the merit of the pair (m, n), based on the difference between their labels. For each n_j, there should exist an m_i such that $\mu(m_i, n_j)$ is high, and such that the arcs (m, m_i) and (n, n_j) have similar labels. If the best choice for m_i still gives low merit to (m_i, n_j), or a large difference between the labels of (m, m_i) and (n, n_j), then $\mu(m, n)$ should be reduced. This process can be iterated, since the reductions of the $\mu(m_i, n_j)$'s can lead to a further reduction of $\mu(m, n)$. After a few iterations, the process tends to stabilize, with good pairings having high merits and all other pairings having low merits [15].

As a simple example, suppose we have segmented a picture of a room into regions, and suppose we have a room model involving objects such as "wall," "door," "light switch," "doorknob," and so on. We can assign initial merits to each region as being any of these objects, based on its size, shape, position, color, etc. If region m is a doorknob, it should be surrounded by a region that is a door; while if m is a light switch, it should be surrounded by a wall region. Thus if no region with high door merit surrounds m, we reduce m's doorknob merit, and similarly for wall and light switch. A system that applies model-based constraints of this type to segmented scenes is described in [5].

It may sometimes be advantageous to transform the graph correspondence problem into another type of graph problem for which efficient heuristics exist. For example, we can transform it into the problem of finding cliques (=maximal complete subgraphs, in which every node is joined to every other node) by proceeding as follows: Let $\overline{G}$ be the graph whose nodes are the pairs (m, n), for all $m \in N_G$ and $n \in N_H$ having compatible labels. Let (m_1, n_1) and (m_2, n_2) be joined by an arc in $\overline{G}$ provided that $(m_1, m_2) \in A_G$ iff $(n_1, n_2) \in A_H$, and that these arcs have compatible labels. Then any correspondence between G and H gives rise to a complete subgraph of $\overline{G}$, and maximal correspondences give rise to cliques. On this approach to correspondence finding see [3, 4].

When symmetries are present in the model (e.g., the two supporting rectangles in the arch), the cost of finding correspondences increases, because the inherent ambiguity implies that fewer combinations will be eliminated. In fact, in such cases one does not really want to find all the correspondences; those which differ only in that symmetric nodes have been interchanged are redundant. The redundancy can be eliminated by representing all model nodes that have the same description by a single node that allows multiple instantiations. The same idea can be used to decrease the cost of finding correspondences between a description and a set of models, by using a single graph to represent all the models, with special labels on the nodes and arcs indicating to which models they belong. A model matching system based on these ideas is described in [33].

b.   Inferring models from sets of examples

The process of inferring a grammar from a set of samples of its language is known as *grammatical inference*. There is considerable literature on the inference of string grammars, and many of the ideas can also be applied to

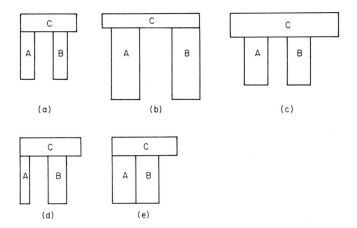

Fig. 8 (a)–(c) Examples of arches. (d) and (e) Examples of nonarches.

picture grammars of various types. A review of grammatical inference can be found in [10]. We will not treat this topic here, nor will we discuss how to infer declarative array or region models from examples. (For some types of models this may be straightforward, e.g., inferring a gray level probability density or a random field from a set of pictures.) In this subsection we briefly discuss the inference of declarative relational models from sets of examples [34].

To illustrate the inference process, let us again consider the arch example of Fig. 5. Suppose that we are given the examples shown in Figs. 8a–8e and are told that examples (a), (b), and (c) are arches, while examples (d) and (e) are "near misses." From these examples we want to be able to deduce that:

(1) An arch consists of three upright rectangles, one of them on top of the other two [(a)–(c)].

(2) The supporting rectangles must be congruent (d) and must not touch (e).

Note the important role played by the "near misses" in helping to clarify the concept of an arch.

In order to make these deductions, we create relational descriptions of the examples and compare them. By doing so, we discover what features the positive examples (a)–(c) have in common, and in what ways the negative examples (d) and (e) differ from them. It is important to use negative examples that are "*near* misses" so that we can find the correct near-matches between them and the positive examples, and thus detect the discrepancies that prevent a complete match.

To see how this process might work, consider the descriptions of examples (a)–(c) shown in Figs. 9a–9c. We show here only a subset of the properties and relations that could have been used; for example, we do not show that the rectangles are elongated, that two of them are congruent, or that one of them is to the left of another. Note that we have not included negative relations such as "nonadjacent," or relations based on relative property values such as "larger than," since such relations would hold between nearly every pair of objects, and so would not be very informative. We will see below, however, how to discover when such relations are needed.

Since the "above" (and "left") relations are asymmetric, there is only one way to construct correspondences between these descriptions. When we do so, we discover that the properties of rectangularity and orientation, and the relations of adjacency and aboveness (i.e., "on top of"), all match, but the size properties differ. This leads us to propose the preliminary model shown in Fig. 10a, where we have preserved the properties and relations that are common to the three descriptions, and ignored those that are not. This model is thus a simple generalization from the three descriptions in Figs. 9a–9c.

Example (d) has the description shown in Fig. 9d. This description matches the model of Fig. 10a, but we know that this example is not an arch. In terms of the properties used in the descriptions, the crucial difference must involve the size information. For example, it might be that the rectangles in the arches all have sizes in certain ranges, while those in the nonarches lie outside these ranges. Another possibility is that the relative sizes within each description are different for the arches than for the nonarches. In fact, the two lower blocks in each arch (a)–(c) have the same height and width, while in (d) their sizes differ. We might be led to discover this crucial discrepancy by applying a "meta-rule" of the following sort: if no crucial differences can be found in the property values, look for such differences in the relationships defined by comparing the property values. Once we have discovered how (d) differs from (a)–(c), we refine our preliminary model by making the crucial difference explicit, i.e., we specify that "equals" relations hold for the heights and widths of the two lower blocks, as shown in Fig. 10b.

Finally, example (e) has the description shown in Fig. 9e, which (nearly) matches the model in Fig. 10b, except for the additional adjacency relation between the two lower blocks. Since this is the only exception, it must be the crucial discrepancy: in other words, in an arch, this relation must *not* hold. This leads us to the final model shown in Fig. 10c.

The illustration given here was greatly simplified in many respects. In particular, we have not described in detail how near-matches are discovered and discrepancies detected. A more thorough treatment of the process of infering relational descriptions from examples can be found in [34].

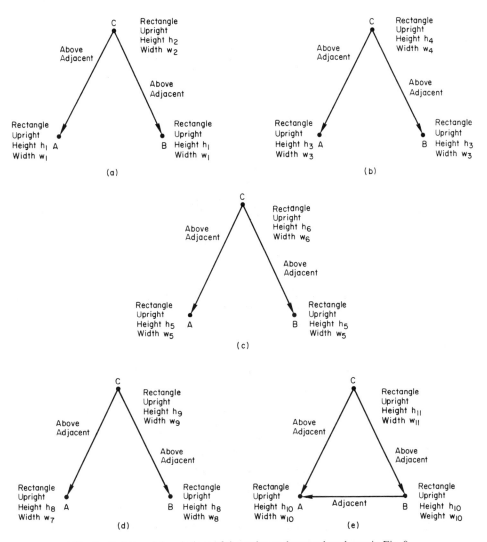

Fig. 9 Relational descriptions of the arches and nonarches shown in Fig. 8.

c. Criteria for constructing models

We assumed in the two preceding subsections that we knew what properties and relations to use for our picture description task, and that models based on these properties and relations could be formulated. In fact, however, defining the appropriate level of description, and constructing models at that level, may be very difficult tasks. In this subsection we discuss various aspects

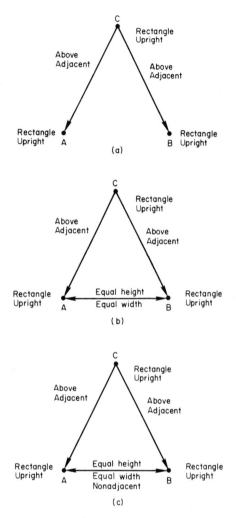

Fig. 10 Arch models inferred from the descriptions in Fig. 9. (a) Model based on generalization from arch examples (a)–(c). (b) Refinement of the model based on nonarch example (d). (c) Final refinement based on example (e).

of the model construction problem, particularly as regards the problem of model complexity.

Many of the properties and relations that we would like to use in describing pictures are not mathematically well defined. We have already discussed the difficulty of defining geometric relations such as "to the left of," particularly for extended objects. As regards properties, there does not seem to be any simple way to characterize classes of natural textures, e.g., grass in a TV picture, cytoplasm in a photomicrograph, lung tissue in a radiograph, woods in an aerial photograph. It is similarly very difficult to characterize many classes of natural shapes, e.g., the shape of a handprinted capital A, a tree crown, a cell nucleus, a human heart, a cumulus cloud. Of course, these are all fuzzy or probabilistic classes, so we cannot expect to define them exactly; but we do not even know how to define their fuzzy membership functions or probability densities. In fact, the allowable variations are likely to depend on the context in which the shapes, textures, etc., occur, and even on the interpretations that we assign to them in terms of real-world entities. It is easy to define classes defined by simple types of transformations, e.g., a specified pattern plus Gaussian noise, or a specified pattern together with (some of) its translations, rotations, rescalings, etc.; but it is much harder to define classes of shapes that involve nontrivial types of geometrical distortions, or classes of textures that involve nontrivial constraints on the arrangement of gray levels. In many cases, we cannot even give verbal, imprecise descriptions of the classes, let alone precise mathematical definitions. Similar remarks apply to nontrivial models on the relational level.

As an alternative to characterizing such classes completely, we can model them in terms of constraints that they must satisfy, e.g., textures in terms of their populations of local property values, shapes in terms of their populations of border curvatures or maximal block sizes, etc. It must be realized, however, that such partial models may be satisfied by pictures that do not belong to the classes, so that model matching becomes unreliable. In designing models, a compromise must be made between fidelity and simplicity; accurate models may be too complex to define completely, while simple models may be too inaccurate. These problems arise for syntactic, as well as for declarative models: a grammar that defines a class accurately may be impractically complex, but a simpler grammar may define too poor an approximation to the class.

When a model is very complex, the process of matching descriptions with it, or parsing descriptions with respect to it, becomes complex too, and hence has high computational cost. This problem is especially acute when it is necessary to try a large number of models. The cost can be reduced by structuring the models so that the matching or parsing process can be broken down into stages. Two ways of doing this will be discussed in the following paragraphs. On the analogous problem of reducing the computational cost of picture matching see Section 9.4.4.

One approach is to define the models hierarchically, as consisting of submodels (and perhaps these in turn of subsubmodels, etc.). The cost of matching a model with a description goes up very rapidly with the size of the model. For example, to find isomorphisms of a model graph H, say having

k nodes, with subgraphs of a description graph G, in the worst case we must examine all k-tuples of nodes in G; the number of these grows essentially exponentially with k. Suppose that we can break H up into parts $H_1, \ldots, H_r$ having $k_1, \ldots, k_r$ nodes, respectively, in such a way that H can be represented by an r-node graph specifying the relations between the H_i's. Then the cost of matching the H_i's with G grows with the sum of $k_1, \ldots, k_r$, rather than with their product; and the additional cost of finding configurations of these matches that match H grows only with r. Similar remarks apply to syntactic models; it is advantageous to break large rules up into small pieces, find matches to the pieces, and then look for the proper pattern of those matches.

Another possibility is to use "nested sets" of models of increasing specificity (and complexity). The idea is to first try simple models that make crude class discriminations, but are inexpensive to test; depending on the outcomes of these tests, more complex models can then be tried; and so on. Thus the complex models need only be used when the simple models have succeeded, so that the average computational cost is reduced. For example, if we are looking for a dark triangle in a picture, we need not apply triangle templates in all positions; rather, we can threshold the picture, i.e., extract all the dark objects, and then test these for triangularity. This idea too applies to both declarative and syntactic models. If we need to find matches to a set of models, we should use simple models that are relevant to as many of them as possible, so that a few simple tests provide a lot of information; this reduces the computational cost still further. The basic segmentation and description techniques described in Chapters 10–12 implicitly define a set of "general-purpose models" that are useful in a wide variety of picture description tasks.

d.   Building descriptions that match models

Even when we have good models, applying them to a picture is not always straightforward, since pictures can have very ambiguous descriptions, and only a few of the possible descriptions may satisfy the model. Thus picture analysis involves not just matching given descriptions to models, but finding descriptions that fit models. For syntactic models, this is one aspect of the parsing problem: there are usually many ways to attempt a parse, and only a few of them may succeed. In this subsection we discuss the problem of finding picture descriptions that match declarative models.

The basic idea is to use the model as a guide in building a description. The model refers to certain types of picture parts, and to certain properties of the relations among these parts; thus we should try to segment the picture so as to extract parts of the desired types, and if necessary, modify the segmentation until we obtain parts that have the desired properties and relations. Typically, we would use the array-level information in the model as a basis for the initial segmentation, e.g., given that the parts are dark or have busy textures, use an appropriate pixel classification technique to extract them. We would use the region-level information as a basis for the next stage of segmentation, e.g., given that the parts have certain sizes or shapes, perform merges or splits on the initial segments to produce such parts. Finally, we would use the relation-level information as a basis for final adjustments to the segmentation to produce parts having descriptions that match those in the relational model. Thus the model is used at every level as a guide in selecting methods of segmentation and correcting the results that they produce. More generally, when we evaluate the segmentation results in the light of the model, we can go back and modify the segmentation criteria; for example, if thresholding produces regions that are too large, we can raise the threshold. Still more generally, when we succeed in extracting some picture parts that conform to the model, we can try to extract the other parts that the model requires by applying fine-tuned segmentation criteria in the appropriate places. These examples illustrate a "top-down" approach to picture analysis in which the analysis steps are selected on the basis of information provided by the model; this contrasts with the "bottom-up" approach in which we first build a description and then compare it with the model. (The terms "top-down" and "bottom-up" are used in the theory of parsing.)

This model-guided, or "knowledge-based," approach to picture analysis certainly makes sense in principle, but it is not always easy to implement in practice. Models are usually oversimplified, and may not be able to provide sufficient information to cover all the situations that can occur in the picture. In particular, in models at the relational level, detailed geometric information has been discarded, and even in region-level models, detailed gray level information has been discarded. Thus it is hard to control segmentation at a given level using models at a higher level. For example, in segmenting at the array level, one can usually make use only of array-level information in the model, but not of geometric information (e.g., there is no obvious way to favor the extraction of convex regions by pixel classification); only after the segmentation has been done can geometric information be used to correct it (e.g., merge concave regions with neighboring regions).

When we use models to control the process of building picture descriptions that match the models, we are blurring the distinction between declarative and procedural models. In fact, given any procedure for constructing descriptions, we can use it to define classes of pictures—namely, the pictures for which it is able to successfully construct descriptions of given types. Thus such a procedure can be regarded as implicitly specifying a model. In cases where explicit formulation of declarative models is too difficult, such "procedurally embedded" models may provide a practical alternative.

It would be very desirable to develop systematic methods of constructing picture models that define a wider variety of real-world classes of pictures. There is also a need for systematic methods of defining optimal picture analysis strategies for given classes of pictures, based on models for the classes. Progress in these areas would help make picture analysis less of an art and more of a science.

Advances in our ability to analyze and interpret pictures by computer have resulted in the development of many practical systems for pictorial pattern recognition. At the same time, these advances bring us closer to the goal of providing computers with the capability of understanding visual input. Computer vision is a major aspect of machine perception, and will be a key factor in achieving machine intelligence.

12.3 BIBLIOGRAPHICAL NOTES

The use of the Riesz representation theorem to characterize linear properties first appears in Rosenfeld [25]. On normalization with respect to groups of transformations see Pitts and McCulloch [22].

Moment properties were introduced in the early 1960s by several investigators; see [25] for detailed references. For a recent review of texture analysis see Haralick [13], where references can be found to the individual approaches mentioned here.

For a collection of papers on image modeling at the array level see Rosenfeld [28]. On the role of image models in image segmentation see Rosenfeld and Davis [29]. Detailed treatments and examples of syntactic models can be found in Fu [8, 9] and Gonzalez and Thomason [11]; see also Rosenfeld [27] on array grammars and acceptors. For a review of grammatical inference see Fu and Booth [10]. A general theory of pattern analysis and synthesis, together with many applications, is presented by Grenander [12].

APPENDIX: ANALYSIS OF THREE-DIMENSIONAL SCENES

The information contained in many types of pictures is basically two-dimensional, and does not need to be interpreted in terms of a three-dimensional scene. For example:

(1)   Pictures of printed or written characters, line drawings, and sketches are inherently two-dimensional (although drawings can be used to represent 3-d information).

(2) Pictures of terrain taken from far away (e.g., from satellite altitudes) are essentially two-dimensional, since the terrain relief is negligible at great distances.

(3) X-rays are projections of a scene onto a plane, i.e., they are two-dimensional "shadows."

(4) Photomicrographs taken at high magnification and shallow depth of field show only a thin slice of the microscope slide scene, since only this slice is in focus.

On the other hand, pictures of scenes taken from relatively close by usually do require three-dimensional interpretation in terms of perspective views of opaque solid objects that may occlude one another.

This appendix briefly discusses some of the basic approaches to the interpretation of pictures as images of three-dimensional scenes. Detailed references will not be given here; for a recent review and bibliography see Shirai [32]. The representation of three-dimensional information about a scene was discussed in Section 11.1.6.

a. Model matching

One approach to extracting three-dimensional information from a picture is to identify a known object in the picture, and determine how the object must be positioned and oriented in space in order to give rise to its observed image in the picture.

A brute-force approach to recognizing the 3-d position of an object from its image in a picture is to store information about a large set of possible projections of the object, and find the one that matches the observed image. For example, we might store a large set of projected object silhouettes, and use them as templates to match against the image. Alternatively, we can store properties of the silhouettes, such as moments (Section 12.1.3) or Fourier descriptors (Section 11.3.3d), and match them with the corresponding properties of the image. In other words, we can represent a three-dimensional object as a set of two-dimensional projections, and identify the object, as well as its position and orientation, by finding a projection that matches a region extracted from the picture.

A better approach can be used if the object contains identifiable local features, some of which are likely to be visible in any projection. We then use a stored three-dimensional model of the object, and try to establish a correspondence between features detected in the picture and the features of the model. As we begin to build this correspondence, the projection of the model onto the picture becomes determined, and we can verify correspondence by checking that other features also correspond as predicted. Note that when we match picture features with projections of model features, we must take into account the fact that only part of the model can be visible in any given projection; determining which parts are not visible is the well-known hidden surface problem in 3-d computer graphics. This approach was used by Roberts, who was the first to make use of model matching to extract three-dimensional information from two-dimensional pictures; he took pictures of polyhedral objects, and used edges and vertices as the features. The feature-based approach has the advantage that if enough features are visible, reliable matches can be obtained even if the object is partly hidden; this would not be true if we used global features such as Fourier descriptors or moments.

b.   Depth cues

When we do not know what objects will be present in a scene, we cannot use specific object models as aids in interpreting pictures of the scene. Nevertheless, it may be possible to derive three-dimensional information about the scene from a picture by making use of various "depth cues" that are present in the picture. On the role of depth cues in visual perception see Section 3.5.

One source of three-dimensional information in a picture is *shading*, i.e., the variation of gray level across a region. If the region represents a uniformly reflective surface, this variation must be due to the changes in slope of the surface relative to the source of illumination and the viewer, so that the gray level variations impose constraints on the three-dimensional shape of the surface. The shape may not be determined unambiguously from the shading information in a single picture, but we can resolve the ambiguity by using pictures of the same scene taken under different conditions of illumination. Extensive work on this approach has been done by Horn. *Shadows* in a picture provide another type of illumination-based depth cue, since they represent parts of surfaces from which a light source cannot be seen because some other surface is interposed. Still another cue is provided by *highlights* that may occur on glossy surfaces. As an example, shadow and highlight cues may help in distinguishing between convex and concave edges in a picture of a polyhedron; we might expect some shadowing on a concave edge, and some highlighting on a convex edge.

Another source of three-dimensional information is provided by local cues to *occlusion*, which indicate, at places where regions meet, which of them may be in front of the other(s). For example, suppose that three regions meet at a "*T*" junction, e.g.,

$$
\begin{array}{c}
A \\
\hline
B \mid C
\end{array}
$$

Then we can conjecture that the vertical edge comes to an end because it is occluded by the horizontal edge—in other words, that region A lies in front of regions B and C. If we are dealing with polyhedral objects, the shape of a junction imposes constraints on which of the regions meeting at the junction belong to the same or different objects, if we assume that the objects are being viewed from a general position, so that improbable alignments of vertices or edges with each other do not occur. A set of rules for linking regions (that appear to belong to the same object) at junctions, based on the work of Guzman, was described in Section 10.4.2e. Rules of this type, and the geometrical and physical constraints that underlie them, have been extensively studied by Huffman, Clowes, and Mackworth, who also made use of duality relationships between points and planes as an aid in finding consistent interpretations of a picture in terms of polyhedra. This approach has also been extended by Waltz to allow cracks and shadows to be present; by Kanade to allow the objects to be two-dimensional (i.e., thin); and by Turner to allow quadric-surface objects.

Still another class of depth cues involve *size, shape,* and *perspective.* If we interpret a pair of convergent lines as being parallel but tilted, or a region as being planar and symmetric but tilted, we can compute their spatial orientation from the angle by which they must be rotated in order to make them parallel or symmetric. Similarly, if we measure the *texture gradient,* i.e., the rate of change of texture coarseness in a region of a picture, and assume that the corresponding surface in the scene is uniformly textured, we can determine the slope of the surface. Given a plane curve in the picture, it is sometimes reasonable to assume that of all possible corresponding space curves in the scene, the one having minimum curvature is the correct one; this idea has been investigated by Barrow, Tenenbaum, and Marr.

c. Range data and stereo

If an array of range values is available, representing the distance from the viewer to each visible point of the scene, the problem of three-dimensional scene analysis is greatly simplified. One can then segment the scene into surfaces of approximately constant range by clustering and thresholding the range values, or by partitioning the array into regions of homogeneous range, using methods analogous to those in Chapter 10; or, more generally, one can find planar surfaces of arbitrary orientation in the scene by fitting planes to the range values. Similarly, one can detect range edges, i.e., places where the range changes abruptly; these correspond to occluding edges where one surface lies behind another. One can also detect various types of local range features, corresponding (in the case of terrain, where knowing the range is essentially equivalent to knowing the terrain elevation) to peaks, pits, ridges, and ravines. Shirai has investigated the constraints defined by various types of region junctions in arrays of range data.

Segmentation techniques can also be applied to arrays of surface slope values; one can detect patches of constant slope, edges where the slope changes abruptly, and so on. In analogy with the Hough transform, one can map each surface element into a point in "slope space" or "gradient space," i.e., if the components of the element's slope are u, v, it maps into the point (u, v); clusters in this space then correspond to parallel or coplanar sets of surface elements, e.g., the faces of a polyhedron.

Range information can be obtained directly from a scene using a variety of range finding devices, which will not be described here. (In the case of terrain, radar provides range information in terms of the time required for the radar reflection to return to the transmitter.) If two pictures of a scene, taken from different, known positions, are available, the three-dimensional positions of points appearing on both pictures can be determined from their *stereoscopic parallax*, i.e., from the relative positions of their images in the pictures, as discussed in Section 9.2.3. The pictures need not be taken at the same time; if we compare successive pictures taken while the camera is moving, we can use *motion parallax* to derive three-dimensional information. The "*optical flow*" of the visual environment that takes place during motion provides powerful cues to the locations and orientations of objects relative to the observer.[§]

REFERENCES

1. H. Abelson, Computational geometry of linear threshold functions, *Informat. Control* **34**, 1977, 66–92.
2. H. Abelson, Towards a theory of local and global in computation, *Theoret. Comput. Sci.* **6**, 1978, 41–67.
3. H. G. Barrow, A. P. Ambler, and R. M. Burstall, Some techniques for recognizing structure in pictures, *in* "Frontiers of Pattern Recognition" (S. Watanabe, ed.), pp. 1–29. Academic Press, New York, 1972.
4. H. G. Barrow and R. J. Popplestone, Relational descriptions in picture processing, *in* "Machine Intelligence" (B. Meltzer and D. Michie, eds.), Vol. 6, pp. 377–396. Edinburgh Univ. Press, Edinburgh, Scotland, 1971.
5. H. G. Barrow and J. M. Tenenbaum, MSYS: A system for reasoning about scenes, Tech. Note 121. Artificial Intelligence Center, SRI International, Menlo Park, California, 1976.

[§] In the case of "scenes" composed of transparent or translucent objects, there are other ways of building up three-dimensional scene models from multiple pictures of the scene. In microscopy, a three-dimensional model can be constructed by "stacking" pictures obtained at different focus settings, corresponding to different cross sections of the scene. In radiology, three-dimensional information can be reconstructed from a sufficiently large set of projections of the scene, as described in Chapter 8.

6. V. Claus, H. Ehrig, and G. Rozenberg (eds.), "Graph-Grammars and their Application to Computer Science and Biology." Springer-Verlag, Berlin and New York, 1979.
7. L. S. Davis, S. Johns, and J. K. Aggarwal, Texture analysis using generalized co-occurrence matrices, *IEEE Trans. Pattern Anal. Machine Intelligence* 1, 1979, 251–259.
8. K. S. Fu, "Syntactic Methods in Pattern Recognition." Academic Press, New York, 1974.
9. K. S. Fu (ed.), "Syntactic Pattern Recognition, Applications." Springer-Verlag, Berlin and New York, 1977.
10. K. S. Fu and T. L. Booth, Grammatical inference: introduction and survey, *IEEE Trans. Systems Man Cybernet.* 5, 1975, 95–111, 409–423.
11. R. C. Gonzalez and M. G. Thomason, "Syntactic Pattern Recognition: An Introduction." Addison-Wesley, Reading, Massachusetts, 1978.
12. U. Grenander, "Lectures in Pattern Theory," Vol. I: Pattern Synthesis, Springer-Verlag, Berlin and N.Y., 1976; Vol. II: Pattern Analysis, 1978.
13. R. M. Haralick, Statistical and structural approaches to texture, *Proc. IEEE* 67, 1979, 786–804.
14. L. Hodes, The logical complexity of geometric properties in the plane, *J. ACM* 17, 1970, 339–347.
15. L. Kitchen, Relaxation applied to matching quantitative relational structures, *IEEE Trans. Systems Man Cybernet.* 10, 1980, 96–101.
16. L. Kitchen and A. Rosenfeld, Discrete relaxation for matching relational structures, *IEEE Trans. Systems Man Cybernet.* 9, 1979, 869–874.
17. J. T. Maleson, C. M. Brown, and J. A. Feldman, Understanding natural texture, *Proc. Image Understanding Workshop*, October 1977, 19–27.
18. M. L. Minsky and S. Papert, "Perceptrons: An Introduction to Computational Geometry." MIT Press, Cambridge, Massachusetts, 1969.
19. M. Nagl, "Graph-Grammatiken: Theorie, Andwendungen, Implementierung." Vieweg, Braunschweig, 1979.
20. G. Nagy and N. Tuong, Normalization techniques for handprinted numerals, *Comm. ACM* 13, 1970, 475–481.
21. K. Paton, Conic sections in chromosome analysis, *Pattern Recognition* 2, 1970, 39–51.
22. W. Pitts and W. S. McCulloch, How we know universals—the perception of auditory and visual forms, *Bull. Math. Biophys.* 9, 1947, 127–147.
23. W. K. Pratt, O. D. Faugeras, and A. Gagalowicz, Visual discrimination of stochastic texture fields, *IEEE Trans. Systems Man Cybernet.* 8, 1978, 796–804.
24. A. Rosenfeld, "Picture Processing by Computer," Sections 7.2 and 7.4. Academic Press, New York, 1969. See also the first edition of this book, Section 10.1.
25. A. Rosenfeld, "Picture Processing by Computer," Section 7.3. Academic Press, New York, 1969. See also the first edition of this book, Section 10.2.
26. A. Rosenfeld, Array and web languages: an overview, *in* "Automata, Languages, Development" (A. Lindenmayer and G. Rozenberg, eds.), pp. 517–529. North-Holland Publ., Amsterdam, 1976.
27. A. Rosenfeld, "Picture Languages." Academic Press, New York, 1979.
28. A. Rosenfeld (ed.), "Image Modeling." Academic Press, New York, 1981.
29. A. Rosenfeld and L. S. Davis, Image segmentation and image models, *Proc. IEEE* 67, 1979, 764–772.
30. B. Schachter and N. Ahuja, Random pattern generation processes, *Comput. Graphics Image Processing* 10, 1979, 95–114.
31. L. G. Shapiro, Data structures for picture processing: a survey, *Comput. Graphics Image Processing* 11, 1979, 162–184.

32.   Y. Shirai, Recent advances in 3-D scene analysis, *Proc. Internat. Joint Conf. on Pattern Recognition, 4th* 1978, 86–94.

33.   M. O. Shneier, A compact relational structure representation, *Proc. Internat. Joint Conf. on Artificial Intelligence, 6th* 1979, 818–826.

34.   P. H. Winston, Learning structural descriptions from examples, *in* "The Psychology of Computer Vision" (P. H. Winston, ed.), pp. 157–209. McGraw-Hill, New York, 1975.

Index